T0305737

Fundamental and Practical Aspects of Tribology

Fundamental and Practical Aspects of Tribology introduces the rudiments of engineering surfaces and teaches the basic phenomena of interacting surfaces in relative motion, major modes of friction and wear, and theories of contact evolution and lubrication. Fundamental topics include friction, wear, and lubrication; surface properties and surface topography; friction of surfaces in contact; wear and surface failures; biotribology; boundary lubrication; fluid properties; hydrodynamic lubrication; bearing selection; and introductory micro- and nanotribology. This book also considers the relationship between nano- and macrotribology, rolling contacts, tribological problems in magnetic recording and electrical contacts, and monitoring and diagnosis of friction and wear.

- Offers a comprehensive review of the fundamentals, providing basic information for scientists and engineers just being introduced to the tribology field
- Teaches tribological methods of measurements and characterization
- Includes examples of real-life tribological problems and case studies of engineering problems and solutions
- Gives an overview of current advancements in the field
- Features end-of-chapter problems and video content for reinforcement of material

This textbook is written for students taking courses in tribology and lubrication, as well as surface engineering. It will also appeal to scientists and engineers who are new to tribology.

The text also offers sample laboratory demonstrations available to qualifying adopting professors.

Fundamental and Practical Aspects of Tribology

Diana Berman, Andreas Rosenkranz, and
Max Marian

CRC Press
Taylor & Francis Group
Boca Raton London New York

CRC Press is an imprint of the
Taylor & Francis Group, an **informa** business

Designed cover image: Shutterstock – Skrypnykov Dmytro

First edition published 2025
by CRC Press
2385 NW Executive Center Drive, Suite 320, Boca Raton FL 33431

and by CRC Press
4 Park Square, Milton Park, Abingdon, Oxon, OX14 4RN

CRC Press is an imprint of Taylor & Francis Group, LLC

ISBN: 978-1-032-50225-0 (hbk)
ISBN: 978-1-032-50236-6 (pbk)
ISBN: 978-1-003-39751-9 (ebk)
ISBN: 978-1-032-80355-5 (eBook+)

DOI: 10.1201/9781003397519

Typeset in Times
by codeMantra

Access the Instructor and Student Resources: https://www.routledge.com/9781032502250

Contents

Preface

Difficult times demand swift and effective solutions to problems. Our world has moved into a series of global challenges. Geopolitical uncertainties persist, and the spectrum of climate change looms large, reminding us of our collective vulnerability. Sustainable energy, resource conservation, and the imperative to forge new paths in industry to combat global warming are now more critical than ever before. In this intricate web of challenges, tribology emerges as a beacon of interdisciplinary science, uniquely positioned to confront these multifaceted issues by unraveling the complexities of friction, lubrication, and wear.

In the pages of this book, we embrace not only the responsibility to disseminate knowledge but also the duty to train young and bright minds that will shape our future. The challenges we face today demand not just solutions but the cultivation of a new generation of thinkers and problem solvers. As authors, educators, and mentors, we recognize the profound impact that nurturing young minds can have on the trajectory of scientific innovation. This book serves as a stepping stone for curious minds eager to contribute to the ever-evolving field of tribology. In each chapter, we strive not only to impart information but also to ignite the spark of curiosity that propels young minds toward exploration and discovery. As we share our insights, experiences, and the joy of unraveling tribological mysteries, we hope to inspire the next wave of scientists and engineers who will carry the torch forward.

In this sense, this book aims at providing a comprehensive understanding of the discipline, bridging the gap between fundamental theory and practical applications. It delves into the fundamentals of engineering surfaces, unraveling the intricacies of interacting surfaces in relative motion. From major modes of friction and wear to theories of contact evolution and lubrication, this book covers essential topics such as surface and fluid properties, mechanics of dry and lubricated contacts, tribological testing and characterization fundamentals on different scales, and surface engineering approaches. Practical case studies offer real-world insights, illustrating the application of tribological principles to engineering problems and beyond.

In the spirit of gratitude, we extend our heartfelt love and thanks to our families whose unwavering support made this endeavor possible. Special acknowledgment also goes to our former supervisors as well as all our past and present colleagues and collaborators, whose guidance, expertise, and knowledge have shaped our understanding of tribology.

On a lighthearted note: To our coffee machines for being the unsung heroes fueling late-night writing sessions and to the resilient keyboards that endured countless keystrokes without a single complaint.

Finally, we would like to take a moment to express our gratitude to the bright minds who have graced our classrooms: Our students – present, past, and future. To the budding tribologists-in-training, your curiosity and enthusiasm have been the driving force behind our commitment to this book. As we share this compilation of tribological wisdom, it is with you in mind – the next generation of thinkers, innovators, and problem solvers.

May the pages of this book spark your imagination, kindle the flame of inquiry (just like the matchstick on the cover of this book, which was ignited by friction), and serve as a compass in navigating the fascinating world of tribology. The future is undoubtedly in capable hands with students and young professionals like you at the helm.

Happy reading!

Diana Berman
Andreas Rosenkranz
Max Marian

PS: Of course, no book is complete without the potential for a few sneaky typos, errors, or "slip-ups." To our vigilant readers, consider yourselves honorary members of the "Error Spotting Squad." Should you encounter any errors, rest assured they were purely intentional – a clever ploy to keep you on your toes. Consider them like little Easter eggs sprinkled throughout the text, waiting to be discovered. After all, who doesn't love a good treasure hunt? If you find any, feel free to imagine a mischievous wink from the authors, acknowledging your sharp eye for detail. Happy hunting!

Authors

Diana Berman is an Associate Professor of Materials Science and Engineering at the University of North Texas. Her research focuses on synthesis and characterization of nanostructures, surfaces, and interfaces of ceramic and carbon-based materials for precise control and improvement of their tribological properties and functionality. She has published more than 80 papers in peer-reviewed journals and holds over 10 patents (both US and international). Among her awards are TechConnect Innovation Awards, Society of Tribologists and Lubrication Engineers Early Career Award, UNT Early Career Professorship Award, UNT Research and Creativity Award, NSF Early Career Award, and Fulbright Scholarship Award. She also serves on the editorial boards for *Nature Scientific Reports* and *Tribology Letters* journals.

Andreas Rosenkranz is a Professor of Materials-Oriented Tribology and New 2D Materials in the Department of Chemical Engineering, Biotechnology, and Materials at the University of Chile. His research focuses on the characterization, chemical functionalization, and application of new 2D materials. His main field of research is related to tribology (friction, wear, and energy efficiency), but he has also expanded his fields toward water purification, catalysis, and biological properties. He has published more than 160 peer-reviewed journal publications, is a fellow of the Alexander von Humboldt Foundation, and acts as a scientific editor for different well-reputed scientific journals, including *Applied Nanoscience* and *Frontiers of Chemistry*.

Max Marian is a Professor of Multiscale Engineering Mechanics at the Department of Mechanical and Metallurgical Engineering of Pontificia Universidad Católica de Chile and Deputy Director of the Institute for Machine Design and Tribology (IMKT) of Leibnitz University Hannover, Germany. His research focuses on energy efficiency and sustainability through tribology, with an emphasis on the modification of surfaces through micro-texturing and coatings. Besides machine elements and engine components, he expanded his research toward biotribology and artificial joints.

His work is particularly related to the development of numerical multiscale tribo-simulation and machine learning approaches. He has published more than 50 peer-reviewed publications in reputed journals, given numerous conferences and invited talks, and been awarded various individual distinctions as well as the best paper and presentation awards. Furthermore, he was listed among the Emerging Leaders of Surface Topography: Metrology and Properties, is on the editorial boards of several recognized journals, and is a member of the Society of Tribologists and Lubrication Engineers (STLE) and the German Society for Tribology (GfT).

1 Fundamentals of Tribology

The reduction of greenhouse gas emissions has meanwhile become a focus of public awareness and the development of energy-efficient as well as sustainable technologies remains a major challenge for the 21st century. While fossil fuels are already increasingly being replaced by renewable energy sources to reduce CO_2 emissions, the influence of tribology on the energy efficiency of a wide range of technical processes has hardly reached public awareness. Yet, tribological processes offer considerable potential for saving CO_2 and resources. In 2017, Holmberg and Erdemir [1] estimated that around 23% of the world's primary energy is consumed to overcome friction and to repair or replace worn components in tribo-technical systems (Figure 1.1a). This corresponds to 119 EJ (= 119 000 000 000 000 000 000 J) of annual energy losses. These hardly comprehensible dimensions can be understood more easily by comparing it to the energy released by 2.8 billion tons of exploding TNT, burning roughly 2.6 billion tons of coal, or the kinetic energy of 119 asteroids, each of the size of the Egyptian Cheops pyramid, hitting earth with 10 km/s (Figure 1.1b). It can also be translated into 8120 Mt of annual CO_2 emissions or €250000 million Euro of total costs for friction and wear. Thereby, the impact of friction has increased in recent years, and, compared to wear, the influence on energy consumption and CO_2 emissions is six times higher, while its effect on the associated costs is three times higher [1]. The authors assumed that these energy losses can be reduced by up to 40% in the long term for various

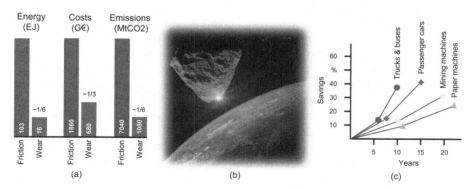

FIGURE 1.1 (a) Global annual energy consumption, costs, and CO_2 emissions due to friction and wear. (Redrawn from Ref. [1] with permission by CC BY 4.0.) (b) Graphical illustration of an asteroid hitting earth, equivalenting a kinematic energy of 1 EJ, i.e., 1/119 of the annual global energy losses due to friction and wear. (Created with the assistance of DALL·E 2.) (c) Estimated potential savings over state-of-the-art through advanced tribological solutions for different case studies and their time to implementation. (Redrawn from Ref. [1] with permission by CC BY 4.0.)

DOI: 10.1201/9781003397519-1

industrial sectors (Figure 1.1c), through tribological advances, such as new lubricants and lubrication concepts, new materials, or new surface technologies [1].

The word "tribology" originates from a study by *H. Peter Jost* on the importance of friction and wear on the national economy of Great Britain in 1966 [2]. The term is composed of the ancient Greek words τριβειυ (tribein = to rub) and λογια (logia = to teach or knowledge of) and literally refers to the theory of friction. Following its original definition from Peter Jost, "tribology is the science and technology of interacting surfaces in relative motion – and of associated subjects and practices" [3]. According to today's understanding, this includes not only friction, lubrication, and wear, but also the interfacial interactions of solids with solids, liquids, or gases. The artificial intelligence chatbot ChatGPT (https://chat.openai.com/chat, queried "define tribology" on March 25, 2023) defines tribology as follows:

> Tribology is the study of surfaces in relative motion, which includes the principles of friction, wear, and lubrication. It is an interdisciplinary field that encompasses various scientific and engineering disciplines such as physics, chemistry, materials science, and mechanical engineering. The goal of tribology is to understand how materials interact with each other when they are in motion, and to develop technologies to optimize wear and friction, which are critical factors that affect the efficiency, durability, and reliability of machines and devices in various industries. By applying tribological principles, researchers and engineers can optimize the design and performance of components such as bearings, gears, engines, or biomedical implants, among others, leading to improved efficiency, reduced energy consumption, and increased lifespan of these critical systems.

Despite the undeniable influence on global energy consumption, costs, and emissions, when asking "people on the street" or even engineering students at the undergrad or graduate level "what tribology is, most of them will not be able to associate this word with any practical object or problem" [4]. According to Popov, the relatively low social and technological awareness of tribology can also be related to the poor availability [4], which also has to do with the fact that the accessibility of a subject is strongly connected to its complexity as well as the existing models and methods to describe the relevant relations and processes. While students of other disciplines, for instance, the basic mechanics of materials, merely need proficiency in the analysis of a few variables and simple differential equations, the most striking deficiency in the current state-of-the-art is that we require partial differential or integral equations with mixed boundary conditions to formulate the most ordinary tribo-contact problems [4]. Moreover, we still cannot reliably predict the coefficient of friction (COF) in practically any material pairing and frequently we still do not even understand the main governing parameters [4]. This certainly is related to the complexity of interfacial phenomena and processes, which can be nicely underpinned with the quote from the eminent Austrian physicist and Nobel Prize winner **Wolfgang Pauli** (1900–1958):

> God made the bulk; surfaces were invented by the devil.

Nevertheless, after years of extensive research and development (partly systematically and partly by trial-and-error), many complex tribological phenomena have been solved/understood or have been met with working practical solutions [4], thus having contributed to the functioning vehicles, machines, and other modern conveniences we are using on a daily base. Therefore, this chapter aims at providing students or engineers new to the subject with the fundamentals of tribology and related knowledge. This covers the historical developments of tribology (1.1), the definition of tribological systems (1.2), tribologically relevant materials and their properties (1.3), surface properties and how to describe them (1.4), contact mechanics (1.5), friction (1.6) and wear (1.7), some hands-on examples of estimating tribological behaviors of different materials (1.8), tribo-effects (1.9), and finally the mechanisms of friction and wear control (1.10).

1.1 HISTORY OF TRIBOLOGY

Even though the term tribology was not coined until 1966, phenomena like friction and wear have concerned mankind ever since as illustrated in Figure 1.2. Even two of the earliest and arguably most significant discoveries – the fire and the wheel – have their origins in friction. The controlled ignition of fire by rubbing two pieces of wood together as practiced already in the prehistoric epoch some ten thousand years ago required friction, while the wheel was a way to reduce it. In this regard, humans soon realized that the transport of goods on rollers had considerable advantages in terms of the required energy over transport on sledges or skids. From the rollers, which were comparatively impractical in their handling, the wheel evolved already around 8000 BC. This can be considered groundbreaking from a technical point of view since there was no example for it in nature. The wheel is characterized by the rolling of its circumference on the surface, which results in lower friction and less wear

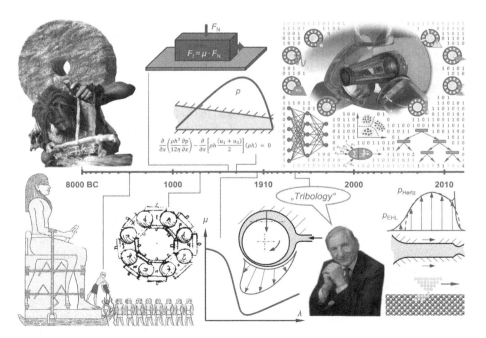

FIGURE 1.2 Timeline of some important tribological advances.

than in sliding motion, which revolutionized transportation as well as warfare, thus considerably influencing the further social development of mankind. The bearing of the wheel on an axle or pivot has been initially realized in the form of (lubricated) plain bearings or bushings. Surprisingly quickly, the simple disc wheel evolved into the lighter spoke wheel by removing areas of the disc between the hub and the rim that were not necessary for the wheel to function. It was also recognized that the rolling resistance of soft materials, e.g., wood, is higher than that of hard materials, e.g., iron, due to the greater deformation during rolling.

As suggested by drawings/engravements or archeological findings, the ancient **Egyptians, Greeks**, and **Romans** were among the first human civilizations to use stone sockets and lubricants to reduce friction and wear between moving parts. The Egyptians employed water, mud, and animal fat to lubricate their carts, while the Greeks and Romans used olive oil and other vegetable oils. The latter also made advancements in early machine elements, including Archimedes gears/screws or ball bearings/joints for ships and rotating platforms. Similarly, lubricants, such as animal fat or vegetable oils, have been used in ancient **China** to lubricate chariots, carriages, and other vehicles.

Besides progress in metalworking as well as materials or lubricants associated with machineries like waterwheels or windmills, few advances were made during the Middle Ages and it was not until 1500 AD when the universal genius **Leonardo da Vinci** (1452–1519) was the first scholar to study rolling bearing designs and to systematically investigate friction and wear [5]. Thereby, he already concluded that friction is influenced by surface roughness and that the frictional force is proportional to the normal force (load) and independent of the nominal contact area [6]. Consequently, he was the first that related the frictional to the normal force, which nowadays is known as "coefficient of friction" (COF), and to experimentally determine its typical values around 0.1–0.5. The proportionality between the frictional and normal forces was re-discovered 200 years later by **Guillaume Amontons** (1663–1705) and today is known as Amontons' law of friction [6,7]. **John Theophilus Desaguliers** (1683–1744) is considered the first to recognize the possible role of cohesion and adhesion in frictional mechanisms [7].

Even though rolling friction was less considered compared to sliding friction due to its much smaller magnitude, **Robert Hooke** (1635–1702) developed the first ideas for rolling friction on plastically deformable bodies, which today are still largely considered to be correct [6].

The term "tribometer" was first mentioned by **Petrus van Musschenbroek** (1692–1761) for his device to measure bearing friction, in which he obtained quantitative results for dry as well as lubricated slider and journal bearings [8].

Leonhard Euler (1707–1783) focused on the mathematical description of friction assuming a proportionality to the gradient of interlocking triangular surface irregularities. Moreover, he distinguished between static and kinetic frictional forces and successfully resolved the issue of rope friction, which may have been the initial contact mechanical problem to be solved analytically. Also, he was the first to introduce the symbol μ for the COF [6].

Later, **Charles Augustin Coulomb** (1736–1806) experimentally confirmed Amontons' results considering the material pairing, surface condition, lubrication, speed, or ambient conditions. Moreover, he found that sliding friction is largely independent of the sliding speed. Coulomb also contributed to Euler's differentiation between kinetic and static frictions and extended the latter by the contribution of adhesion, thus explaining why static friction increases with the time an object has stayed stationary [6,9].

Ultimately, the industrial revolution marked a turning point since the development of steam engines and other machines required new approaches for reducing friction and wear. Engineers began to search for the best material pairings for engine components or machine elements such as bearings, gears, and other components.

Understanding that there are always no-slip as well as slip areas in rolling contacts of a driven wheel and attributing rolling friction to local sliding, **Osborne Reynolds** (1842–1912) was to first to attempt putting tribological contacts under the magnifying glass and questioning the differentiation into static and dynamic friction [6]. Employing the fundamentals for the mechanical contact between two curved, elastic bodies from **Heinrich Rudolf Hertz** (1857–1894) [10], **Frederick William Carter** (1870–1952) later was able to derive a quantitative theory for partial slip in rolling contacts [6].

In experiments on journal bearings of railway axle boxes with immersed shafts, **Beauchamp Tower** (1845–1904) observed oil rising upward the oil feed holes, thus discovering the hydrodynamic lubrication phenomena [11]. Based on further experiments on the fluid flow, **Osborne Reynolds** derived a dimensionless quantity to express the ratio of inertial forces to viscous forces (today known as Reynolds number), thus differentiating between turbulent and laminar flow as well as the theory of hydrodynamic lubrication expressed by the "Reynolds equation." This holds true for the complete separation of both rubbing bodies without contacting roughness and allows relating the film thickness and the length of a lubricated contact with the COF [6,12].

The transition of full film separation to mixed lubrication, i.e., contacting roughness features, was studied in detail by **Richard Stribeck** (1861–1950), who expressed the dependence of friction on the sliding speed (or normal load or lubricant viscosity) in the so-called "Stribeck curve" [10]. Although these findings are usually attributed to Stribeck, **Robert H. Thurston** (1839–1903) and **Adolf Mertens** (1850–1914), who developed testing machines to study the frictional forces of different materials and lubricants, came to similar conclusions several years earlier [13].

William Bate Hardy (1864–1934) examined processes occurring in the absence of a sufficient lubricant film and showed that even molecular grease layers (boundary lubrication) have a substantial influence on tribological behavior. By studying the lubricants' and surfaces' molecular weights, he was able to trace this back to polymer chains of the grease adhering to the metal surfaces [6,14].

Frank Philip Bowden (1903–1968) and **David Tabor** (1913–2005) further contributed to our understanding of contact mechanics and dry friction, highlighting the role of roughness as well as the real area of contact, which typically is some orders of magnitude smaller than the nominal one. Furthermore, they suggested that the formation and shearing of cold weld junctions at contacting asperities of metallic surfaces contributed to their sliding friction. According to them, the friction coefficient for isotropic, plastic materials approximates 1/6, corresponding to the ratio of critical shear stress to hardness. This approximation comes very close to measured COFs for steel/steel, steel/bronze, or steel/iron pairings under dry conditions [6].

These findings inspired a whole line of research on the contact mechanics of rough surfaces as well as associated adhesive and abrasive wear phenomena, whereby the names of John Frederick Archard (1918–1989), Ernest Rabinowicz (1927–2006), James Anthony Greenwood and Brian Williamson, and Bo Persson are to be mentioned [6].

The mechanisms underlying the fictional behavior of elastomers, i.e., energy dissipation through deformation and thus the dependence on the rheological behavior, can largely be traced back to **Karl Grosch** (1923–2012) [6].

Duncan Dowson's (1928–2020) extensive research on the behavior of lubricated contacts under high pressures led to the development of theories and equations that describe the lubrication regimes and film thickness in the area of elastohydrodynamic lubrication, which have been crucial in designing efficient and durable machine components or joint replacements. Additionally, Dowson was instrumental in advancing the understanding of surface roughness effects on lubrication, friction, and wear as well as in establishing experimental techniques for tribological studies.

Much of the progress and current knowledge can certainly be attributed to focused, methodical investigations. In this context, the name of **Horst Czichos** (1937) is to be mentioned, who suggested a systems approach for tribology, which will be highlighted in more detail in the following chapter.

Over the years, tribology has witnessed several remarkable developments that have expanded its horizons and practical applications. Advanced materials and coatings have emerged, engineered to enhance wear resistance and reduce friction in various industrial sectors. Nanotechnology has played a pivotal role, enabling the design of novel lubricants and additives at the nanoscale to optimize surface interactions. Furthermore, the integration of data-driven approaches and machine learning has revolutionized tribological research, allowing for predictive modeling of friction and wear based on real-time operating conditions. In this regard, the concept of tribology 4.0 aligned with the fourth industrial revolution envisions a future, where smart technologies and connectivity are seamlessly integrated into tribological systems. This includes the implementation of sensors, IoT (Internet of Things) devices, and data analytics to monitor and optimize machinery performance in real time. Proactive maintenance strategies can be developed, thus minimizing downtime and maximizing efficiency. Additionally, additive manufacturing and 3D printing transform the fabrication of complex components with tailored surface features, influencing friction and wear characteristics. As industries move toward sustainable practices, tribology also contributes by fostering eco-friendly lubricants and energy-efficient designs. These recent developments and the paradigm of tribology 4.0 collectively underscore the field's evolution from fundamental studies to cutting-edge technologies that impact diverse sectors, from manufacturing to transportation and beyond.

1.2 TRIBOLOGICAL SYSTEMS

The base for the definition of tribological systems, according to Czichos and Habig [15], is the general function of technical systems. In this context, tribo-systems can be divided into mainly energy-, substance-, and information-determined functional classes with regard to their technical function, namely the conversion of mechanical quantities, substances, or information. In its simplest form, this corresponds to an abstract black box representation as schematically depicted in Figure 1.3. A technical system therefore can be described by

- system elements,
- input quantities,
- output quantities,
- a functional translation of input quantities into output quantities, and
- a system boundary enclosing the system.

The system function translates the input variables into output variables and is supported by the structure of the system. The overall system structure, in turn, consists of individual elements and their properties and interactions. Moreover, losses and disturbance variables occur, which influence the system function.

Transferred to a tribo-technical system, this results in the representation shown in Figure 1.4. The system structure consists of the pair of interacting surfaces of the **base body** and the **counter-body**, the **intermediate medium** (e.g., a lubricant), and the **ambient medium** (e.g., air). The assignment of base body and counter-body can

FIGURE 1.3   Black box representation for a target function.

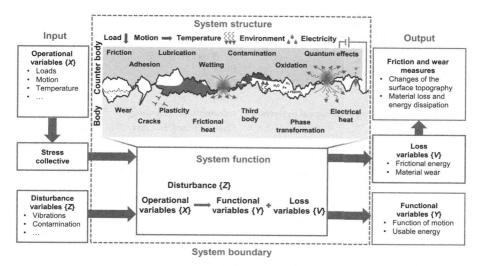

FIGURE 1.4   General system representation of a tribological system, its target function as well as interactions in tribological contacts. (Redrawn and modified from Refs. [16–18] with permission.)

be made as desired. In most cases, the one that is more important for the system function from the user's point of view is designated as the base body. A distinction can be made between open and closed systems. **Open systems**, on the one hand, are characterized by a constant flow of material into and out of the system or by stressing of the main body by constantly new areas of the counter-body (e.g., in manufacturing processes, wheel/rail-contacts). Frequently, the wear (material loss) of the base body is important for the main function. **Closed systems**, on the other hand, experience a permanent, sometimes interrupted tribological stress on the elements (e.g., rolling bearings, cam/followers).

The **system function** of the tribo-system is formed by the interacting surface pair. Thereby, the active surfaces are subjected to external forces as well as relative movements and temperatures for different stress durations. This is referred to as tribological **stress collective** and corresponds to the input variables of the system function. Microscopically, the elementary physical and chemical interaction mechanisms take place at locally and temporally varying **points of action**. If the sum of all points of contact is combined to form a **real area of contact**, this is usually substantially smaller than the geometric contact area formed by the nominal component dimensions. This results in **loss variables** such as friction and wear, which result in a change in the surface topography, a loss of material and energy dissipation, and the actual **functional variables** of the tribological system. Examples for energy, material, or information-converting tribo-technical systems are summarized in Table 1.1.

The mechanisms and applications of tribological systems span over several scales, which are illustrated in Figure 1.5 with the corresponding characteristic length and

---

TABLE 1.1
Functions of Tribological Systems [15]

Energy-Converting Tribological Systems	Mechanical Engineering, Precision Engineering
Transmission of motion	Guides, joints, bearings
Inhibition of motion	Brakes
Transmission of power	Couplings
Transmission of energy	Gears
Information-Converting Tribological Systems	**Information Technology**
Storage technologies	Computer hard disk drive, CD, DVD
Signal transfer	Cam/followers, switching circuits
Signal output	Inkjet printer
Material-Converting Tribological Systems	**Production Engineering, Transportation**
Original forming	Casting, pressing, and extruding tools
Forming	Bending, rolling, forging, and drawing tools
Separating	Drilling, turning, milling, and grinding tools
Joining	Fittings, friction welding
Coating	Surface technologies
Changing material properties	EDM processes, lithography
Sealing	Seals, valves, piston/cylinder liner
Material transport	Conveying systems, pipeline, fluidics
Goods transport	Tire/road, wheel/rail

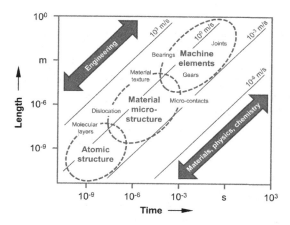

FIGURE 1.5 Characteristic dimensional scales in tribological systems. (Redrawn and modified from Ref. [15].)

velocity ranges. This extends from processes at the **nano-** or **micro-level** in the field of physics, chemistry, and materials science, such as the influence of basal orientation of molybdenum disulfide (MoS_2) with regard to the sensitivity of coatings to humidity. It ends with machine elements and assemblies containing multiple tribological contacts in engineering sciences on the **macro-level**, for example, large-diameter bearings in the wind energy industry with outer diameters of more than three meters or large gearboxes with gears of similar dimensions.

EXERCISE 1.1

For the following tribo-technical systems, identify the base body and counter-body, the intermediate and environmental medium, and the system type (open or closed system):

- Gear transmission
- Wheel/rail
- Journal bearing
- Excavator
- Crushing plant

1.3 SOLID ENGINEERING MATERIALS AND THEIR BULK PROPERTIES

Solid engineering materials can be broadly categorized into

- metals (alloys),
- polymers,
- ceramics, and
- composites.

Each category possesses unique properties and applications that cater to specific engineering needs. **Metals** and **alloys** are renowned for their ductility, high strength, and load-bearing capacity, making them suitable for applications like bearings, gears, and sliding components. Some common metals/alloys used in engineering include steel, aluminum, and copper. **Polymers** offer lightweight and elasticity, while featuring low friction and self-lubricating properties, making them ideal for applications where reduced wear and noise are essential. They exist in various forms, ranging from thermoplastics to thermosetting polymers. Tribological applications of polymers include seals, bushings, and sliding components in biomedical devices. **Ceramics** exhibit excellent thermal and electrical insulating properties, a high hardness and wear resistance, making them suitable for tribological applications under high-load and high-temperature conditions. Examples include cutting tools, bearings for high-speed machinery, and brake pads. **Composites** combine at least two different classes of materials, offering tailored properties for specific engineering requirements. Composite materials can be engineered to have high strength-to-weight ratios, making them ideal for aerospace components, sporting goods, and advanced automotive parts.

One fundamental aspect of material science worth highlighting is the types of atomic and molecular bonds that hold materials together since this is essential for all mechanical, electrical, thermal, and chemical properties of materials, thus notably affecting their tribological behavior. A summary of primary and secondary bonds as well as their associated material properties is presented in Table 1.2. Please note that many materials exhibit combinations of different types of bonding, which further influences their characteristics.

For most engineering materials, the **elastic** behavior is observed when the **deformation** (or strain) is small (Figure 1.6). In a uniaxial stress test of a sample with cross-sectional area A, as the load F gradually increases, the material initially falls within its elastic region. This results in a (almost) linear relationship between **stress** $\sigma = F/A$ and **strain** $\varepsilon = \Delta l/l_0$. This is known as Hooke's law, and the slope of the straight line represents **Young's modulus**, also referred to as elastic modulus, which can be defined by

$$E = \frac{\partial \sigma}{\partial \varepsilon}. \tag{1.1}$$

At the end of this linear region, the proportional limit is reached. If the load is further increased, the stress–strain relationship may slightly deviate from the straight line, but the deformation remains elastic. In this region, if the load is slowly released, the material will revert to its original shape with no significant residual deformation.

The transverse contraction coefficient or **Poisson's ratio** v is another important property of solid materials, defined as the ratio of lateral strain to longitudinal strain under uniaxial loading. An incompressible material has a Poisson's ratio of 0.5. Likewise, the **shear modulus** is defined as the proportionality coefficient between shear stress and the resulting shear deformation and is connected to the elasticity coefficient and Poisson's ratio in the following way:

$$G = \frac{E}{2(1+v)}. \tag{1.2}$$

TABLE 1.2

Summary of Different Bond Types with a Brief Description, the Resulting Properties, and Some Common Examples

Type	Description	Typical For	Bond Energy	Properties	Examples
Primary: Ionic	Complete transfer of electrons between elements leading to electrostatic forces	Ceramics and glasses	Large	High T_{melt}, large E, small thermal expansion, low thermal and electrical conductivities	NaCl, MgO, CaF$_2$
Primary: Covalent	Sharing of electrons between two elements	Semiconductors, ceramics, polymer chains	Variable: large for diamond, small for bismuth	Moderate T_{melt}, moderate E, variable thermal expansion, low thermal and electrical conductivities	SiC, GaAs, C (diamond)
Primary: Metallic	Interactions between ion cores and sea of valence electrons	Metals and alloys	Variable: large for tungsten and small for mercury	Moderate T_{melt}, moderate E, moderate thermal expansion, high thermal and electrical conductivities	Cu, Ag, Ni, W
Secondary: Van der Waals, hydrogen, dipole/dipole	Interaction between permanent and/or temporary atomic or molecular dipoles	Inter-chain in polymers, inter-layers in 2D materials, inter-molecular	Small	Low T_{melt}, small E, large thermal expansion, low thermal and electrical conductivities	C (graphite), H$_2$O, Ar

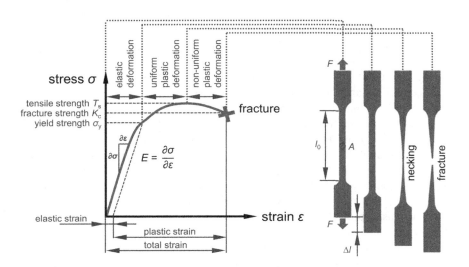

FIGURE 1.6 Typical stress–strain relationships obtained from tensile testing.

For brittle materials like cast iron or ceramics, rupture may occur relatively early without any significant **plastic deformation**. However, as the applied load increases in a typical ductile material such as steel, the material eventually reaches its initial yielding limit, resulting in noticeable plastic and permanent deformation. This initial limit is referred to as the **yield strength** and reflects a critical property of ductile materials. This value is commonly defined at 0.2% plastic strain based on the stress–strain curve under uniaxial stress conditions. Notably, the yield strengths under tension and compression are generally similar for ductile materials. Beyond the yield point, the plastic strain can increase substantially, and the material may reach its ultimate strength before eventually breaking. Once the material enters the plastic zone beyond the yielding limit, its behavior becomes more intricate. If the load is released in the plastic zone, the stress returns to zero or very low levels, but some strain remains as plastic deformation. The slope of the stress–strain curve in the plastic zone, which is known as the **tangent modulus**, may or may not be constant and arises from the combined effects of both elastic and plastic strains.

A typical technique to determine the yield point of an elastically perfectly-plastic material connects with hardness testing, which involves indenting a rigid pyramid (**Vickers hardness**; HV) or ball (**Brinell hardness**; HB) into the surface to be examined. The indentation hardness is only slightly affected by the shape of the indenter, and, in a first-order approximation, can be obtained by calculating the ratio of the normal force applied to the area of the indent, which is known as the indentation hardness. In contrast, the **Rockwell hardness** (HRA, HRB, HRC) is determined by the indentation depth, which has advantages in terms of speed and costs. In this context, either hardened steel balls or spherical diamond cones are used, while the last letter (A, B, C) specifies the load and indenter geometry. Concerning the tribological characterization of materials, hardness measurement plays a crucial role as tribological processes are mainly influenced by micro-asperities and interactions similar to the hardness test.

Some elastomers and rubbers exhibit a non-linear stress–strain relationship and can undergo large deformations, while maintaining their ability to return to their original shape upon unloading. Unlike linear-elastic materials, these **hyperelastic** materials do not follow Hooke's law and display instead a strain-dependent stress response. Consequently, the deformation is governed by strain energy functions, which describe how the stored elastic energy changes with deformation. Moreover, **viscoelastic** materials, such as human skin, wood, and some polymers, unify both elastic and viscous behaviors, which implies that they do not only store energy when deformed but also dissipate energy over time due to internal friction and molecular relaxation. This results in time-dependent responses, for which the material's behavior is influenced not just by the applied stress but also the rate at which the stress is applied or removed. This material response can be described by relaxation modulus and creep compliance functions, which illustrate the time-dependent nature of the material's response.

In the realm of tribology, relevant material characteristics extend beyond classic elastic or plastic properties, encompassing a diverse range of properties crucial to understand and manage surface interactions. Just to mention a few, fatigue resistance becomes paramount when considering cyclic loading scenarios, in which repeated stresses can lead to material degradation over time (linear reciprocating sliding motion). Thermal conductivity is significant, impacting the ability of a material to dissipate heat generated during sliding. Chemical reactivity and corrosion resistance are vital in determining a material's compatibility with its environment, particularly in aggressive or corrosive conditions. Moreover, the ability of a material to attract lubricants, known as wettability or surface energy, influences the effectiveness of lubrication strategies. A selection of relevant material properties, microstructural characteristics, and their application-related effects are presented in Table 1.3. The comprehensive understanding of these material characteristics enhances the design of components and systems with the overall goal to improve the performance and durability of various tribological applications. A qualitative comparison of metallic (M), polymer (P), and ceramic (C) materials is as follows [15]:

- Mass forces $\quad\quad\quad\quad\quad\quad\quad\quad$ $P < C < M$
- Hertzian pressure $\quad\quad\quad\quad\quad$ $P < M < C$
- Friction-induced temperature rice $\quad$ $M < P < C$
- Surface tension (adhesion energy) $\quad$ $P < M < C$
- Abrasive wear $\quad\quad\quad\quad\quad\quad$ $C < M < P$
- Chemical reactivity $\quad\quad\quad\quad\quad$ $P < C < M$

while some tribologically important characteristics as well as typical value ranges for different materials are provided in Figure 1.7.

It is essential to acknowledge that while solid materials are traditionally assumed to be homogeneous continua, they often contain various types of inhomogeneities, such as impurities, inclusions, voids, microstructural defects, and non-uniform

TABLE 1.3
Material Properties, Microstructural Characteristics, and Their Related Effects in Tribo-Applications [15]

Material Characteristic/Property	Effect On
Elastic modulus and Poisson's ratio	Hertzian shear stresses
Hardness, yield strength, tensile strength	Permissible material strain, size, and nature (elastic/ plastic) of the real contact area
Fracture toughness	Crack propagation, separation of wear products
Fatigue strength	Surface spallation
Residual stresses	Surface spallation
Density	Mass forces, material strain
Thermal conductivity, specific heat capacity	Temperature increase due to friction
Coefficient of thermal expansion	Thermally induced residual stresses, dimensional changes, distortions (e.g., bearing clearance)
Surface energy	Friction coefficient, wettability, adhesion
Gibbs' free energy	Reaction kinetics of tribo-chemical reactions
Arrhenius constant	(thermodynamic equilibrium and rate constants)
Phases (type, number, size, distribution)	Wear mechanisms
Grid structure	Adhesion, friction coefficient, abrasion
Texture	

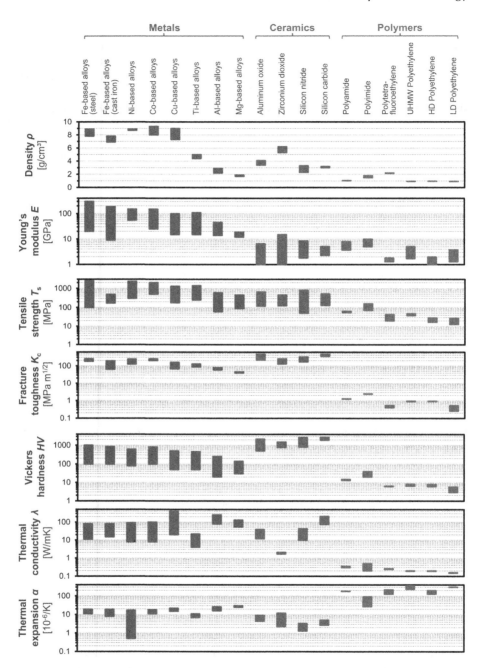

FIGURE 1.7 Overview of ranges for some tribologically relevant parameters of various materials.

surface layers, which can significantly influence interfacial characteristics. In modern times, diverse material and surface treatment technologies have been developed and commonly used, some of which may leverage favorable material inhomogeneities.

1.4 TECHNICAL SURFACES

1.4.1 SURFACE STRUCTURE AND PROPERTIES

The journey from the material core to its surface is a fascinating exploration of how **properties** evolve and adapt in response to changing environments. While bulk properties represent the characteristics of the material's interior (Section 1.3), the **surface structure** and the arrangement of molecules/atoms near the surface often deviate from the well-ordered lattice structure (in the case of metals or alloys), which is typically present in the bulk. Surface atoms may exhibit a lower degree of coordination, causing distortions, vacancies, and even reconstructions of the crystal lattice. These structural alterations introduce new electronic states within the band structure, potentially leading the modification of various properties (Figure 1.8). Through interactions of the material with the surrounding media, changes in surface composition and the incorporation of components from the environment can occur. Depending on the reactivity of the base material, **physisorption** can take place via van der Waals' forces (reversible processes, where an adsorbed substance is bound to the surface layer of the material by physical forces without chemical changes and with only minor changes due to relaxation), or a stronger **chemisorption** requiring higher activation energy and involving covalent or ionic bonding fractions, which results in a stronger and irreversible bond between the adsorptive and the adsorbent. In the case of metallic materials, oxide layers are generally present due to exposure to atmospheric oxygen, and gaseous or liquid impurities are adsorbed onto these layers through physical or chemical means. Furthermore, the influences of the manufacturing processes are to

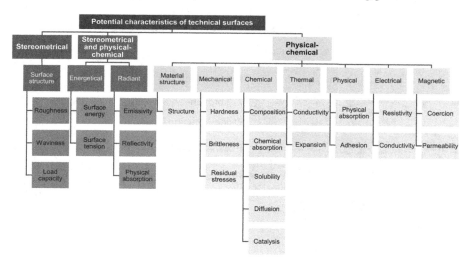

FIGURE 1.8 Summary of potential characteristics of technical surfaces.

be considered. Surfaces that have undergone machining and deformation exhibit different solidification, residual stresses, and texture inhomogeneities between the edge zone and the bulk of the material. When characterizing technical surfaces, three regions are distinguished from the innermost to the outermost layers:

- **Bulk material**
- **Inner boundary layer**
- **Outer boundary layer**

As illustrated in Figure 1.9, the inner boundary layer consists of a deformation or consolidation zone adjacent to the bulk material, depending on the manufacturing/forming process. The outer boundary layer typically exhibits a composition distinct from the base material and can include oxide layers, adsorption layers, and impurities. In the context of tribology, the differences between the bulk material and the surfaces, i.e., the boundary layer regions of technical surfaces, are crucial to understand how surfaces contact with each other.

The **chemical composition** of surfaces can significantly differ from that of the bulk due to the incorporation of components from the surrounding environment. It should be noted that in addition to the influences of the ambient medium, there can also be an enrichment of alloying elements from the interior of the material surface in alloys (surface segregation). These aspects imply that, in interpreting tribological processes on material surfaces, one cannot generally rely on the usual volume-based chemical composition. Instead, the actual chemical composition within the outer boundary layers has to be considered.

Within the material's **microstructure**, differences can exist not only in the grain sizes of crystallites at the surface and in the bulk, but also in aspects such as the density of vacancies and dislocations. This, in turn, holds fundamental importance for the strength properties of surface areas. Due to the discussed differences in chemical composition and microstructure, the surface **hardness** can also vary significantly from that of the interior of the material. In this context, **surface engineering techniques** – such

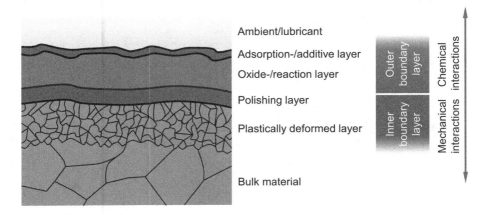

FIGURE 1.9   Schematic representation of a technical surface in cross-section.

as hardfacing (e.g., fusion hardfacing or thermal spray), the deposition of coatings (e.g., by physical or chemical vapor deposition), diffusion processes (e.g., carburizing, nitriding, and boriding), selective hardening methods (e.g., flame, induction, and laser hardening), ion implantation, or techniques like laser cladding and weld overlay – are nowadays extensively utilized to bolster surface strength and improve friction. The predominant advantage of **surface hardening** lies in its capacity to enhance strength solely within a specific case depth (few mm), leading to substantial improvements, for instance, in rolling contact fatigue, while preserving the integrity and ductility of the material's bulk as well as keeping costs comparatively low.

1.4.2 SURFACE TENSION AND ENERGY

At the interface between a liquid and a solid, the cohesive forces between the liquid's molecules result in the formation of a meniscus subject to interfacial tension. This so-called **surface tension** characterizes the behavior of liquids at their interfaces with other substances and is defined as the tangential force F per unit length L acting along the interface between the liquid and a solid surface:

$$\gamma = \frac{F}{L}. \tag{1.3}$$

This phenomenon plays a significant role in processes like capillary action, for which liquids rise or fall in narrow tubes. For technical surfaces, surface tension influences how liquids, such as water or lubricants, spread and interact with the surface, which is crucial in processes involving wetting, adhesion, cleaning, painting, or coating. Consequently, the **contact angle** is a fundamental concept that characterizes the wetting behavior of a liquid on a solid surface. It is the angle formed between the tangent to the liquid–vapor interface at the three-phase contact point and the solid surface. The contact angle θ is an indicator of the degree of **wetting** and provides insights into the interaction between the liquid, solid, and vapor phases. Young's equation relates the static contact angle, the surface tension of the liquid γ_L, the surface tension of the solid γ_S, and the interfacial tension between the liquid and solid γ_{LS}:

$$\gamma_L \cos\theta = \gamma_S - \gamma_{LS}. \tag{1.4}$$

The contact angle is driven by the balance between adhesive forces (liquid–solid interaction) and cohesive forces (liquid–liquid interaction). A small contact angle indicates good wetting, i.e., the liquid spreads readily on the surface. In contrast, a large contact angle resembles poor wetting, for which the liquid tends to form droplets on the surface. By carefully placing droplets and measuring the contact angle (Figure 1.10) of a set of selected liquids covering a range of known surface tensions (e.g., deionized water, diiodomethane, ethylene glycol, among others) onto the solid surface, the solid surface energy can be determined by re-arranging the above-mentioned Young's equation. It is important to note that this method assumes that the solid surface is homogeneous and the liquid droplets maintain their shape without any penetration into the surface. In this regard, the static contact angle can be distinguished from the dynamic contact angle, whereby the contact angle hysteresis is

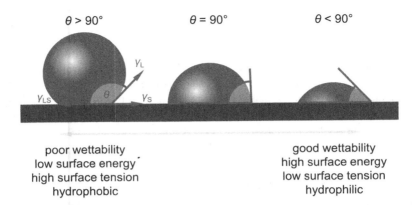

FIGURE 1.10 Schematic representation of different wettability states of the underlying surface as indicated by the measured contact angle.

determined at discrete time steps when the three boundary phases are in motion due to the solution impurities absorbing on the surface. Additionally, **contact angle measurements** can be influenced by factors such as contamination, environmental conditions, and the surfaces' micro-geometry (roughness). The latter will be addressed in more detail in the following.

1.4.3 SURFACE MICRO-GEOMETRY

Besides the structural and physical–chemical properties, technical surfaces feature micro-geometric deviations from the ideal macroscopic geometry of components. This roughness is influenced by manufacturing or processing and manifests as a three-dimensional stochastic distribution of "peaks" and "valleys" at different scales (Figure 1.11). On a coarser scale, **waviness** refers to longer wavelength deviations from the nominal surface profile, i.e., an overall pattern or undulation of the surface. It is often caused by factors such as machine tool vibrations, tool deflection, or the inherent nature of the

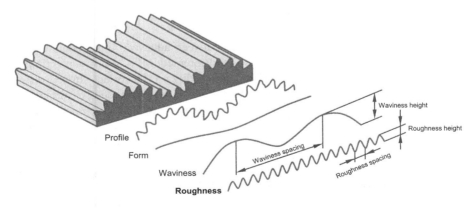

FIGURE 1.11 Surface profile composed of form deviation, waviness, and roughness.

manufacturing process itself. In other words, it represents the broader trends or variations on the surface that can affect the overall fit and function of components.

In contrast, **roughness** refers to the finer-scale irregularities (so-called **asperities**) present on a surface. These irregularities are characterized by short wavelengths and relatively small amplitudes. Roughness is a result of the machining or finishing process and can greatly influence properties such as friction, wear, sealing, and appearance. Even though it is necessary to simplify the specification of surface properties through means such as drawings, blueprints, and technical documents in engineering practice, the topographic features of surfaces exhibit a complex nature, often displaying a significant randomness. Considering a surface of 100×100 mm² and the size of one atom to be 1 A (10^{-10} m), a surface would consist of 10^{18} atoms. To avoid dealing with such large numbers, various statistical roughness parameters are used, which differ in the nature of the determination (and measurement) and the types of surfaces they are best suited to describe. Within this chapter, we focus on some chosen parameters, which are more frequently employed in practice. These refer to profile or area sections of the technical surface under consideration as determined from contact- or non-contact-based methods as discussed in Chapter 2.

Line surface roughness ("R") parameters are used to quantify the deviations of a two-dimensional surface profile from a reference line. They are typically measured along a single line or profile on the surface (see Figure 1.12, bottom left) as typically obtained by contact-based measurement methods or in a three-dimensional surface measured by non-contact methods. These parameters are well-suited to capture the variations along a specific path and are commonly used in industries, where the surface's performance is primarily affected by directional features. **Area surface roughness ("S") parameters**, in turn, focus on characterizing the overall texture of a three-dimensional surface across a defined area. These parameters are measured across a space, capturing the variations in height over an entire surface region (see Figure 1.12, top), as directly obtained by non-contact methods. Alternatively, multiple individual line contact-based measurements need to be combined into a surface. Area parameters are particularly useful for describing surfaces with complex and intricate patterns that might not be adequately represented by a single-line profile.

The respective roughness parameters are typically calculated by the height distance (z-axis) from a **reference line** in x-axis (R parameters), or **surface** (over x- and y-axes; S parameters), which has to be determined in the first place either by just averaging the height data or fit a linear or higher order mathematical function in the case of slightly tilted or curved surfaces. Afterward, these corrected values can be used to determine statistical characteristics reflecting the height distribution. The most frequently reported parameter is the **average surface roughness**

$$R_{\mathrm{a}} = \frac{1}{L} \int_0^L |z(x)| \, \mathrm{d}x, \qquad (1.5)$$

$$S_{\mathrm{a}} = \frac{1}{A} \int_0^A |z(x)| \, \mathrm{d}x \, \mathrm{d}y, \qquad (1.6)$$

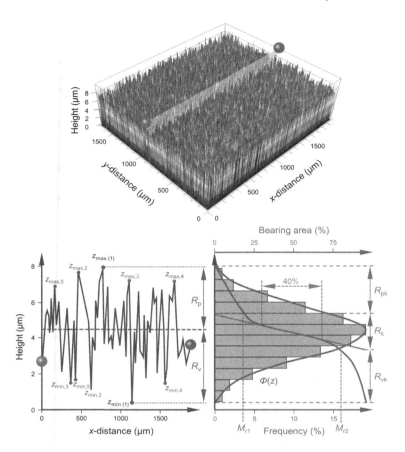

FIGURE 1.12 Representative 3D rough surface (top) and 2D line section of the height versus z-distance with the indication of various relevant roughness parameters (bottom left), which is analyzed using a histogram, probability function as well as the bearing area (bottom right). Please note that a much larger magnification is typically applied for the height in the height (z-direction) compared to the lateral x- and y-directions.

within the evaluation length L or area A. In the case of N discrete measurement points z_i (line) or z_{ij} (area), these equations can be expressed as

$$R_a = \frac{1}{N} \sum_{i=1}^{N} |z_i|, \tag{1.7}$$

$$S_a = \frac{1}{N} \sum_{i=1,\, j=1}^{N} |z_{ij}|. \tag{1.8}$$

Since the S parameters can be adopted from the R formulations in the same way by considering the additional y-axis as shown in the previous example, this will not be explicitly written in the following.

To solve this issue of the roughness characterization, another quantitative parameter, the **root mean square (RMS) roughness**

$$R_q = \sqrt{\frac{1}{L} \int_0^L z(x)^2 \mathrm{d}x} = \sqrt{\frac{1}{N} \sum_{i=1, j=1}^{N} z_i^2} \qquad (1.9)$$

is used. Since tribological contacts typically consist of the pairing of two surfaces with roughness each, a **composite average roughness**

$$R_a = R_{a1} + R_{a2} \qquad (1.10)$$

with the individual average roughness R_{a1} and R_{a2} as well as a **composite RMS roughness**

$$R_q = \sqrt{R_{q1}^2 + R_{q2}^2} \qquad (1.11)$$

with the individual RMS roughness R_{q1} and R_{q2} can be calculated.

Apart from these averaged values, there are some parameters relying on extreme values, namely the **maximum profile peak height** R_p, which is equal to the distance z_{max} between the highest peak and the reference line, and the maximum profile valley depth R_v, which is equal to the distance $|z_{min}|$ between the deepest valley and the reference line, see Figure 1.12. The **mean maximum height** is usually defined[1] as the average distance between the five highest peaks $z_{max,i}$ and five deepest valleys $z_{min,i}$:

$$R_z = \frac{1}{5} \sum_{i=1}^{5} \left(z_{max,i} - |z_{min,i}| \right) . \qquad (1.12)$$

While most other parameters correlate between line and area values, the mean height differs from this rule since it describes the entire peak-to-valley height, i.e., the difference between the single, highest measured peak z_{max} and the single, lowest measured valley z_{min}:[2]

$$S_z = z_{max} - |z_{min}| = S_p + S_v. \qquad (1.13)$$

The analog line value is referred to as **maximum height of the profile**

$$R_t = R_p + R_v. \qquad (1.14)$$

Evidently, these parameters are substantially influenced by scratches, contamination, and measurement noise due to their reliance on peak values. An overview of typical R_a and R_z value ranges for various manufacturing and surface finishing methods is summarized in Figure 1.13. Importantly, different surfaces can have the same R_a (Figure 1.14), therefore, it is important to include several roughness parameters.

---

[1] Some other definitions might be encountered in literature.
[2] Imagine measuring millions of points and then basing a parameter on only two of them... That is S_z!

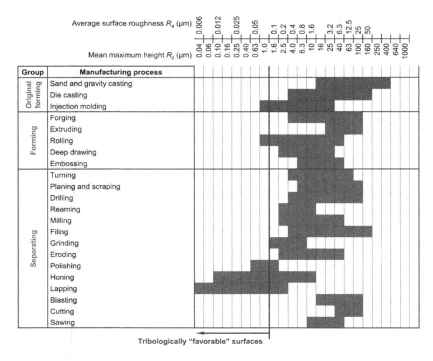

FIGURE 1.13 Typical surface roughness values achievable with various manufacturing processes.

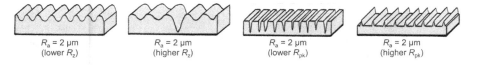

FIGURE 1.14 Schematic illustration of surfaces with the same averaged roughness but different maximum heights and different peak heights.

EXERCISE 1.2

To get a better feeling for roughness parameters, let us calculate the roughness for several representative roughness profiles. Given are the following surface line profiles:

1.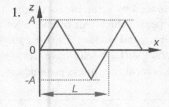

2.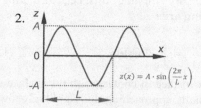

$$z(x) = A \cdot \sin\left(\frac{2\pi}{L} x\right)$$

3.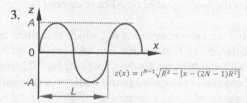

$$z(x) = i^{N-1}\sqrt{R^2 - [x - (2N-1)R^2]}$$

For each of them, determine:
- The maximum height of the profile R_t
- The averaged roughness R_a
- The RMS roughness R_q

In fact, two surfaces with the same averages, RMS or maximum height profile values still can exhibit a very distinct contact mechanical or tribological behavior due to differences in the distributions of their peaks and valleys. To obtain more information on the statistical distribution of surface heights, the so-called **probability density function (PDF)** describes the probability that the height of a randomly chosen position on the surface falls within a small histogram interval, i.e., it quantifies how densely the height values are distributed over a range of values (Figure 1.12, bottom right). In the case of a normal (Gaussian) distribution, which is frequently the case for freshly machined surfaces, the PDF can be expressed as follows:

$$\Phi(z) = \frac{1}{R_q\sqrt{2\pi}} e^{-\frac{z^2}{2R_q^2}}. \tag{1.15}$$

The probability for a roughness height to be within the values z_1 and z_2

$$P_r(z_1 \leq z \leq z_2) = \int_{z_1}^{z_2} \Phi(z)\,dz, \tag{1.16}$$

or to be below a value of a certain threshold z_0

$$F(z \leq z_0) = \int_{-\infty}^{z_0} \Phi(z)\,dz \tag{1.17}$$

can be determined by integrating a given PDF. By "flipping" this, i.e., calculating the area from z_0 to infinity, we obtain the **bearing area**

$$\text{BA}(z_0) = 1 - F(z \le z_0) = \int_{z_0}^{\infty} \Phi(z)\,\mathrm{d}z, \tag{1.18}$$

which represents the percentage of a rough surface in contact with the surface of a flat, rigid body at a given distance. By virtually approaching this rigid body toward the rough surface, we can construct the bearing area or **Abbott curve** (Figure 1.12, bottom right) [19].

Based on it, it is possible to derive the **core roughness R_k**, which is defined as the height profile along the z-axis of the BA curve between the intersections of a 40% line placed at the minimum slop point with the 0% and 100% points (Figure 1.12, bottom right). In other words, it measures the depth of the lowest points (valleys) on the surface, considering only the core material's topography and ignoring the influence of peaks. The **peak height R_{pk}** and the **valley depth R_{vk}** (Figure 1.12, bottom right) are the distances between highest profile peak and the intersection with the line of the first material ratio point M_{r1} (i.e., the upper limit of the core roughness) as well as the deepest profile valley and the intersection with the line of the second material ratio point M_{r2} (i.e., the lower limit of the core roughness), respectively. As a measure for peak heights above the core roughness, greater R_{pk} values imply that the surface is composed of high peaks, which might lead to a small initial contact area and thus high local contact stresses (see Section 1.5) and vice versa. In turn, R_{vk} measures the valley depths below the core roughness and may be related to lubricant retention and wear debris entrapment.

Furthermore, the **skewness**

$$R_{sk} = \frac{1}{R_q^3 L} \int_0^L z(x)^3 \,\mathrm{d}z = \frac{1}{R_q^3 L} \sum_1^N z_i^3 \tag{1.19}$$

and the **kurtosis**

$$R_{ku} = \frac{1}{R_q^4 L} \int_0^L z(x)^4 \,\mathrm{d}z = \frac{1}{R_q^4 L} \sum_1^N z_i^4, \tag{1.20}$$

which are the third and fourth statistical moments of the height distribution, respectively, can be employed to describe to shape of the distribution function plot [19]. Thereby, the skewness refers to the degree of asymmetry in the distribution of height values or elevations across the surface. A positively skewed surface would have a longer tail on the bottom side when looking at a profile plot, indicating that there are more deeper valleys and fewer high peaks (Figure 1.15). In contrast, a negatively skewed surface would have a longer tail on the upper side, implying more high peaks and fewer deep valleys. A perfectly symmetric rough surface would have a skewness close to zero, corresponding to a Gaussian normal distribution.

The kurtosis, in turn, indicates the concentration of extreme values (both high and low) in comparison to a normal distribution. A high kurtosis implies that the surface has more relatively sharp peaks or valleys in its profile. A low kurtosis indicates a

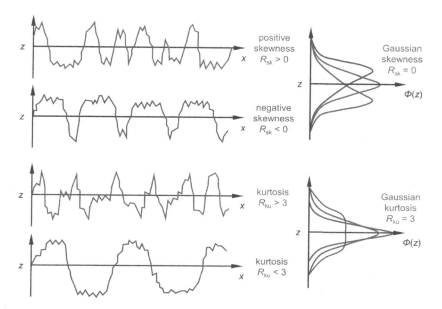

FIGURE 1.15 Schematic illustrations of line profiles (left) and corresponding probability density functions (right) for surfaces with different skewness and kurtosis values.

flatter distribution with broader peaks and valleys. A normal distribution has a kurtosis of 3, and any deviation from this indicates the degree of deviation from a normal distribution.

Moreover, the **autocorrelation function** characterizes the spatial arrangement of surface features by measuring the similarity between a given point on the surface and other points at varying distances. It quantifies how the height of the surface at one point correlates with the height at other points. Mathematically, the autocorrelation function for a shift l across the x-direction is the integration of the product of surface height by its shifted form over the entire domain:

$$R(l) = \lim_{L \to \infty} \frac{1}{L} \int_0^L \left[z(x)z(x+l) \right] dx = \frac{1}{N - N_1} \sum_1^{N-N_1} \left[z_i(x) z_i(x+l) \right], \quad (1.21)$$

with N_1 being the amount of discrete data within l. Apparently, the autocorrelation function becomes R_q^2 when $l = 0$. Therefore, the autocorrelation function can also be normalized by the square of the RMS roughness:

$$r(l) = \frac{R(l)}{R_q^2}, \quad (1.22)$$

becoming one in the case of $l = 0$ [19].

Through mechanical processing of surfaces, a preferential roughness orientation often emerges, which holds particularly true for turning or milling grooves. In such cases, these surfaces are referred to as anisotropic. The dimensionless **Peklenik number**

$$\gamma = \frac{l_x^*}{l_y^*} \qquad (1.23)$$

can be used to quantify the spatial alignment of surface texture and represents the ratio of the characteristic correlation lengths along two mutually orthogonal measurement directions l_x^* and l_y^* [20]. The correlation lengths tell us how far apart two measurement points need to be before the roughness at those points becomes essentially uncorrelated or independent of each other. In other words, if the distance between two points is much larger than the correlation length, the roughness values at those points are likely to be unrelated. In the case of $\gamma > 1$, the roughness features have a higher degree of correlation along the longitudinal measurement direction (x-axis), while $\gamma < 1$ implies a higher degree of correlation along the transversal measurement direction (y-axis). In practical terms, this signifies that the surface exhibits anisotropy, and the roughness characteristics are more pronounced in one direction as exemplarily demonstrated in Figure 1.16. This might occur, for instance, in surfaces that have been machined or processed using techniques that create grooves or patterns aligned in a certain direction. The anisotropy can be desired in certain applications since it affects material properties, friction, wear, and how light or fluids interact with the surface. In the case of $\gamma = 1$, the surface roughness features are equally correlated in both orthogonal measurement directions (Figure 1.16). This implies isotropy, meaning there is no preferential alignment of roughness features along any particular direction. Isotropic surfaces are often preferential in applications where uniformity of friction, wear resistance, or other material properties is important. Many naturally occurring surfaces tend to be isotropic.

Actually, tribological processes frequently modify the surface topography of both involved rubbing surfaces. For instance, during running-in, surfaces can become smoother, or excessive wear and pitting caused by contact fatigue can make them even rougher (see Section 1.8). Moreover, the approaches introduced so far can, however, still not cover the **multiscale** nature of surface roughness, i.e., the presence of distinct features and variations in surface topography at different length scales. To this end, fractal analysis characterizes the self-similarity and complexity of rough surfaces across scales as you zoom in or out. Higher fractal dimensions indicate rougher surfaces with more intricate features. By Fourier transformation, the surface profile can be broken into its frequency components (frequency decomposition), which helps to identify different spatial wavelengths present in the roughness. A power spectrum reveals

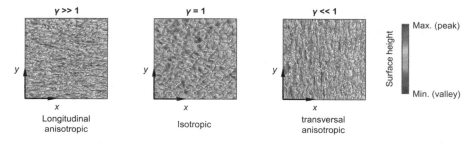

FIGURE 1.16 Various surface roughness orientations in dependency of the Peklenik number.

the contribution of each frequency component, helping to understand the distribution of features across scales. Wavelet transform can be used for analyzing localized features as it breaks down the surface profile into different wavelet functions with varying scales, allowing to identify features that are prominent at specific scales while providing both frequency and spatial information. However, these methods are complex in implementation and beyond the scope of this book.

1.5 TRIBO-CONTACTS AND CONTACT MECHANICS

The contact between the main body and the counter-body of a tribological system encompasses a wide array of scenarios and the principles of contact mechanics govern the performance, longevity, and efficiency of numerous systems encountered in both everyday life and industrial applications. In the following chapters, we will present and summarize the theoretical foundations to understand tribo-contact mechanics.

1.5.1 TYPES OF CONTACTS

A contact generally occurs when the surfaces of two bodies come into physical proximity. The area, where this touch occurs, is referred to as the contact zone or **contact area**. This contact is typically induced by external forces acting on the contacting partners. According to Newton's third law, this leads to reaction forces on the contact area, which manifest as distributed contact loads on the contact surface. A distributed load acting perpendicular to the contact surface is also termed **surface pressure**. In the case of elastic bodies, these surface pressures result in a displacement of the surfaces, leading to **strains and deformations** within the bodies. According to the extended Hooke's law, this generates a **multi-axial stress state** within the bodies in contact.

Depending on the **direction** of the **reaction forces** on the contact surface, we distinguish between normal and tangential contacts as outlined in Figure 1.17. **Normal contacts** are characterized by contact forces acting solely perpendicular (normal) to the contact surface. Under the assumption of frictionless contact surfaces, normal contacts pertain to static problems, as relative motion between the contacting partners always triggers tangential force components. **Tangential contacts**, in addition

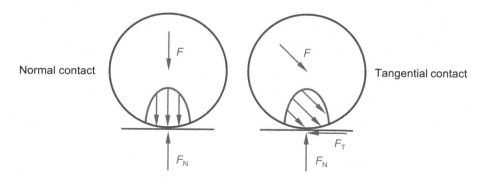

FIGURE 1.17 Classification of contacts according to the force conditions into normal and tangential contacts.

to normal forces, exhibit contact forces parallel (tangential) to the contact surface, which can include frictional or shear forces. Dynamic contacts between bodies with differing relative velocities typically fall under the category of tangential contacts.

Depending on the **geometry** of the contacting bodies, i.e., how well the contacting surfaces match, contact mechanics distinguish between conformal and non-conformal contacts (Figure 1.18). **Conformal contacts** are recognized by the fact that the surface geometries fit together, resulting in a relatively large, shared nominal contact area. Contacts between a shaft and a ring/bore are typical examples of conformal contacts. **Non-conformal contacts**, also known as concentrated or counter-formal contacts, encompass all types of contacts with a small, locally limited contact zone. Typical examples can be found in the contact between two gear teeth or rolling elements. Differently shaped contact areas are formed based on the contour of the surfaces of the contacting partners. Essentially, three different types of contacts can be distinguished:

- **Area contacts** represent conformal contacts and emerge when two flat surfaces come into contact, such as between a block and a plane or between two conformally curved bodies with identical radii of curvature, as experienced between the outer bearing ring and its housing block.
- **Point contacts** are a form of non-conformal contact and arise when a sphere, tip, or corner makes stress-free contact with a counter-body, such as another sphere, a cylinder, or a plane.
- **Line contacts** are also a form of non-conformal contact and occur when a rotationally symmetrical body or an edge makes stress-free contact with another rotationally symmetrical counter-body or a flat surface.

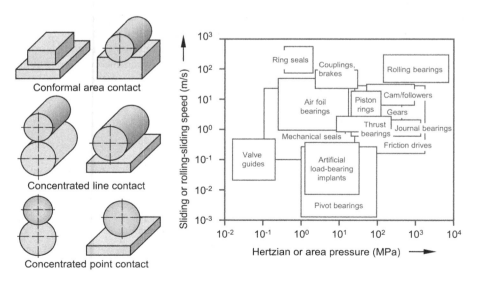

FIGURE 1.18 Classification of contacts into conformal area, concentrated line, and point contacts as well as characteristic dimensional ranges of velocities and pressures in various tribological systems. (Adapted and extended from Refs. [15,17].)

In conformal contacts, such as thrust slider or journal bearings, relatively low nominal contact pressures up to a few tens of MPa are present due to the relatively large contact areas, and the contacting bodies are frequently considered rigid. Strictly speaking, in the case of point and line contacts, they are only ideal point or line contacts when there are no applied loads. Nonetheless, the term "point contact" remains applicable as long as the radius of the contact circle is small compared to the sphere's radius. In this context, a "line contact" is considered for contacts, for which the width of the contact area is small compared to its length. In the general case of surfaces with all-sided curvature pairs, the elastically deformed contact areas can be represented by an ellipse, which transforms into an extremal strip (line contact, curvature in only one spatial axis) or a circle (point contact, curvature direction-dependent). As observed for cam-follower arrangements, gear teeth contacts or rolling bearings, nominal pressures of several hundred MPa or even a few GPa can be reached, resulting in localized elastic deformations of the contacting partners. For these elastic deformations, significantly lower contact pressures are sufficient, especially when one of the contacting partners is relatively soft, such as in seals or artificial knee implants.

Contacts can also be distinguished by means of their surfaces:

- **Smooth contacts** designate contacts of two bodies with ideally smooth surfaces, which are more theoretical in nature (see Section 1.4).
- **Rough contacts** are contacts between contact partners with rough surfaces, which also include finely machined surfaces, which are produced by hollowing or lapping.

The **material properties** of the involved contact partners play also a role in categorizing various types of contacts (Figure 1.19):

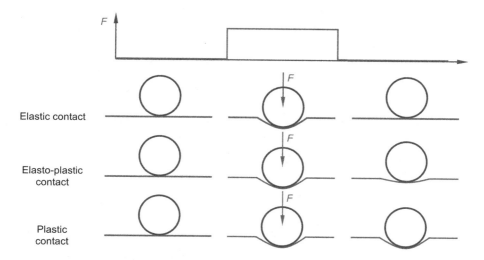

FIGURE 1.19 Classification of contacts according to the material behavior into elastic, elasto-plastic, or plastic contacts. Please note that here only one body is assumed to deform, while the counter-body remains rigid. In reality, frequently both bodies will undergo deformation.

- **Elastic contacts** occur whenever the strains resulting from contact forces completely disappear upon unloading. A distinction is made between linear and non-linear elasticity, while the former can be described by Hooke's law. A typical elastic contact connects with the interaction between two bodies made of rubber or steel under moderate stress. Rubber displays a non-linear-elastic behavior, while steel exhibits a linear-elastic behavior.
- **Plastic contacts** exhibit deformation of the contact partners under load, which remains after unloading. The relationship between strains and stresses cannot be described by Hooke's law. The increase in stresses within the material in relation to strains generally occurs more gradually than in elastic contacts and cannot become arbitrarily large. An example of a plastic contact is the interaction between two bodies made of modeling clay.
- **Elasto-plastic contacts** represent a combination of elastic and plastic contacts. In this case, the material of the bodies initially exhibits a purely elastic behavior, often following a linear relationship between strains and stresses according to Hooke's law. After surpassing a defined stress value known as the yield point, the material begins to behave plastically. From this point onward, stresses increase only moderately with increasing strain. Upon unloading, only the elastic portion of the strains reverts, while the plastic portion remains. Elasto-plastic contacts arise, for instance, in contact partners made of steel under high-stress conditions.
- **Rigid contacts** do not show any deformation, regardless of the load. However, this material behavior is not observed in practice and is only of theoretical significance.

Furthermore, contacts can be categorized based on their **kinematics** into the following four types (see also Figure 1.20):

- **Sliding contacts** occur whenever a pure sliding motion is present in the contact area. An important prerequisite is that the velocity vector of the rolling element remains parallel to the contact surface. The relative motion between both bodies can thus occur in translational, rotational, or a combination of both modes of motion. Sliding contacts are fundamentally possible between bodies of any geometry.
- **Spinning contacts** represent a special form of sliding contacts, where at least one body undergoes a purely rotational relative motion around its own axis, which is perpendicular to the contact surface.

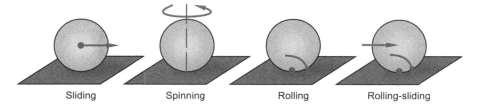

| Sliding | Spinning | Rolling | Rolling-sliding |

FIGURE 1.20 Classification of contacts according to their kinematics into sliding, spinning, rolling, and rolling-sliding contacts.

- **Rolling contacts** refer to contacts between bodies, where the condition of pure rolling is met during their motion, i.e., the body has no relative velocity to the counter-body at the point of contact and covers a distance equal to its circumference during the rotation. Rolling contacts are essentially possible only when at least one body is singly or doubly curved.
- **Rolling-sliding contacts** arise when the relatively moving body undergoes a rolling motion on the counter-body, while simultaneously experiencing a macro- or microscopic sliding motion. Therefore, for rolling-sliding contacts, it is also necessary that at least one body is singly or doubly curved. The ratio between rolling and sliding components can vary freely.

Characteristic values for the conditions of movement are often related to the circumferential speeds u_1 (base body) and u_2 (counter-body). Thereby, the **sum velocity**

$$u_s = |u_1 + u_2|,$$ (1.24)

the **relative velocity**

$$u_r = |u_1 - u_2|,$$ (1.25)

the **mean** or **effective velocity**

$$u_m = \frac{u_1 + u_2}{2},$$ (1.26)

the **slip**

$$s = 2\frac{u_1 - u_2}{u_1 + u_2},$$ (1.27)

as well as the **slide-to-roll ratio**

$$SRR = 2\frac{|u_1 - u_2|}{u_1 + u_2}$$ (1.28)

are of particular importance. The latter two refer to the ratio between translational/ sliding and rotational/rolling components of motion. For pure sliding contacts, which implies that one body has a surface velocity of zero, the absolute value of the slide-to-roll becomes $|SRR| = 2$. For opposing sliding velocities, slip conditions can occur in the range $2 < |SRR| < \infty$. For pure rolling contacts, SRR equals to 0, and for rolling-sliding contacts with the same direction of motion, $0 < |SRR| < 2$. The temporal evolution of motion can either be continuous or intermittent. While model experiments (see Chapter 2) often focus on constant motion conditions, real-world tribological applications frequently involve temporally variable motion conditions.

EXERCISE 1.3

Two bodies 1 and 2, which are assumed to be rigid, are contacting and moving relatively to each other according to the following sketches:

1.
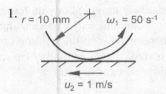

2.
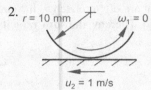

3.
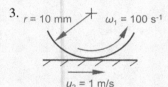

4.
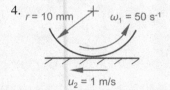

Determine:

- Sum velocity
- Relative velocity
- Slide-to-roll ratio

What type of kinematic contacts are these?

Regarding lubrication, contacts can be categorized into two groups:

- **Dry contacts** involve interactions between two solid bodies without the presence of an intermediate substance.
- **Lubricated contacts** have an intermediate substance (e.g., lubricating oils, greases or solid lubricants; see Section 1.6, Chapters 3 and 4 for more details) between the two solid contact partners to separate the surfaces as well as to reduce friction and wear.

1.5.2 NORMAL CONTACTS (HERTZIAN THEORY)

For the calculation of pressures in normal contacts, an approximation based on the force per unit area

$$p = \frac{F}{A} \tag{1.29}$$

or force per projected area

$$p = \frac{F}{2Rl} \tag{1.30}$$

might be accurate enough for parallel and conformal contacts, respectively (Figure 1.21). However, this approach represents an oversimplification of concentrated contacts.

A satisfactory mathematical formulation for various non-conformal contact problems was first postulated by Heinrich Hertz [21] in 1881 and has been repeatedly confirmed in its accuracy since then. However, the mathematical derivation of the theory results in a series of assumptions that must be at least approximately fulfilled:

- The material of both contact partners is isotropic, homogeneous, and behaves purely elastically.
- The contact being calculated is a dry normal contact.
- The surfaces of the contact partners are ideally smooth and frictionless.
- Both contact partners are free from residual stresses.
- The contact area should be as flat as possible and small in its dimensions compared to the radii of curvature of the contact bodies.

EXERCISE 1.4

Is the Hertzian theory applicable to the contact between two gear teeth?

In general, the **Hertzian theory** deals with contacts between two surfaces that are curved on all sides and pressed against each other with a force F (Figure 1.22). The curvature of a surface can be described by its maximum and minimum curvature radii r_{ij}, which lie in two mutually perpendicular planes (principal curvature planes). The curvature radii are positive for convex (center of curvature inside the body) and

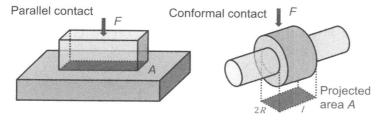

FIGURE 1.21   (Projected) contact area in parallel and conformal contacts.

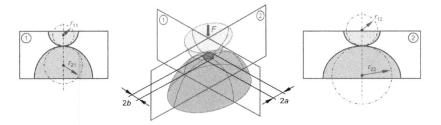

FIGURE 1.22   Concentrated contact between two ellipsoids (center) as well as principles planes of curvature (left and right). (Redrawn and adapted from Ref. [22].)

negative for concave surfaces (center of curvature inside the body), see Figure 1.22. The smaller the curvature radius is, the more pronounced the curvature of the surface is. The principal **curvatures** ρ_{ij} along these planes can be described by

$$\rho_{ij} = \frac{1}{r_{ij}} \text{ for } i,j = 1,2. \tag{1.31}$$

EXERCISE 1.5

Which basic geometries are behind the following paired curvatures?

1. $\rho_{11} = \rho_{12}$ and $\rho_{21} = \rho_{22}$
2. $\rho_{11} = \rho_{12}$ and $\rho_{21} = \rho_{22} = 0$
3. $\rho_{11} = 0$ and $\rho_{21} = \rho_{22} = 0$
4. $\rho_{11} = 0$ and $\rho_{21} = 0$

The size of the elliptical contact zone in contacts between surfaces curved on all sides, and the resulting contact pressures, stresses, and deformations strongly depend on the respective principal curvatures ρ_{ij} of the bodies. Therefore, Hertz defined an auxiliary value

$$\cos \tau = \frac{\rho_{11} - \rho_{12} + \rho_{21} - \rho_{22}}{\rho^*}, \tag{1.32}$$

with the cumulative curvature

$$\rho^* = \sum_{i,j=1}^{2} \rho_{ij} \tag{1.33}$$

Furthermore, Hertz introduced the auxiliary values ξ, η, and ψ, which represent transcendental functions of the auxiliary value $\cos \tau$, involve elliptic integrals, and are necessary to determine the semi-axes of the contact ellipse. A variety of values are provided in Table 1.4. In the simplest case ($\cos \tau = 0$), the coefficients are $\xi = \eta = \psi = 1$. For approximation of the coefficients featuring errors below 1% and without the necessity to solve the elliptical integrals, two Hertzian coefficients can be estimated from

$$\ln(\xi) \approx \frac{\ln(1-\cos\tau)}{-1.53 + 0.333 \cdot \ln(1-\cos\tau) + 0.0467 \cdot \ln(1-\cos\tau)^2}, \tag{1.34}$$

$$\ln(\eta) \approx \frac{\ln(1-\cos\tau)}{1.525 - 0.86 \cdot \ln(1-\cos\tau) - 0.0933 \cdot \ln(1-\cos\tau)^2} \tag{1.35}$$

for $\cos \tau < 0.949$ and from

$$\ln(\xi) \approx \sqrt{-0.4567 - 0.4446 \cdot \ln(1-\cos\tau) + 0.1238 \cdot \ln(1-\cos\tau)^2}, \tag{1.36}$$

$$\ln(\eta) \approx -0.333 + 0.2037 \cdot \ln(1-\cos\tau) + 0.0012 \cdot \ln(1-\cos\tau)^2 \tag{1.37}$$

for $0.949 < \cos \tau < 1$ [23], while the last coefficient depends on the former [24]:

$$\psi = \frac{\eta^2 + \xi^2}{2} + \cos\tau \frac{\eta^2 - \xi^2}{2}. \tag{1.38}$$

TABLE 1.4
Hertzian Coefficients following [22]

$\cos\tau$	ξ	η	ψ	$\cos\tau$	ξ	η	ψ	$\cos\tau$	ξ	η	ψ
0.9995	23.95	0.163	0.171	0.9770	5.63	0.338	0.476	0.9280	3.55	0.428	0.630
0.9990	18.53	0.185	0.207	0.9765	5.58	0.339	0.478	0.9260	3.51	0.431	0.634
0.9985	15.77	0.201	0.230	0.9760	5.53	0.34	0.481	0.9240	3.47	0.433	0.638
0.9980	14.25	0.212	0.249	0.9755	5.49	0.342	0.483	0.9220	3.43	0.436	0.642
0.9975	13.15	0.022	0.266	0.9750	5.44	0.343	0.486	0.9200	3.40	0.438	0.646
0.9970	12.26	0.228	0.279	0.9745	5.39	0.345	0.489	0.9180	3.36	0.441	0.650
0.9965	11.58	0.235	0.291	0.9740	5.35	0.346	0.491	0.9160	3.33	0.443	0.653
0.9960	11.02	0.241	0.302	0.9735	5.32	0.347	0.493	0.9140	3.30	0.445	0.657
0.9955	10.53	0.246	0.311	0.9730	5.28	0.349	0.495	0.9120	3.27	0.448	0.660
0.9950	10.15	0.251	0.320	0.9725	5.24	0.35	0.498	0.9100	3.23	0.45	0.664
0.9945	9.77	0.256	0.328	0.9720	5.20	0.351	0.500	0.9080	3.20	0.452	0.667
0.9940	9.46	0.26	0.336	0.9715	5.16	0.353	0.502	0.9060	3.17	0.454	0.671
0.9935	9.17	0.264	0.343	0.9710	5.13	0.354	0.505	0.9040	3.15	0.456	0.674
0.9930	8.92	0.268	0.350	0.9705	5.09	3.555	0.507	0.9020	3.12	0.459	0.677
0.9925	8.68	0.271	0.356	0.9700	5.05	0.357	0.509	0.9000	3.09	0.461	0.680
0.9920	8.47	0.275	0.362	0.9690	4.98	0.359	0.513	0.8950	3.03	0.466	0.688

(Continued)

TABLE 1.4 (*Continued*)
Hertzian Coefficients following [22]

cosτ	ξ	η	ψ	cosτ	ξ	η	ψ	cosτ	ξ	η	ψ
0.9915	8.27	0.278	0.368	0.9680	4.92	0.361	0.518	0.8900	2.97	0.471	0.695
0.9910	8.10	0.281	0.373	0.9670	4.86	0.363	0.522	0.8850	2.92	0.476	0.702
0.9905	7.93	0.284	0.379	0.9660	4.81	0.365	0.526	0.8800	2.86	0.481	0.709
0.9900	7.76	0.287	0.384	0.9650	4.76	0.367	0.530	0.8750	2.82	0.485	0.715
0.9895	7.62	0.289	0.388	0.9640	4.70	0.369	0.533	0.8700	2.77	0.49	0.721
0.9890	7.49	0.292	0.393	0.963	4.65	0.371	0.536	0.8650	2.72	0.494	0.727
0.9885	7.37	0.294	0.398	0.962	4.61	0.374	0.540	0.8600	2.68	0.498	0.733
0.9880	7.25	0.297	0.402	0.961	4.56	0.376	0.543	0.8550	2.64	0.502	0.739
0.9875	7.13	0.299	0.407	0.96	4.51	0.378	0.546	0.8500	2.60	0.507	0.745
0.9870	7.02	0.301	0.411	0.959	4.47	0.38	0.550	0.8400	2.53	0.515	0.755
0.9865	6.93	0.303	0.416	0.958	4.42	0.382	0.553	0.8300	2.46	0.523	0.765
0.9860	6.84	0.305	0.420	0.957	4.38	0.384	0.556	0.8200	2.40	0.53	0.774
0.9855	6.74	0.307	0.423	0.956	4.34	0.386	0.559	0.8100	2.35	0.537	0.783
0.9850	6.64	0.31	0.427	0.955	4.30	0.388	0.562	0.8000	2.30	0.544	0.792
0.9845	6.55	0.312	0.430	0.954	4.26	0.39	0.565	0.7500	2.07	0.577	0.829
0.9840	6.47	0.314	0.433	0.953	4.22	0.391	0.568	0.7000	1.91	0.607	0.859
0.9835	6.40	0.316	0.437	0.952	4.19	0.393	0.571	0.6500	1.77	0.637	0.884
0.9830	6.33	0.317	0.440	0.951	4.15	0.394	0.574	0.6000	1.66	0.664	0.904
0.9825	6.26	0.319	0.444	0.95	4.12	0.396	0.577	0.5500	1.57	0.69	0.922
0.9820	6.19	0.321	0.447	0.948	4.05	0.699	0.583	0.5000	1.48	0.718	0.938
0.9815	6.12	0.323	0.450	0.946	3.99	0.403	0.588	0.4500	1.41	0.745	0.951
0.9810	6.06	0.325	0.453	0.944	3.94	0.406	0.593	0.4000	1.35	0.771	0.962
0.9805	6.00	0.327	0.456	0.942	3.88	0.409	0.598	0.3500	1.29	0.796	0.971
0.9800	5.94	0.328	0.459	0.94	3.83	0.412	0.603	0.3000	1.24	0.825	0.979
0.9795	5.89	0.33	0.462	0.938	3.78	0.415	0.508	0.2500	1.19	0.85	0.986
0.9790	5.83	0.332	0.465	0.936	3.73	0.418	0.613	0.2000	1.15	0.879	0.991
0.9785	5.78	0.333	0.468	0.934	3.68	0.42	0.618	0.1500	1.11	0.908	0.994
0.9780	5.72	0.335	0.470	0.932	3.63	0.423	0.622	0.1000	1.07	0.938	0.997
0.9775	5.67	0.336	0.473	0.93	3.59	0.426	0.626	0.0500	1.03	0.969	0.999

The behavior of two bodies in contact is influenced by their material properties. For instance, soft bodies deform much more under an applied load than hard bodies, which affects the maximum pressure, the size of the contact area, or the stresses beneath the surfaces. In this case, to calculate contacts between bodies of different materials, it is essential to consider the **reduced elastic modulus**

$$E' = \frac{\dfrac{1 - v_1^2}{E_1} + \dfrac{1 - v_2^2}{E_2}}{2} , \tag{1.39}$$

which considers the elastic properties of both materials. In the case where both contact partners have the same material properties, the equation simplifies to

$$E' = \frac{1-v^2}{E}. \tag{1.40}$$

Quite frequently one can also find the **equivalent elastic modulus** in literature:

$$E^* = \frac{1}{2E'}, \tag{1.41}$$

which (in contrast to the reduced elastic modulus) gives the same unit as the conventional Young's modulus of a material (see Section 1.3).

The shape of the contact area depends mainly on the geometry of the contact partners. For point contacts between two all-around curved bodies, an elliptical contact area is formed, which can be described using the major semi-axis a and the minor semi-axis b. The **contact area** of the resulting ellipse can then be calculated as follows:

$$A = a \cdot b \cdot \pi. \tag{1.42}$$

If the principal curvature planes, which include the minimum and maximum curvature radii, are identical for both bodies, the **semi-axes** are given as follows:

$$a = \xi \cdot \sqrt[3]{\frac{3 \cdot F \cdot E'}{\rho^*}}, \tag{1.43}$$

$$b = \eta \cdot \sqrt[3]{\frac{3 \cdot F \cdot E'}{\rho^*}}. \tag{1.44}$$

Thereby, the **pressure** acting on the contact area is not constant but can be described for any point $P(x,y)$ by

$$p(x,y) = f(x) = \begin{cases} p_{max} \cdot \sqrt{1-\left(\frac{x}{a}\right)^2 - \left(\frac{y}{b}\right)^2}, & \text{for } P(x,y) \in A \\ 0, & \text{for } P(x,y) \notin A \end{cases}, \tag{1.45}$$

with the **maximum (Hertzian) pressure**

$$p_{max} = \frac{3 \cdot F}{2 \cdot a \cdot b \cdot \pi} = \frac{\sqrt[3]{\frac{3 \cdot F \cdot \rho^{*2}}{E'^2}}}{2 \cdot \xi \cdot \eta \cdot \pi}. \tag{1.46}$$

The **elastic deformation**

$$\delta = \psi \cdot \sqrt[3]{\frac{9 \cdot F^2 \cdot E'^2}{8}} \tag{1.47}$$

describes the degree both bodies are approaching each other due to the applied loads. It represents the total indentation of both bodies at the contact point in load direction. Note that all aforementioned equations can be adapted to include the equivalent Young's modulus instead of the reduced one.

EXERCISE 1.6

Let's assume the contact between a sphere with diameter $d_s = 10$ mm and the grooved raceway on the outer jacket surface of a ball bearing's inner ring:

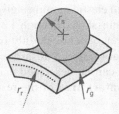

Both sphere and ring are made from steel ($E = 210$ GPa, $\nu = 0.3$). At the point of contact, the ring has a radius of $r_r = 100$ mm and its groove has a radius of $r_g = -6.2$ mm. The contact is loaded with 100 N.

- Calculate the size of the resulting contact area as well as the maximum Hertzian pressure.
- How will the contact area and contact pressure change if the groove radius is reduced to -5.2 mm?

For the **contact between two spheres** of diameter d_1 and d_2, the aforementioned equations simplify to

$$a = b = \sqrt[3]{\frac{3 \cdot F \cdot E'}{4\left(\dfrac{1}{d_1} + \dfrac{1}{d_2}\right)}} \, , \tag{1.48}$$

$$p_{\max} = \frac{1}{\pi} \sqrt[3]{\frac{6 \cdot F}{E'^2}\left(\frac{1}{d_1} + \frac{1}{d_2}\right)^2} \, , \tag{1.49}$$

$$\delta = \sqrt[3]{\frac{9 \cdot F^2 \cdot E'^2}{2}\left(\frac{1}{d_1} + \frac{1}{d_1}\right)} \, , \tag{1.50}$$

whereby the sphere-on-sphere contact with radii $r_1 = 0.5 \cdot d_1$ and $r_2 = 0.5 \cdot d_2$ can be converted into an equivalent sphere-on-plane contact using the equivalent radius

$$R = \frac{1}{r_1} + \frac{1}{r_2} \, . \tag{1.51}$$

For the **contact between a sphere** of radius R **and a plane**, the equations become

$$a = b = \sqrt[3]{\frac{3 \cdot F \cdot E' \cdot R}{2}}, \tag{1.52}$$

$$p_{max} = \frac{1}{\pi} \sqrt[3]{\frac{6 \cdot F}{4 \cdot E'^2} \left(\frac{1}{R}\right)^2}, \tag{1.53}$$

$$\delta = \frac{3.97 \cdot F^{0.9}}{10^5 \cdot l^{0.8}} \tag{1.54}$$

Both sphere-on-sphere and sphere-on-plane yield perfect point contacts (circular-shaped contact area), which are rarely found in actual applications. However, the majority of fundamental tribological experiments are based on ball-on-plane experiments (see Section 2.1.3). The aforementioned expressions nicely demonstrate that:

- If $F\uparrow$, then $a\uparrow$, $p_{max}\uparrow$ and $\delta\uparrow$
- If $R\uparrow$, then $a\uparrow$, $p_{max}\downarrow$, and $\delta\downarrow$
- If $E\uparrow$, i.e. E'$\downarrow$, then $a\downarrow$, $p_{max}\uparrow$ and $\delta\downarrow$

EXERCISE 1.7

Two steel spheres ($E = 210\,GPa$, $v = 0.3$) are pressed against each other with the force 10 N. The radius of both spheres is 10 mm.

- Find the radius of the resulting contact area as well as the maximum Hertzian pressure.
- How will the contact radius and contact pressure change if the radius of the spheres is reduced by a factor of 2 (to 5 mm)?

EXERCISE 1.8

Adapt/simplify the aforementioned formulas to calculate the dimensions of the contact area and maximum pressure for the following pairings:

1. Cylinder with length l and diameter d on plane.
2. Cylinder on cylinder with length l and diameters d_1 and d_2, respectively.

The elastic deformations of line contacts cannot be calculated with the Hertzian theory. Using an empirically derived method by Lundberg [25], the deformation for the contact between a cylindrical steel body and a flat surface can be approximated as

$$\delta = \frac{3.97 \cdot F^{0.9}}{10^5 \cdot l^{0.8}} \tag{1.55}$$

Please note that the elastic deformation in this case does not depend on the curvature of the body. This approach is well documented and aligns with modern numerical simulations, which is why it has also found its way into international standardization (e.g., in the rolling bearing industry).

When considering the contact between two parallel cylinders, the elastic deformation that occurs can only be approximated using the following empirically determined equation [26]:

$$\delta = \frac{2 \cdot F}{\pi \cdot l}\left[\frac{1-v_1^2}{E_1}\left(\ln\frac{d_1}{b} + 0.407\right) + \frac{1-v_2^2}{E_2}\left(\ln\frac{d_2}{b} + 0.407\right)\right]. \tag{1.56}$$

The pressures determined according to the Hertzian theory only reflect the normal stress state at the surface of the bodies. However, fatigue damage typically occurs underneath the surface and relates to **sub-surface stresses**. For point contacts (sphere-on-plane), the stress component acting parallel to the external load can be calculated as follows [22]:

$$\sigma_z = \frac{p_{max}}{\left(\dfrac{z}{a}\right)^2 + 1}. \tag{1.57}$$

Please note that in the case of $z = 0$, we obtain $\sigma_z = p_{max}$. The stress component acting parallel to the contact surface in all radial directions (assuming rotational symmetry) can be determined using

$$\sigma_r = p_{max}\left\{\frac{1}{2\left(\dfrac{z}{a}\right)^2 + 1} - (1+v)\left[1 - \frac{z}{a}\tan^{-1}\left(\frac{a}{z}\right)\right]\right\}, \tag{1.58}$$

while the principal shear stress can be calculated using Mohr's stress circle or the relationship $(\sigma_z - \sigma_r)/2$ [22]:

$$\tau = p_{max}\left\{\frac{(1+v)}{2}\left[1 - \frac{z}{a}\tan^{-1}\left(\frac{a}{z}\right)\right] - \frac{3}{4\left(\dfrac{z}{a}\right)^2 + 4}\right\}. \tag{1.59}$$

The stress distributions between a cylinder and a plane can be analytically determined below the central axis of the contact zone ($x = 0$). Since theoretically it is an infinitely long cylinder, it can be assumed that $\sigma_y = 0$. Therefore, a plane stress state is established, while allowing the determination of the principal stress in normal depth direction based on

$$\sigma_z = \frac{p_{max}}{\sqrt{\left(\dfrac{z}{b}\right)^2 + 1}}. \tag{1.60}$$

This equation also yields $\sigma_z = p_{max}$ for $z = 0$. The stress component acting parallel to the contact surface is

$$\sigma_y = p_{max}\left[\frac{2z}{b} - \frac{1 + 2\left(\dfrac{z}{b}\right)^2}{\sqrt{\left(\dfrac{z}{b}\right)^2 + 1}}\right], \tag{1.61}$$

while the principal shear stress component becomes

$$\tau = p_{max}\left[\frac{z}{b} - \frac{\left(\dfrac{z}{b}\right)^2}{\sqrt{\left(\dfrac{z}{b}\right)^2 + 1}}\right]. \tag{1.62}$$

As can be seen from Figure 1.23, the two principal stress components, σ_z and $\sigma_{r,x}$, exhibit their maximum values at the surface and decrease with increasing depth z.

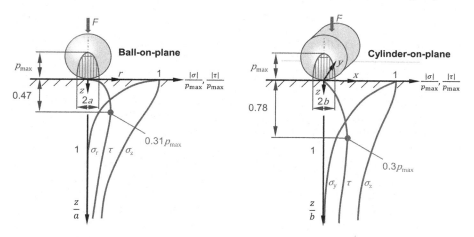

FIGURE 1.23 Evolution of sub-surface tensions in the center line of the contact for a ball-on-plane (left) as well as cylinder-on-plane contact (right). (Redrawn and adapted from Ref. [22].)

In contrast, the maximum orthogonal shear stress does not occur at the surface but rather at $0.47 \cdot b$ and at $0.78 \cdot b$ below the surface for point and line, respectively, whereby their maxima at these positions are $0.31 \cdot p_{max}$ and $0.30 \cdot p_{max}$, respectively [22]. This aligns well with studies on material fatigue, which indicate that cracks usually originate in the interior and then propagate toward the surface. In a frictionless sliding or rolling process of a sphere or cylinder over a plane (high-cycle loading conditions), stresses within the bodies increase, reaching their maximum value directly beneath the point of contact and then gradually decrease.

In the case of uniaxial tensile or compressive loading, it is experimentally feasible to determine the stress value at which yielding begins (yield strength σ_y, see Section 1.3). For pure shear stressing, experiments indicate that the material begins to yield already at shear stresses substantially below the yield strength. However, multi-axial stress states are challenging, since the estimation of yield strength for each individual component of the stress tensor would theoretically require an infinite number of experiments due to the infinite number of potential stress conditions. Therefore, a transfer function can be used to convert the multi-axial stress state into an equivalent uniaxial stress state, allowing to compare this with the material properties determined from uniaxial tests and providing insights into whether plastic yielding or fracture occurs under the respective stress conditions. To this end, various strength or stress hypotheses are used to determine the equivalent stress, which are mostly based on the plasticity theory. The validity of these hypotheses depends on the material of the component, the mode of its failure, the time course of the loading, the temperature of the component, among others. In practice, several **strength hypotheses** have proven their effectiveness:

- The **maximum shear stress (Tresca) criterion** can be directly derived from the mechanical concept that the material's stress can be determined by the largest occurring shear stress, which is half of the greatest principal normal stress difference [22]:

$$\sigma_{Tresca} = 2\tau_{max} = \sigma_{max} - \sigma_{min}, \tag{1.63}$$

which changes to the following for a plane stress condition:

$$\sigma_{Tresca} = \sqrt{\left(\sigma_x - \sigma_y\right)^2 + 4\tau_{xy}^2} . \tag{1.64}$$

Therefore, a hydrostatic stress state would not lead to failure since only the deviatoric part of the stress state is relevant to damage. The maximum shear stress hypothesis assesses failure by yielding, which occurs when the largest occurring shear stress reaches a critical value (in the case of static loading, the shear yield strength) and is generally applicable to ductile materials that tend to fail due to yielding or sliding fractures, while brittle materials exhibit deformation-free sliding fractures under compressive loading. According to the maximum shear stress hypothesis, the maximum equivalent stress within the material is $\sigma_{Tresca,max} = 0.6 \cdot p_{max}$ at a depth of $z = 0.78 \cdot b$ [22]. Furthermore, it is possible to consider the influence of residual stresses by superimposing them with the applied stresses and attributing plastic deformations to shearing caused by shear stresses. It has been shown that the maximum shear stress criterion is a good approximation to determine stresses in rolling contacts. However, other approaches might be more accurate and simpler in handling.

- The **alternating shear stress criterion** also describes failure in ductile materials through yielding or sliding fractures. However, it considers the most significant coordinate shear stress, τ_{xz}, acting in the plane parallel to the surface, as the most relevant parameter:

$$\sigma_{alt} = 2\left|\tau_{xy\ max}\right|. \tag{1.65}$$

When considering the rolling axis as the time axis in rolling contact, each material element at a certain depth below the surface undergoes the entire shear stress profile, which alternates between positive and negative values during rolling. The alternating shear stress criterion gives good agreement between the depth of pits occurring and the location of the maximum alternating shear stress in Hertzian contacts. This is why several standardized calculation methods, e.g., calculation of the life of rolling element bearings (DIN ISO 76, DIN ISO 281), are based on the alternating shear stress hypothesis. A drawback of using this hypothesis is that it cannot account for residual stresses or frictional influences. Moreover, certain changes in structural directions or location of fatigue cracks cannot be explained by this approach.

- The **maximum distortion energy (von Mises) criterion** can be physically interpreted in such a way that when a body is subjected to hydrostatic loading, it undergoes purely elastic deformation, changing only its volume and not its shape. Therefore, for the material to fail, the specific stored plastic deformation energy (not the volume change) must exceed a certain threshold. The term distortion energy refers to the work required to induce deformation through slip within the crystal lattice. In a generalized three-axial stress state, this can be expressed by

$$\sigma_{von\ Mises} = \sqrt{\frac{1}{2}\left[\left(\sigma_y - \sigma_x\right)^2 + \left(\sigma_x - \sigma_z\right)^2 + \left(\sigma_z - \sigma_y\right)^2\right]}, \tag{1.66}$$

while the following expression holds true for bi-axial stresses

$$\sigma_{von\ Mises} = \sqrt{\sigma_x^2 + \sigma_y^2 - \sigma_x\sigma_y + 3\tau_{xy}^2}. \tag{1.67}$$

The maximum distortion energy theory can generally be applied to ductile materials under both static and dynamic loading conditions as well as to estimate rolling fatigue in Hertzian contacts, demonstrating good agreement with experimental results. According to the maximum distortion energy hypothesis, the maximum equivalent stress within the material is $\sigma_{von\ Mises,max} = 0.56 \cdot p_{max}$ at a depth of $z = 0.71 \cdot b$ and many authors consider the von Mises theory as the most suitable approach to describe rolling fatigue [22]. The equivalent von Mises stress distribution below the surface of a frictionless, normal Hertzian contact is illustrated in Figure 1.24.

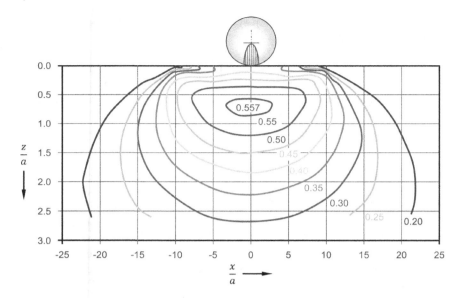

FIGURE 1.24 von Mises stress distribution normalized on the Hertzian pressure ($\sigma_{\text{von Mises}}/p_{\text{max}}$) for a frictionless normal contact.

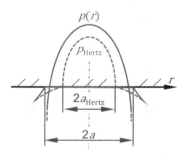

FIGURE 1.25 Deformation and pressure distribution in an adhesive normal contact compared to the Hertzian theory.

1.5.3 Adhesive Contacts

As indicated, the Hertzian theory covers several assumptions and neglects the forces of **adhesion** (Section 1.6.2). As such, for example, a rubber ball brought close to a surface is pulled spontaneously into contact in a "neck-shape," giving a pronounced contact area even with zero nominal normal force (Figure 1.25). As with Hertz, the contact area increases with force. However, to remove the rubber ball, we have to apply a negative force. The **Johnson-Kendall-Roberts (JKR) model** [27] takes into account the surface energy (Section 1.4.2) and the contact width can be approximated by

$$a = \sqrt[3]{\frac{3 \cdot R}{4 \cdot E^*}\left(F + 2 \cdot F_a + 2\sqrt{F_a^2 + F_a \cdot F}\right)}, \tag{1.68}$$

with the adhesive force

$$F_a = 3 \cdot \gamma \cdot \pi \cdot R. \tag{1.69}$$

This approach is particularly suitable for relatively large, flexible spheres where the deformation is dominated by elastic effects, i.e., soft samples with high adhesion, but neglecting long-range interactions outside the contact area.

In contrast, the **Derjaguin–Muller–Toporov (DMT) model** [28] assumes that the adhesive forces are primarily electrostatic and includes van der Waals' forces outside the contact area. This model approximates the contact width by

$$a = \sqrt[3]{\frac{3 \cdot R}{4 \cdot E^*}(F + F_a)}. \tag{1.70}$$

Both models are useful tools to understand adhesion in normal contacts. The choice between these models depends on the specific conditions of the contact, the materials involved, and the relative contributions of van der Waals' and electrostatic forces. In practice, the actual behavior of adhesive contacts often lies somewhere between these idealized models, and modifications or more advanced models may be needed to accurately describe complex adhesion phenomena [29]. These models have been instrumental in fields like microelectromechanical systems (MEMS), nanotechnology, and materials science, where adhesion plays a crucial role in device performance and reliability. For more information, the interested reader may refer to Ref. [6].

1.5.4 TANGENTIAL CONTACTS

Pure normal contacts are typically not encountered in tribological systems. In most cases, normal forces are accompanied by a component of tangential forces, leading to the presence of both normal and shear stresses within the contact zone. Assuming that the COF is constant within the contact region (Section 1.6), μ represents the ratio between the normal and tangential stress components

$$\mu = \frac{\tau_{zy}}{\sigma_z}. \tag{1.71}$$

This interplay between normal and tangential forces is a critical aspect in tribo-systems, impacting the performance and durability of concentrated contacts. The equivalent von Mises stress distribution below the surface of normal contact subjected to a COF of 0.25 is illustrated in Figure 1.26. Apparently, the location of the highest stress, i.e., the maximum equivalent stress, moves closer to the surface (Figure 1.27) and deviates from the symmetric position with increasing tangential stress component. Furthermore, the value of the maximum equivalent stress rises. Additionally, a second, initially local stress maximum forms directly at the surface. For COFs of about 0.3, the maximum of the equivalent stress reaches the surface as illustrated in Figure 1.27.

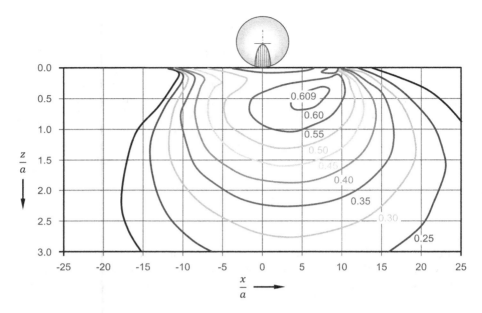

FIGURE 1.26   von Mises stress distribution normalized on the Hertzian pressure ($\sigma_{\text{von Mises}}/p_{\text{max}}$) for a tangential contact with a COF of 0.25.

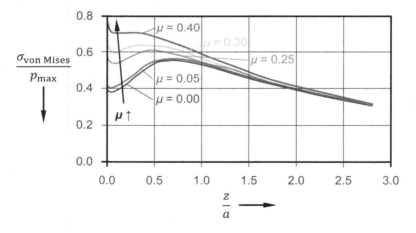

FIGURE 1.27   Shift of the von Mises stress distribution normalized on the Hertzian pressure ($\sigma_{\text{von Mises}}/p_{\text{max}}$) in the contact center dependent on the assumed COF.

1.5.5   Contacts with Residual Stresses

Residual stresses refer to internal stress components that remain within a material after the release of all external effects (load, temperature, among others). They arise from mechanical impact during manufacturing or local plastic deformations during operation, e.g., due to particle over-rolling or fatigue processes. Thermal effects induced by heat treatments also generate residual stresses. Residual stresses can occur in the immediate vicinity of foreign inclusions. The generation and occurrence of residual stresses in tribo-contacts cannot be entirely avoided due to the thermomechanical loading situation, nor can they be precisely predicted or easily considered in calculations. Residual stress fields overlay the applied load stress fields. Depending on whether these residual stresses are tensile or compressive, they can either increase or decrease the overall stress state. Generally, it is assumed that tensile residual stresses have a negative impact, while not excessively large compressive residual stresses may positively impact the overall stress condition, including the fatigue strength or fatigue life. For instance, shot peening (bombarding a material's surface with small spherical particles at high velocities) is employed to induce plastic deformation in the material, which introduces compressive sub-surface residual stresses.

1.5.6 Rough Contacts (Real Area of Contact)

The contact between two perfectly smooth surfaces represents an idealized scenario and is often employed in introductory contact mechanics. Thereby, the contact area is well-defined and can be calculated using analytical approaches introduced in Section 1.5.2 for simple geometries or using numerical simulations. However, in reality, smooth contact models fall short of representing the complexities introduced by surface roughness and asperities (Section 1.4.2). When two rough surfaces come into contact, the actual area of contact is much smaller than the **nominal contact area** that would be predicted by smooth contact models. This **real area of contact** (Figure 1.28) is influenced by the topography of the surfaces.

Modeling rough contacts requires considering the statistical distribution of asperities and their interactions can become complex due to necessity of an extremely fine discretization. Therefore, various stochastic approaches have been developed to calculate the real area of contact. In the most simplified form, a rough surface can be imagined as a regular pattern of half-sphere-shaped asperities with equal curvature and height. In this idealized scenario, the treatment of a contact problem is straightforward.

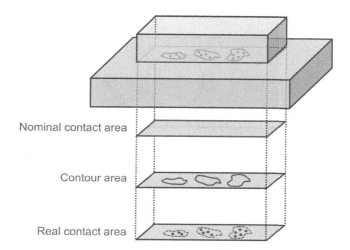

Nominal contact area

Contour area

Real contact area

FIGURE 1.28 From the nominal to the real contact area of two contacting surfaces. (Redrawn and adapted from Ref. [15].)

The total force results from summing all equal asperity forces, which can be calculated using Hertzian contact theory. Therefore, the individual "micro"-contact areas and the total contact area would scale proportionally to $F^{2/3}$. However, this contradicts empirical observations as well as Coulomb's law of friction (Section 1.6), from which one expects a more linear relationship between contact area and normal force. This changes when considering that real surfaces are typically stochastically rough.

1.5.6.1 Archard Model

In 1957, Archard [30] suggested a model for transitioning from a single Hertzian sphere-on-plane contact with radius R_1 (see Section 1.5.2) to a multiple asperity contact by introducing an additional roughness with the radius R_2 to the initial sphere as depicted in Figure 1.29. Assuming that the density of the additional radii is η_2, for any differential ring at distance r with the area $dA = 2 \cdot \pi \cdot r \cdot dr$, the number of asperities in contact is $dN_{2,r} = \eta_2 \cdot 2 \cdot \pi \cdot r \cdot dr$. Thus, the force carried by the area dA is

$$dF = p_{max} \cdot \sqrt{1 - \left(\frac{r}{a}\right)^2}\, 2\pi r dr, \tag{1.72}$$

and the load carried by each asperity is

$$F_{a,2} = \frac{dF}{dN_{2,r}} = p_{max} \cdot \frac{\sqrt{1 - \left(\frac{r}{a}\right)^2}}{\eta_2} = C_2 F^{\frac{1}{3}} \sqrt{1 - \left(\frac{r}{a}\right)^2}, \tag{1.73}$$

making use of the proportionality between p_{max} and $F^{1/3}$. In the following, C_i is the family of the constants used for simplicity of the formula deriving. From the Hertzian theory, we know that the contact for each asperity with a given radius R_2 is

$$A_{a,2} = C_3 F_{a,2}^{\frac{2}{3}} = C_4 F^{\frac{2}{9}} \left(1 - \left(\frac{r}{a}\right)^2\right)^{\frac{1}{3}}, \tag{1.74}$$

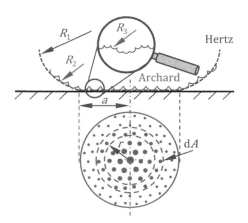

FIGURE 1.29　Archard model for the contact of a rough sphere with a smooth plane.

and the total contact area becomes

$$A_c = \int_0^a A_{a,2} \eta_2 \, 2\pi r \, dr = \int_0^a C_4 F^{\frac{2}{9}} \left(1 - \left(\frac{r}{a}\right)^2\right)^{\frac{1}{3}} \eta_2 \, 2\pi r \, dr. \tag{1.75}$$

By introducing $x = r/a$, we obtain $2 \cdot \pi \cdot r \cdot dr = 2 \cdot \pi \cdot a^2 \cdot x \, dx$ (remembering that $\pi a^2 = C_1 \cdot F^{2/3}$), the total contact area becomes

$$A_c = C_4 F^{\frac{2}{9}} \int_0^1 \left(1 - (x)^2\right)^{\frac{1}{3}} \eta_2 \, 2\pi a^2 x \, dx = C_5 F^{\frac{8}{9}}, \tag{1.76}$$

i.e., is proportional to $F^{8/9}$. By further roughening the sphere with asperities of even smaller radii, the total contact area becomes proportional to $F^{26/27}$, thus suggesting that each step of the contact refinement leads closer to a linear proportionality between the contact area and the applied force.

Since the effective contact area of the multitude micro-contacts is way smaller than the nominal contact area according to the Hertzian theory, local pressures substantially exceed the Hertzian pressure, which modifies sub-surface stress fields. However, the influence on the stress distribution by each asperity contact is very local and rapidly diminishes, while the "global" stress distribution corresponds to that of the Hertzian contact (Figure 1.24). Only toward the surface of the contact area, increasing and partially significant local deviations from the conditions of the smooth contact become apparent.

1.5.6.2 Greenwood–Williamson Model

Greenwood and Williamson [31] introduced a fundamental method for modeling irregular surfaces, today known as the **Greenwood–Williamson (GW) model**. Thereby, all asperities were assumed to have the same radius of curvature β and to be sufficiently spaced apart so that the deformations can be treated individually. However, their heights were stochastically distributed around an averaged value following the probability density function as illustrated in Figure 1.30.

Consequently, the surfaces are separated by the averaged distance h_0, and the overall contact is formed by the asperities exceeding this value ($z > h_0$). Hence, the actual contact area results from the accumulation of these individual asperity contacts, while the total applied load is the sum of the loads carried by each asperity

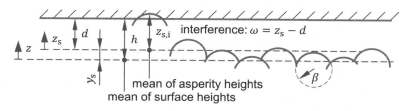

FIGURE 1.30 Simplified model of stochastic surfaces underlying the GW model. (Reprinted and adapted from Ref. [32].)

contact. The penetration depth of a single asperity at a height $d = z - h_0$ and the contact area of a single asperity is $A_a = \pi \cdot d \cdot \beta$. Following the Hertzian theory (Section 1.5.2) in slightly modified form, the single force of an asperity is

$$F_a = \frac{4}{3} \cdot E^* \cdot \beta^{\frac{1}{2}} d^{\frac{3}{2}}. \tag{1.77}$$

Using the asperity density η in the contact area A (the product of both yields the total number of asperities in the contact area) as well as integration between h_0 and ∞ over the probability density function $\Phi(z)$, we obtain the number of asperities in contact

$$N_c = \eta \cdot A \int_{h_0}^{\infty} \phi(z) \mathrm{d}z \tag{1.78}$$

and the total asperity contact area

$$A_c = \pi \cdot \beta \cdot \eta \cdot A \int_{h_0}^{\infty} (z - h_0) \cdot \phi(z) \, \mathrm{d}z. \tag{1.79}$$

Accordingly, the total contact load is given by

$$F_c = \frac{4}{3} \cdot E^* \cdot \beta^{\frac{1}{2}} \cdot \eta \cdot A \int_{h_0}^{\infty} (z - h_0)^{\frac{3}{2}} \cdot \phi(z) \, \mathrm{d}z. \tag{1.80}$$

As the bodies are brought closer together (decreasing separation distance h_0), the number of contacts, total contact area, and total applied force experience an exponential increase, while the proportional relationships remain remarkably consistent. The averaged pressure in the area becomes

$$\overline{p}_c = \frac{F_c}{A} = \frac{4}{3} \cdot E^* \cdot \beta^{\frac{1}{2}} \int_{h_0}^{\infty} (z - h_0)^{\frac{3}{2}} \cdot \phi(z) \, \mathrm{d}z. \tag{1.81}$$

Note that these formulations appear to be independent of the externally applied load to the contact. However, the number of asperities that come into contact as well as how they deform and how much individual load they bear is driven by the surfaces' distance. In a tribo-contact, the bodies' distance would rather be the response to the external load, so that the sum of contacting asperities can carry the load, which would require an iterative solution until the mentioned equilibrium is reached.

Assuming an exponential PDF of the form

$$\Phi(z) = \frac{2}{R_q} e^{-\frac{2z}{R_q}} \tag{1.82}$$

for $z > 0$, we can further modify the aforementioned equations (note: this is not a Gaussian distribution). The number of asperities in contact becomes

$$N_c = \eta \cdot A \int_{h_0}^{\infty} \frac{2}{R_q} e^{-\frac{2z}{R_q}} \, dz = \frac{2 \cdot \eta \cdot A}{R_q} \left[-\frac{R_q}{2} \cdot e^{-\frac{2z}{R_q}} \right]_{h_0}^{\infty} + \frac{R_q}{2} \cdot e^{-\frac{2z}{R_q}} = \eta \cdot A \cdot e^{-\frac{2z}{R_q}}. \quad (1.83)$$

Therefore, by introducing some curve fitting using the power law for the relation between asperity contact width and solid body approach (see Ref. [33]), the total force carried by all asperities in contact is

$$F_c = \frac{4}{3} \cdot E^* \cdot \beta^{\frac{1}{2}} \cdot \eta \cdot A \int_{h_0}^{\infty} (z - h_0)^{\frac{3}{2}} \cdot \frac{2}{R_q} \cdot e^{-\frac{2z}{R_q}} \, dz \approx \frac{\sqrt{2\pi}}{4} \cdot E^* \cdot \beta^{\frac{1}{2}} \cdot \eta \cdot A \cdot R_q^{\frac{3}{2}} \cdot e^{-\frac{2z}{R_q}},$$
$$(1.84)$$

and the averaged load per asperity becomes

$$\bar{f}_a = \frac{F_c}{N_c} = \frac{\sqrt{2\pi}}{4} \cdot E^* \cdot \beta^{\frac{1}{2}} \cdot R_q^{\frac{3}{2}}. \quad (1.85)$$

Note that this does not depend on the total applied force. The averaged contact area of all asperities can then be estimated as

$$\bar{a}_a = \pi \cdot \left(\frac{3 \cdot \bar{f}_a \cdot \beta}{4 \cdot E^*} \right)^{\frac{3}{2}} = \pi \cdot \left(\frac{3}{16} \sqrt{2\pi} \right)^{\frac{3}{2}} \cdot \beta \cdot R_q, \quad (1.86)$$

which depends mainly on the size of the asperities.

Restructuring these equations, we can also obtain the number of asperities in contact

$$N_c = \frac{4 \cdot F_N}{\sqrt{2\pi} \cdot E^* \cdot \beta^{\frac{1}{2}} R_q^{\frac{3}{2}}}, \quad (1.87)$$

as well as the total real area of contact

$$A_c = N_c \cdot \bar{a}_a \approx \frac{3 \cdot F_N}{E^*} \cdot \sqrt{\frac{\beta}{R_q}} \quad (1.88)$$

for a given normal load, which shows a linear proportionality.

Basically, a contact model involving two perfectly aligned rough surfaces would produce results that are practically identical to those obtained using a mode, in which a rough surface is arithmetically combined with a smooth, flat surface. Therefore, Greenwood and Tripp [34] extended the single rough surface contact model to scenarios involving two rough surfaces simply using the RMS roughness R_q. These approaches have later been expanded to consider a variety of scenarios, including but not limited to

- Asperities with elliptical shapes [35–37],
- Varying radii of curvature at asperity peaks [38,39],
- Interactions between individual asperities [40,41],
- Non-Gaussian surfaces [42–45], and
- Elastic-plastic materials [46–48].

In addition to these asperity-based contact models, alternative stochastic contact models have been developed for applications to fractal surfaces [49–51]. For a more comprehensive understanding of the underlying principles and mathematical equations, readers interested in further details are encouraged to consult [52].

The usage of statistical methods to describe rough surfaces and conduct contact analyses is quite simple and transparent way. It greatly facilitates the derivation of closed-form contact equations and the development of efficient contact modeling techniques. Nevertheless, it is essential to stress that the fundamental assumption of a simplified asperity summit shapes contradicts the diverse and realistic range of engineering surface topographies. Furthermore, it is worth emphasizing that analyzing each isolated individual asperity contact using Hertzian theory while disregarding interactions among neighboring asperities can pose challenges and limitations, especially in heavily loaded contacts. In these cases, substantial mutual influence among asperities and deep penetration may occur, and neighboring asperity contact regions frequently tend to merge. In reality, the stochastic models mentioned earlier are best suited addressing lightly loaded rough surface contacts. In recent years, rough surface contact problems have increasingly been tackled through numerical simulation methods, e.g., using the elastic half-space theory (BEM) or Finite Element Modeling (FEM). An example of the locally resolved surface contact area and pressure of two rough surfaces as well as the averaged pressure as a function of the distance is shown in Figure 1.31.

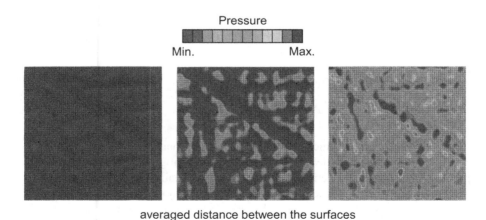

averaged distance between the surfaces

FIGURE 1.31   Pressure distribution and real area of contact between two rough surfaces at different averaged distances. (Reprinted and adapted from Ref. [32].)

1.6 FRICTION

Friction is a fundamental force that opposes relative motion between two surfaces in contact. It plays a crucial role in our everyday lives, affecting everything from walking to the operation of machinery. Amonton's law of friction states that the friction force between two surfaces is directly proportional to the normal force and a material pairing constant. Therefore, the COF

$$\mu = \frac{F_F}{F_N} \tag{1.89}$$

is a dimensionless number that quantifies the level of friction between two surfaces in contact as the ratio of frictional force F_F to the normal force F_N and is independent of the nominal contact area (Section 1.5.3). Friction is a vector quantity, which implies that it has both magnitude and direction. The COF, as a scalar value, does not include any information about the acting direction. The direction of the frictional force opposes the relative motion or the tendency for motion (Figure 1.32).

Fundamentally, based on the state of motion, static and dynamic kinetic frictions can be distinguished as shown in Figure 1.33. **Static adhesive friction** refers to the resistance encountered when an object initially attempts to move while in contact with another object. It involves the attraction between the molecules on the surfaces in contact, which holds the objects together and resists attempts to initiate motion. Static friction is a self-adjusting force, which means that it can vary in magnitude depending on the applied force trying to initiate motion and also the time the surface has been in contact. When the applied force is less than the maximum static frictional force, the object remains at rest. In this context, the static COF measures the ratio of the maximum force required to overcome this initial resistance with the normal force.

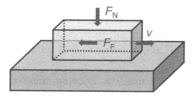

FIGURE 1.32 Friction force resulting from block-on-plane contact subjected to normal load and motion.

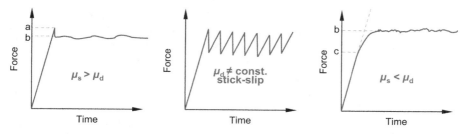

FIGURE 1.33 Force required to drag an object over time in dependency of characteristics regarding static (μ_s) and dynamic (μ_d) COFs.

Once an object overcomes the initial resistance and starts moving relative to the other surface, the frictional force transitions from static to **dynamic kinetic friction**. Dynamic friction is typically weaker than static friction and arises due to the irregularities on the surfaces interacting with each other as they slide past one another. The kinetic COF measures the frictional force that acts on the object while moving. Typically, the COF generally depends on the materials as well as their surface conditions, temperature, or the presence of a lubricant and can range from 0 (no friction) to 1 (high friction) and can even exceed 1 for extreme cases. It is important to note that friction is never a constant for a tribo-pair, but always underlies more or less pronounced fluctuations over time or sliding distance.

In this regard, the **stick-slip** phenomenon is a fascinating behavior observed in dynamic friction scenarios, see Figure 1.33 (middle). When an object is in motion, it may not move smoothly but rather exhibit an intermittent jerking or slipping motion. This irregular motion results from alternating periods of sticking and slipping that occur as the dynamic friction force overcomes the static friction force. This stick-slip phenomenon has practical implications in various fields. For instance, it can affect the precision of machinery, causing vibrations and noise. In geological contexts, stick-slip motion can lead to earthquakes when accumulated stress is suddenly released. A detailed understanding and control of the underlying stick-slip behavior is crucial for optimizing mechanical systems and mitigating its potentially disruptive consequences.

Generally, **friction processes** can be divided into the following three phases:

- Energy introduction
- Energy conversion
- Energy dissipation

The **energy introduction** in a tribo-technical system is directly driven by the tribological stresses acting upon the system (Section 1.2) and results from the formation of a contact area between the contacting bodies, micro-contact area enlargement due to normal forces as well as the formation of adhesive bonds depending on the interfacial energy.

The **energy conversion** is caused by **friction mechanisms**, which refer to the motion-restricting, energy-dissipating elementary processes of friction occurring in the contact area of a tribological system. They originate from the locally and temporally stochastically distributed micro-contacts in the contact area. According to the type of motion on a macroscopic level (Section 1.5.1) as illustrated in Figure 1.34, rolling and sliding friction can further be differentiated. **Rolling friction** is composed of the deformation and elastic hysteresis of the contacting partners as well as the compression of the lubricant. The **hysteresis** effect depends on the specific material but also occurs with the assumption of a linear-elastic material behavior, attributed to material damping. A portion of the deformation energy imparted during a loading cycle is converted into heat and is irreversibly lost as kinetic energy. The use of solid, liquid, or gaseous lubricants represents an effective approach in nature and technology to reduce friction in highly stressed tribological contacts through the partial or complete separation of the contacting partners through an intermediate medium with low shear resistance. Depending on the macro-geometry, arrangement, kinematics, and surface topography of the contacting bodies, as well as the

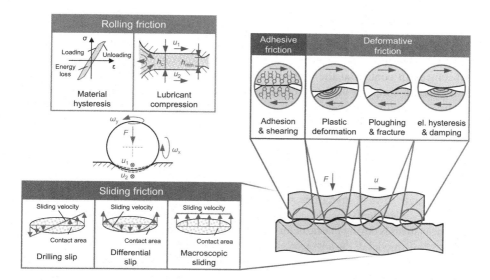

FIGURE 1.34 Macroscopic friction components in rolling-sliding contacts (left) and microscopic sliding friction mechanisms (right). (Adapted from Ref. [17] with permission.)

properties of the lubricant, various lubrication and friction conditions can be established. In the case of lubricated contacts, **rolling friction** can also be attributed to resistance of a lubricant against entering the contact zone and compression due to prevailing contact pressures.

Rolling friction occurs primarily in concentrated contacts and, in many applications, can be practically neglected compared to the dominant **sliding friction**. The latter occurs due to macroscopic sliding, drilling, or rolling movements as well as microscopic sliding motion in the contact area resulting from elastic deformation of the contacting partners. In lubricated contacts, various lubrication and friction conditions can arise, ranging from solid to mixed and fluid friction. The **hydrodynamic friction** component results from the fluid's internal resistance to shear or velocity gradients in the lubricating gap at the hydrodynamic (HD) or elastohydrodynamic (EHD) contact, which will be discussed in detail in Chapter 3. The **contact mechanical sliding friction** component results from the interaction of the surface pairing and the energy balance of the friction process. It should be noted that, even for nominally pure rolling Hertzian contacts of curved bodies, we observe microscopic slip zones with local sliding motion. Thereby, we can distinguish between

- **Reynolds Slip**: Pertains to Hertzian rolling contact between two bodies with different elasticity moduli. During the rolling motion of both bodies, the surface elements of the two contact partners are stretched differently in the tangential direction, resulting in a slip between them.
- **Heathcote Slip**: When the two rolling bodies have different radii of curvature, a curved (Hertzian) contact surface is formed. Due to the unequal distances of individual surface elements from the rolling axis, slip is induced during motion.

1.6.1 ADHESIVE FRICTION

The **adhesive friction** component of sliding friction can be attributed to the formation and subsequent shearing/breaking of physical and chemical bonds between the substances in contact at the atomic and molecular levels. Bowden and Tabor [53] introduced a simple model for the adhesion component of solid friction in metals, relating friction to the shear strength at the interface τ_s and the true contact area A_c during plastic deformation

$$F_F = \tau_s \cdot A_c. \tag{1.90}$$

Indeed, various studies, especially for metals and alloys, have shown that the ratio of shear strength to yield strength results in COFs of about 0.2. However, typical measurements often yield higher values around 1, indicating the complex interaction of normal and shear stresses during plastic deformation in forming the true contact area. Boden and Tabor's basic model of the adhesion component of friction has been expanded by various other approaches, including theories involving surface energy and fracture mechanics. Factors influencing the adhesion component include the deformability of contact partners, the electron structure, existing surface layers, and the presence of intermediate substances and environmental conditions. For instance, metallic materials with hexagonal close-packed structures may exhibit lower adhesion components due to more favorable conditions for the formation of true contact areas. The electron structure plays an important role, while high-density metals with mobile electrons, such as noble metals, tend to exhibit more metallic adhesion compared to metals with lower electron density, such as transition metals. Surface layers are known to significantly affect the adhesion component. For instance, ultra-high vacuum conditions have demonstrated that the formation of thin surface layers can alter the free surface energy and, consequently, friction. Apparently, the presence of intermediate substances and environmental conditions also affects the adhesion component, with conditions like humidity influencing the friction behavior of materials. Therefore, the trick to reduce the adhesive friction component is to lower the shearing strength while maintaining a high hardness of the substrate [15].

1.6.2 DEFORMATIVE AND ABRASIVE FRICTION

The **deformative friction** component is caused by microscopic interactions between the roughness asperities of the surfaces. This includes elastic deformation (hysteresis) or plastic flattening of roughness peaks. The **abrasive friction** (also called plowing friction) component is caused by microscopic interactions between the roughness asperities of the surfaces. This includes the elastic deformation (hysteresis) or plastic flattening of roughness peaks. Furthermore, when two bodies with different hardness come into contact, the harder surface asperities can penetrate into the softer counter-body. During tangential displacement, this results in a frictional component due to the material's resistance to deformation caused by the harder counterpart. Two fundamental possibilities for a frictional component due to deformation arise: Firstly, due to deformation by surface asperities of the counter-body (counter-body deformation) and, secondly, due to deformation by embedded wear particles

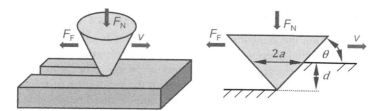

FIGURE 1.35 Model of a single asperity plowing through the softer material.

("particle deformation"). A simple model involves a tangentially moving conical surface asperity (Figure 1.35), whereby the COF depends on the tangent of the asperity's inclination. For a single asperity, $\tan \theta = d/a$ and the forces experienced by the asperities originate from the response of the material to the indentation (hardness of the material H resists the penetration of the asperity):

$$F_F = a \cdot d \cdot H \tag{1.91}$$

and

$$F_N = \frac{1}{2}\pi \cdot a^2 \cdot H. \tag{1.92}$$

By combining both equations, we obtain

$$\mu = \frac{2}{\pi}\tan\theta, \tag{1.93}$$

which indicates that the abrasive friction does not depend on the hardness of the material, though the resulting scratch mark will be different for soft and hard substrates. Since engineering surfaces have asperities with an inclination of only about 4°–6°, the model predicts COF of around 0.05–0.07. Yet, this value can be considered as a rather lower limit as the model neglects phenomena like material pile-up in front of the deforming asperity. Since micro-cracking can occur in brittle materials during deformation, an extended model for the deformation component of friction was proposed by Zum Gahr [54], whereby fracture toughness, elastic modulus, and hardness play an influencing role. Another form of deformation can be caused by embedded wear particles. This is highly sensitive to the ratios of the curvature radii of the wear particles and the penetration depth. It is estimated that COFs of about 0.2 can be derived for the deformation component, emphasizing the significance of embedded wear particles (third bodies) in addition to the material properties of the base body and counter-body, which are crucial for deformation behavior [15].

It is important to emphasize that, in real applications, there is a superposition of the above-mentioned friction mechanisms in the contact area, which are uncertain in time and space. Therefore, the resulting friction behavior can hardly be characterized theoretically or by numerical modeling but has to be determined experimentally, which will be further addressed in Chapter 2.

1.6.3 FRICTIONAL HEATING

The effect of friction mechanisms can be macroscopically characterized by a frictional force or frictional energy, resulting in **energy dissipation**, i.e., the conversion of mechanical kinetic energy into other forms of energy. The primary process of friction-induced energy dissipation connects with the generation of **frictional heat** (mechanical heat equivalent) in the contacting components. The physical processes involved in friction-induced energy dissipation are highly complex. When viewed at a macroscopic level, in a sliding contact area (A) subjected to normal and frictional tangential forces (F_N, F_R), there is a friction-induced average energy

$$E_F = F_F \cdot s = F_N \cdot \mu \cdot s \tag{1.94}$$

or power consumption

$$P_F = \frac{E_F}{t} = F_N \cdot \mu \cdot v, \tag{1.95}$$

with the sliding distance s and the velocity v. With respect to the nominal contact area, assuming that the frictional energy is entirely converted into heat, this results in a surface-specific frictional power that can be equated to a specific thermal load or heat density:

$$Q_F = \frac{F_N \cdot \mu \cdot v}{A} = p \cdot \mu \cdot v. \tag{1.96}$$

The friction-induced energy dissipation in tribological contacts is associated with heat flows in the contact area and leads, due to the micro-geometry of the contacting surfaces, to temperature distributions characterized by the **average (volume-specific) temperature increases** of both contacting bodies and temperature increases in temporally and spatially static distributed micro-contacts (referred to as "**flash temperatures**"). Determining or calculating friction-induced temperature increases presents a significant challenge, both in experimental terms due to the inaccessibility of the contact boundary and theoretically due to the complexity of the elastic-plastic contact deformation processes, the stochastic nature of energy dissipation mechanisms, and the temperature dependence on relevant properties of the contacting bodies [15].

To theoretically estimate the friction-induced temperature increases occurring in tribological contacts depending on the stress collective and system structure, various "flash temperature hypotheses" have been developed. Due to the stochastic nature of dissipative micro-contact processes leading to "flash temperatures," the occurrence of chemical boundary layers, the fluctuation of local frictional forces, and the often not precisely known dependency of material data on temperature, calculating friction-induced temperature increases is exceedingly difficult. Simplifying theories are based on the following main assumptions [15]:

- The existing micro-contact points are conceptually combined into a resulting contact area where frictional energy is converted into heat (flat heat source).
- The generated heat is dissipated by both contacting bodies.

- The surfaces of both bodies at the contact points have the same temperature.
- The calculated temperature represents an estimation of the temperature increase above the average temperature of the surface.

To find out the temperature increase as a result of the friction-induced heat, let us consider a simple half-space model. In this model, the point heat flux $d\dot{Q}$ contributes to the frictional heat at a distance r (considering the equilibrium heat profile):

$$dT = \frac{d\dot{Q}}{2 \cdot \pi \cdot k \cdot r}, \tag{1.97}$$

with the thermal conductivity k. If the heat source $\dot{q}$ is presented by the small area dA at distance s from the center (non-point source), then

$$d\dot{Q} = \dot{q}\ dA = \dot{q} \cdot s\ ds\ d\varphi \tag{1.98}$$

and temperature increase for the center ($r = s$) is

$$dT = \frac{\dot{q} \cdot s\ ds d\varphi}{2 \cdot \pi \cdot k \cdot s}. \tag{1.99}$$

The total temperature rise in the steady-state regime is then

$$\Delta T = \int_0^a \int_0^{2\pi} dT = \frac{1}{2 \cdot \pi \cdot k} \int_0^a \int_0^{2\pi} \frac{\dot{q}}{s} s\ ds\ d\varphi = \frac{\dot{q} \cdot a}{k}. \tag{1.100}$$

In the case of the tribologically induced heating, we can write

$$\Delta T = p \cdot \mu \cdot \frac{v \cdot a}{k}. \tag{1.101}$$

EXERCISE 1.9

Let's assume a steel-on-wood contact with a contact radius is 5 mm, a sliding speed is 1 m/s, a contact pressure is 1 MPa, and a friction coefficient is 0.2. The thermal conductivity of steel is 20 W/m·K, and we assume that there's no heat dissipation in the wood.

- Estimate the temperature increase as a result of frictional sliding.
- How would the result change if copper with a thermal conductivity of 400 W/m·K is used instead of steel?

For this example, we only considered one heat-conducting surface. In tribological systems, for which by definition two surfaces come into contact, the heat is usually dissipated in both directions. Still, the temperature increase in the center of the contact should be the same for both surfaces, i.e.,

$$\Delta T_1 = \frac{\dot{q}_1 \cdot a}{k_1} = \Delta T_2 = \frac{\dot{q}_2 \cdot a}{k_2}. \tag{1.102}$$

For the frictional contact, we obtain

$$\dot{q}_F = p \cdot \mu \cdot v = \dot{q}_1 + \dot{q}_2 = \dot{q}_1 + \frac{k_2}{k_1}\dot{q}_1 \tag{1.103}$$

and

$$\dot{q}_1 = p \cdot \mu \cdot v \frac{k_1}{k_1 + k_2}, \tag{1.104}$$

suggesting that in the case of $k_1 = 0$ (non-conductive material), all the heat dissipates in the opposite surface with conductivity k_2, and vice versa. Then, the temperature increase is

$$\Delta T_1 = \Delta T = \frac{\dot{q}_1 a}{k_1} = \frac{p \cdot \mu \cdot v \cdot a}{k_1 + k_2}. \tag{1.105}$$

In the case of the self-mated surfaces, the temperature increase is

$$\Delta T = \frac{p \cdot \mu \cdot v \cdot a}{2k}. \tag{1.106}$$

EXERCISE 1.10

For exercise 9 (contact radius 5 mm, sliding speed 1 m/s, contact pressure 1 MPa, friction coefficient 0.2), we now assume a steel-on-steel pairing (thermal conductivity 20 W/m·K).

- Recalculate the temperature increase as a result of frictional sliding.
- How would the result change if the sliding speed increases 10 times (to 10 m/s) or 100 times (to 100 m/s)?

In previous examples, we considered the equilibrium heat flux for the whole half space. However, in the case of a tribological contact under pure sliding conditions, while one surface might remain in contact, a fresh surface is constantly exposed on the counter-body side. This effect becomes even more pronounced as the sliding speed increases, making a constant heat flux impossible. Instead, the heat is also dissipated through the heat convection by heating the new, cold mass of the surface coming in contact. To address this situation, we introduce the so-called Peclet number:

$$Pe = \frac{a \cdot \rho \cdot c}{k}V, \tag{1.107}$$

where ρ is the density and c is the specific heat capacity of the cold material that comes into contact.

If *Pe* is larger than 10, the temperature increase as a result of frictional sliding should be modified as

$$\Delta T = 1.6 \; Pe^{-\frac{1}{2}} \frac{\dot{q}_1 \cdot a}{k}. \qquad (1.108)$$

When *Pe* becomes very large, instead of equilibrating the heat flux inside the body, all the generated frictional heat is dumped in a large mass that always comes with a colder contact.

EXERCISE 1.11

For exercise 10 (contact radius 5 mm, sliding speed 100 m/s, contact pressure 1 MPa, friction coefficient 0.2), we now assume new steel (thermal conductivity 20 W/m·K) coming into contact as well as $\frac{\rho \cdot c}{k} = 0.2 \frac{s}{mm^2}$. Recalculate the temperature increase as a result of frictional sliding.

In addition to the transfer of frictional heat and friction-induced mechanical vibrations through components and materials connected to the immediate frictional bodies, both energy absorption and emission occur. **Energy absorption** can be related to phonons/electron excitation, elastic hysteresis, lattice deformations, generation and transformation of defects and dislocations, phase transformations, formation of residual stresses, or micro-fracture processes. Regarding **energy emission**, frictional electricity, i.e., the generation and transmission of electrical charge through frictional processes, as well as friction-induced sound emission, photon emission (triboluminescence), and ion and electron emission, are to be mentioned [15].

1.7 WEAR

Similar to friction, wear, which is defined as the progressive material loss from the surface of a solid body[3] due to contact and relative motion with a counter-body (solid, liquid, or gaseous), can also be attributed to elementary physical and chemical interactions in the contact area. Wear under dry conditions (solid body wear) occurs in the case of direct solid/solid contact, while it can be also induced under boundary and mixed lubrication. The **quantitative assessment** of wear and its outcomes primarily rely on two key concepts:

- **Wear Measures**: These metrics, represented as numerical values, quantify how the shape or mass of an object changes due to wear.
- **Wear Mechanism Descriptions**: These describe the alterations that occur on the surfaces of materials or components under tribological stress due to wear. This includes changes in chemical composition, microstructure, and the nature as well as the form of the wear particles generated.

---

[3] In general, value-adding machining processes (e.g., cutting) are not considered as wear with regard to the workpiece being produced, despite the similar tribological elementary processes in the contact area.

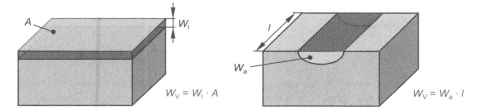

FIGURE 1.36 Schematic illustration of linear and planimetric wear measures.

1.7.1 Wear Measures

As shown in Figure 1.36, measures can be expressed in various dimensions and can be categorized as follows:

- **Linear Wear Length** W_l: One-dimensional changes in the geometry of tribologically stressed materials or components, perpendicular to their common contact surface (linear wear).
- **Planimetric Wear Area** W_a: Two-dimensional changes in the cross-sectional areas of tribologically stressed materials or components, perpendicular to their common contact surface (planimetric wear).
- **Wear Volume** W_v: Three-dimensional changes in the geometric regions of tribologically stressed materials or components within their common contact area (volumetric wear). Wear volumes are linked to the wear mass by the density or specific gravity of the wearing materials or components.
- **Wear Mass Loss** W_m: Changes in the gravimetric mass of the stressed materials or components.

Sometimes, the wear resistance is used as the inverse of a wear measure $(1/W)$. In addition to these "direct" wear measurements, there are also "indirect" wear measures, for which wear is compared to a reference value. These are commonly referred to as "wear rates" and include

- **Wear Velocity**: Measuring wear in relation to the effective duration of the tribological stress during the wear process.
- **Wear-Path Ratio** or **Wear Intensity**: Measuring wear in relation to the distance traveled.
- **Wear-Throughput Ratio**: Measuring wear in relation to the throughput.

Internationally, the **wear coefficient**, also known as the **specific wear rate**, or simply **wear rate** or **wear factor**, serves as a reference value for wear. This coefficient initially introduced by Archard in 1953 represents a normalized wear rate, which is the wear volume per sliding distance s, divided by the normal force:

$$k = \frac{W_V}{s \cdot F_N}. \tag{1.109}$$

This macroscopic parameter represents the volume loss per unit load in a tribo-system over a specific sliding distance (neither rolling nor rolling-sliding) under constant load conditions. It assumes a proportional relationship between the wear volume and these factors. This volumetric wear coefficient is commonly used because it approximates wear results, making them comparable across various geometries, materials, test durations, and loads. It is applicable under dry and lubricated conditions (boundary and mixed lubrication) and ranges from 10^{-10} to $10^{-2}\,\text{mm}^3/\text{N·m}$, whereby $10^{-2}\,\text{mm}^3/\text{N·m}$ is typically seen as a transition from mild to severe wear. However, it is crucial to ensure that the compared values were obtained under "similar" conditions, including pressure, load, temperature, friction, and contact severity. The wear coefficient does not provide information about the wear mechanism and is not a materials' constant since wear results from interaction processes between contacting bodies or substances. Therefore, a wear measurement does not describe the property of an individual material but should be always considered in the context of the specific tribological system studied.

1.7.2 Wear Mechanisms

Wear can be attributed to elementary physical and chemical interactions that occur in the tribological contact area. Although the elementary processes of wear are partially linked to friction mechanisms, high friction does not necessarily coincide with high wear. For sliding contacts, the basic wear processes are illustrated in Figure 1.37. However, under real conditions, these mechanisms usually do not operate independently but interact and are superimposed in a rather complex manner. Thus, wear is often initiated by one mechanism and continued by others. Given the intricacies of wear, it is typically not feasible to calculate wear parameters theoretically.

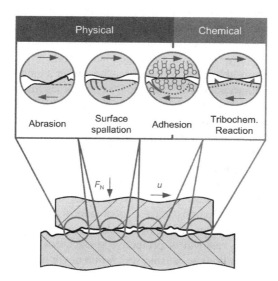

FIGURE 1.37 Schematic representation of different wear mechanisms in sliding contacts. (Adapted from Ref. [17] with permission.)

Instead, they need to be experimentally determined using appropriate measurement and testing techniques (Chapter 3).

The wear mechanism of **abrasion** occurs when one of the two contact bodies is rough and significantly harder than the other, or when hard particles are present in the contact area. Abrasion directly leads to material loss as fine chips or fragments are separated from the (locally) softer contact partner due to the relative motion of the two bodies. This process is associated with an increased abrasive friction. Abrasive wear also occurs in lubricated contacts when the roughness peaks of the two bodies are not completely separated or when particles are present that are larger than the formed lubricating gap. Assuming a single asperity (Section 1.6.2, Figure 1.35) sliding along the distance L ($\gg a$), this creates a grove with a triangular cross-section and volume given as

$$W_{V,i} = a \cdot d \cdot L. \tag{1.110}$$

From the previously used expression for the normal force acting as a resistance of the material to penetration, we can re-write the wear volume as

$$W_{V,i} = a^2 \cdot \tan\theta \cdot L = \frac{2 \cdot F_N}{\pi \cdot H} \tan\theta \cdot L \tag{1.111}$$

as well as

$$W_V = \sum_{i=1}^{N} W_{V,i} = \sum_{i=1}^{N} \frac{2 \cdot F_N}{\pi \cdot H} \tan\theta \cdot L = \frac{2}{\pi \cdot H} \tan\theta \cdot F_N \cdot L = k_{abr} \cdot F_N \cdot L \tag{1.112}$$

in the case of multiple asperities.

Surface spallation or fatigue in sliding contacts is primarily caused by periodic stress at the micro-contact level (microscopic sliding fatigue wear). Similar to material fatigue in cyclically stressed components, surface spallation is a multi-stage process that initially only results in a changed sub-surface microstructure without measurable wear. Noticeable wear only occurs when, after a long phase of formation and propagation of microcracks, the damaged surface areas finally break due to the fusion of surface-parallel cracks, leading to the delamination of plate-like wear particles. Rapid progression to this end stage is facilitated by material defects, existing microcracks, and crack propagation during adhesive contacts.

Adhesive wear, also referred to as cold welding in metals, is caused by the same material interactions of the contact partners responsible for the adhesive component of friction. Usually, the chemical and physical bonds forming in micro-contacts between the contact partners are significantly weaker than the cohesive forces in the surrounding material. In most cases, local adhesive contacts are, therefore, dissolved by shearing the interfacial bonds. Under certain circumstances, such as immediately after the removal of protective surface coatings, strong bonds can form between the contact bodies, causing separation not at the original interface but within the body of lower strength. The material fragments transferred can undergo chemical changes, such as oxidation, during continued sliding, under the influence of other wear mechanisms, and can eventually be detached as wear particles, either wholly or in part, from the contact surfaces.

Let us assume the same model as for the adhesive friction calculations from Section 1.6.1, whereby during sliding, we assume a certain probability P to extract material from the counter-body. Then, we can write the wear volume as

$$W_{V,i} = P \cdot \frac{1}{2} \left(\frac{4}{3} \cdot \pi \cdot a^3 \right) = P \cdot \frac{2}{3} \cdot \pi \cdot a^3. \tag{1.113}$$

This area of the bottom surface remains in contact for distance $\Delta L_i = 2a$. Therefore, the wear volume per unit sliding distance is

$$\frac{W_{V,i}}{\Delta L_i} = \frac{P \cdot \frac{2}{3} \cdot \pi \cdot a^3}{2 \cdot a} = \frac{1}{3} \cdot P \cdot \pi \cdot a^2 = \frac{1}{3} \cdot P \cdot \pi \cdot \frac{F_N}{H}, \tag{1.114}$$

and the total wear becomes

$$W_V = \int_0^L \frac{1}{3} \cdot P \cdot \pi \cdot \frac{F_N}{H} dx = \frac{1}{3} \cdot P \cdot \pi \cdot \frac{F_N}{H} \cdot L = k_{adh} \cdot \frac{F_N}{H} \cdot L. \tag{1.115}$$

The latter is also known as Archard's wear law [55]. Please note that in both abrasive and adhesive wear models, the resulting total wear is directly proportional to the product of applied load and sliding distance, while the coefficient of the proportionally represents the resistance of the material to wear.

Tribo-chemical reactions refer to all chemical reactions of one or both contact bodies with components of the intermediate substance and/or the ambient medium that are favored by friction-induced (local) temperature and pressure increases or the removal of protective coatings by other wear mechanisms that would not occur outside the tribological contact. Strictly speaking, they are not an independent wear mechanism, as they do not lead to material loss but only change the chemical and mechanical behavior, such as adhesion, and potentially the strength of the affected surface areas. In conjunction with other wear mechanisms, they can significantly influence the wear behavior of a tribological system. In sliding contacts, they often lead, in conjunction with separation processes, to the formation of powdery wear particles. However, in some cases, tribological chemical reactions can also have a positive impact on the wear behavior of a tribological system, if they result in reduced interaction between the contact partners.

The occurring wear mechanisms and their interactions lead to characteristic changes in the surface condition (Figure 1.38) of the contact bodies and the formation of wear particles, which depends on the type of wear present due to the system's structure and kinematics. Since the wear mechanisms are usually not directly observable, their identification is mostly possible based on these **wear phenomena**:

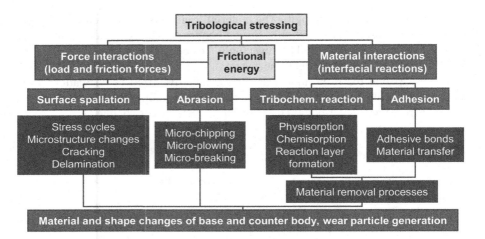

FIGURE 1.38 Material and deformation processes under tribological stress and the effects of wear mechanisms.

- Abrasion → Scratches, grooves, depressions, waves
- Surface spallation/fatigue → Cracks, pits
- Adhesion → Fretting, holes, crests, scales, material transfer
- Tribo-chemical reactions → Reaction products (layers, particles)

This can, for instance, be classified by a wear atlas as found in Ref. [15] or types of expectable wear in accordance with the tribological system and the underlying kinematics as shown in Figure 1.39.

An additional mechanism worth mentioning is associated with melting-induced wear as a result of friction-induced overheating at the contact asperities. Since friction is always accompanied by heating, it has its own restriction imposed by the characteristics of the materials in contact. Specifically, the temperature increase can only reach up to the lowest melting temperature T_{melt} of the contacting materials. Then, the energy produced during sliding is dissipated as

$$
\begin{aligned}
\dot{Q}_{\text{total}} &= \mu \cdot F \cdot V = \dot{Q}_1 + \dot{Q}_2 + \dot{Q}_{\text{melt}} \\
&= \frac{\Delta T_{\text{melt}} \cdot K_1}{\lambda_1 \cdot a} \cdot \pi \cdot a^2 + \frac{\Delta T_{\text{melt}} \cdot K_2}{\lambda_2 \cdot a} \cdot \pi \cdot a^2 + H_{\text{f}} \cdot \dot{W}_{\text{V,melt}},
\end{aligned}
\tag{1.116}
$$

where H_{f} is the heat of fusion of the material with the lower melting temperature. Thus, the melt wear is represented by the molten material that is removed from the sliding contact:

$$
W_{\text{V,melt}} = \dot{W}_{\text{V,melt}} \cdot t \left(\mu \cdot F \cdot L - \frac{\Delta T_{\text{melt}} \cdot K_1}{\lambda_1 \cdot a} \cdot \pi \cdot \frac{a^2 \cdot L}{V} - \frac{\Delta T_{\text{melt}} \cdot K_2}{\lambda_2 \cdot a} \cdot \pi \cdot \frac{a^2 \cdot L}{V} \right) \frac{1}{H_{\text{f}}},
\tag{1.117}
$$

indicating a linear dependence of the melt wear on the sliding distance, while the applied load only affects the first term. The phenomenon of melt wear is commonly encountered in various industrial applications and is particularly relevant in fields such as metallurgy, manufacturing, and aerospace, as well as in winter sports.

System elements	Tribological stressing	Wear type	Prevailing wear mechanisms			
			Adhesion	Abrasion	Surface spallation	Tribochem. reaction
Solid body (1) Intermed. medium (3) Solid body (2) (HD or EHD)	Sliding Rolling Rolling-Sliding Impact	Sliding wear Rolling wear Rolling-sliding-wear Impact wear			X	X
Solid body (1) Solid body (2) (solid, boundary or mixed friction)	Sliding	Sliding wear	X	X	X	X
	Rolling Rolling-sliding	Rolling wear Rolling-sliding wear	X	X	X	X
	Impact	Impact wear	X	X	X	X
	Oscillating	Oscillation/vibration wear	X	X	X	X
Solid body (1) Solid particle (2)	Sliding	Ploughing wear Erosion		X		X
Solid body (1) Solid particle (3) Solid body (2)	Sliding	Grain sliding wear Three body wear	X	X		X
	Rolling Rolling-sliding	Grain rolling wear	X	X		X
Solid body (1) Fluid with solid particles (2)	Flowing	Erosion wear	X	X		X
Solid body (1) Gas with solid particles (2)	Flowing	Erosion wear	X	X		X
	Impact	Impingement Jet wear	X	X		X
Solid body (1) Fluid or gas (2)	Flowing Vibration	Cavitation wear			X	X
	Impact	Drop impact wear			X	X
	Separated wear products	Particle form				
		Particle characteristics	Spherical	Spiral- or chip-shaped	Flaky or lamellar or splintery	Powdery or amorphous

FIGURE 1.39 Prevailing wear mechanisms as well as wear products dependent on the system elements, the acting tribological stress, and wear type.

1.7.3   WEAR EVOLUTION

Since wear is system-dependent, it is necessary to distinctly specify the wear data from both contact partners and potentially the wear data of the entire system individually:

- **Component Wear**: Individual wear measurements for the base body and the counter-body.
- **System Wear**: The cumulative wear measurements of the base body and the counter-body.

The chronological sequence, in which the mechanisms occur, and the nature of their interactions can only be determined through the analysis of wear manifestations at various stages. Generally, the following stages of wear over time can be classified:

- **Initial Stage – Running-In Period**
  - When two surfaces first come into contact, there is typically an initial running-in period.
  - During this phase, surface asperities (microscopic irregularities) on the contacting bodies begin to conform to each other.
  - Initially, this can lead to increased friction and localized wear as high peaks on the surfaces come into contact.
- **Steady-State Stage**
  - After running-in, the system often enters into a steady-state stage.
  - During this phase, wear rates become relatively constant and predictable.
  - The dominant wear mechanisms that occur during this stage depend on factors such as load, speed, temperature, and lubrication.
- **Accelerated Wear Stage**
  - As components continue to operate, wear may accelerate due to various factors. Such as changes in operating conditions (e.g., increased load or temperature), the accumulation of wear debris that acts as abrasive particles, surface fatigue, or microcrack propagation.
  - Accelerated wear can lead to increased friction, reduced performance, and a decreased component lifespan.
- **Failure Stage**
  - In the final stage, wear progresses to a point where it significantly affects the component's functionality.
  - This stage can manifest as catastrophic failure, reduced efficiency, or increased maintenance requirements.
  - It may involve the formation of cracks, delamination, or material loss to a degree that the component can no longer perform its intended function.

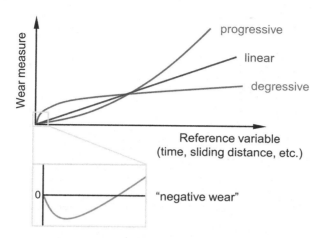

FIGURE 1.40   Different behavior of wear measure progression versus reference variable.

It is important to note that the progression of wear is not linear (Figure 1.40), and different wear mechanisms can coexist and interact. Additionally, the specific stages and their durations can vary widely depending on factors such as the materials involved, operating conditions, and the presence of lubrication. Also, it is possible to observe "negative" wear amounts or rates for some time if, for example, the material is transferred from the counter- to the main body by adhesion processes.

Understanding the progression of wear in a tribological system is crucial for designing effective mitigation strategies and optimizing component longevity. Different wear mechanisms, such as abrasion, surface spallation, and adhesive wear, often coexist and evolve over time. By carefully examining the surface changes and wear particle formation, engineers and researchers can gain insights into the dominant wear mechanisms and their interplay. This knowledge can inform the selection of materials, lubrication methods, and maintenance practices to minimize wear and extend the service life of mechanical systems. The ability to identify and analyze wear patterns is an essential aspect of tribology, contributing to the development of more durable and efficient machinery and equipment.

Wear, at its core, represents the gradual loss of material from a component's surface due to a complex interplay of mechanical, physical, and chemical interactions, but should not be confused with reliability or service life. **Reliability** delves into the realm of a system or component's ability to consistently perform its designated function under specified conditions, devoid of failure, for a predefined duration. It is essentially a measure of the probability that a system or component will operate without faltering over a given timeframe. Reliability, therefore, plays a pivotal role in ensuring that a component or system not only functions as expected but does so consistently, providing both performance and safety. **Service life**, sometimes referred to as lifespan or operational life, extends its focus to the temporal aspects of the components. It signifies the duration, for which a component or system remains functional and meets its performance requirements before it reaches a state of non-usability or obsolescence. **Failure rate** is a pivotal concept that complements reliability and

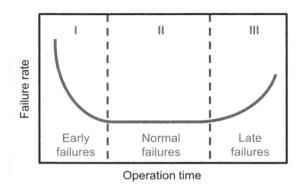

FIGURE 1.41 Failure rate as a function of operating time (bathtub curve).

service life. It represents the likelihood or probability of a system or component failing within a specific time frame. A low failure rate indicates a high level of reliability, as the component is less likely to fail within the given timeframe. When the wear-related failure rate is plotted as a function of the operating time of a tribological system, it often results in a bathtub curve (Figure 1.41). The regime of decreasing failure (I) rate corresponds to the early failures in the system's life. For instance, in tribological systems, this area can be influenced by successful running-in behavior, which helps reduce the rate of early failures. The regime of constant failure rate (II) is typically the domain of normal operating conditions. Failures are generally a consequence of statistically independent factors. In the progressive regime (III), the failure rate increases over time due to the accumulation of damage from acting wear mechanisms. Consequently, this region is particularly characteristic of wear-related failures in tribo-technical systems. Understanding these distinct regimes is crucial for assessing the reliability and longevity of tribological systems. It allows engineers and maintenance experts to develop strategies that address wear-related challenges at different stages of the system's life cycle.

1.8 TRIBOLOGICAL BEHAVIOR OF DIFFERENT MATERIALS

A variety of different materials are ubiquitous in the realm of engineering, offering a diverse array of properties and applications. Understanding their tribological behavior – how they perform in terms of friction and wear – is essential to optimize their performance and ensure the longevity of mechanical systems. In this chapter, we explore the intriguing world of tribology within the context of metallic, ceramic, and polymeric materials, delving into some simplified examples and influencing factors.

1.8.1 METALS

Metals are widely used in various applications, where they may encounter different tribological conditions. The tribological properties of metals are influenced by many factors such as the chemical composition, microstructure, mechanical properties, and surface roughness. **Fully soluble metals** sliding against each other, e.g., Cu-Ni alloy pairings, in a clean environment (vacuum, no surface oxidation/contamination) under high pressure and temperature along with the experienced plastic deformation will facilitate the formation of intermixed phases. The notable strength of the metallic bond in material junctions might result in pronounced adhesive wear. Concerning metals/alloys with **limited solubility**, e.g., tin-lead, iron-copper, steel-gold, and brass-steel, intermixing is less pronounced, thus creating a little better configuration. However, with the limited solubility, one of the materials usually shows much higher hardness than the other one. In this case, although adhesive wear may be quite low, the softer material starts to experience notable abrasive wear. Therefore, metals generally have moderate to high COF values, which implies that they can experience significant frictional forces during sliding or rolling. Similarly, metals are generally susceptible to wear, and their wear resistance depends on many factors, such as hardness, strength, and ductility. In general, harder and stronger metals tend to have better wear resistance, while more ductile metals may exhibit better resistance to adhesive wear.

The cases mentioned above are mainly relevant for the regime of mild wear experienced by materials at low contact pressure and velocity conditions. As outlined in Section 1.7, there are many more wear mechanisms than just abrasive and adhesive wear. The most probable mechanisms experienced by metals are summarized in Figure 1.42.

Upon transitioning to higher contact pressures, materials will start to experience delamination wear. This wear does not depend on the sliding velocity but rather the result of yielding-induced propagation of the stresses in the metals, which induces the fracturing of the large platelets. The underling mechanism bases on the assumption that sliding occurs in an ambient environment inevitably resulting in the formation of a native oxide layer. During sliding under high pressures, dislocations are generated inside the metal and their propagation is initiated. The presence of the oxide layer prevents the release of the internal stresses and causes their propagation over large distances. Once the stress is finally released, large platelets of the material are removed leading to the delamination of them from the surface. Once the pressure exceeds ~0.3–0.5 Y_S, the tribo-contact can experience the seizure or fusion of the whole contact area as a result of the contacting surfaces being buried into each other. Please note that in this case, the real contact area becomes the nominal contact area.

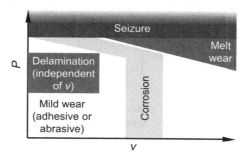

FIGURE 1.42 Exemplarily wear map for metallic materials dependent on the acting contact pressure and sliding velocity.

If the pressure increase is accompanied by an increase in the sliding velocity, metals are prone to corrosion wear. The major contributor for the corrosion activation connects with frictional heating induced at the contact by the flash temperature increase (the temperature increase is proportional to $\mu \cdot P \cdot v$ as outlined in chapter 1.6.3). In this regard, two parameters trade for each other, which occurs only till certain velocity. If the sliding becomes too fast, there is no time for the heat to flow and equilibrate. At the same time, the heating-activated corrosion is possible only till a certain degree. Once the heating at the contact interface becomes too large, the materials start to experience melt wear. Melt wear also usually does not depend on the pressure as the driving force for the material adaptation to the stresses is the hardness of the solid materials.

1.8.2 CERAMICS

Ceramics offer a high potential to develop advanced tribological materials that can withstand harsh environments and provide high performance with minimal maintenance. In general, ceramics feature lower COFs compared to most metals and are characterized by a high wear resistance due to their high hardness, toughness, strength, and chemical inertness resulting in a notable resistance against oxidation and corrosion. Ceramics are frequently employed for extreme conditions (high temperatures, high loads, high speeds). Thereby, ceramics, such as alumina (aluminum oxide), titania (titanium oxide), silicon nitride, titanium nitride, and silicon carbide, are formed from multiple grains sintered together. The binding of the ceramic powders can be enhanced by the introduction of glassy binders or waxes, which is usually pressure-assisted to increase the density of the resulting material. Focusing on a single grain (half of a sphere) of the ceramic sitting on the surface of the material, the energy needed to remove the whole grain is

$$\Delta E_{\text{whole}} = \left(\frac{4\pi R^2}{2} \right) \Delta \gamma_{\text{s}} \qquad (1.118)$$

with the surface energy $\Delta \gamma_{\text{s}}$. Instead of removing the entire particle, wear can be originated from fracturing the grain and its partial removal. Let us consider the situation when the crack is introduced at half of the grain and only half of the grain is removed (i.e., the quarter of the sphere). In this case, the energy needed is

$$E_{\text{half}} = \left(\frac{\pi R^2}{2} \right) \Delta \gamma_{\text{a}} + \left(\frac{2\pi R^2}{2} \right) \Delta \gamma_{\text{s}}, \qquad (1.119)$$

where $\Delta \gamma_{\text{a}}$ is the energy of adhesion inside the grain. These two possible wear mechanisms compete with each other in defining the most favorable route for wear to occur. If $\Delta \gamma_{\text{s}} < 0.5 \Delta \gamma_{\text{a}}$, the whole grain extraction dominates, while otherwise grain refinement dominates. Therefore, to reduce the wear of ceramics, it is important to increase $\Delta \gamma_{\text{s}}$. Possible approaches to achieve this include the use of stronger binders and changing the shape of the grains, e.g., needle-like grains. In the case of the

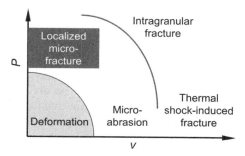

FIGURE 1.43 Exemplarily wear map for ceramic materials dependent on the acting contact pressure and sliding velocity.

latter, the energy of the grain refinement and extraction of half of the grain becomes smaller compared to the removal of the whole grain. The energy needed to wear ceramic materials can be expressed in terms of

$$\frac{\Delta E}{V} \sim \Delta\gamma \cdot \frac{R^2}{R^3} = \frac{\gamma}{R}. \tag{1.120}$$

This expression suggests that to decrease wear, the energy per unit volume of wear should be increased. This can be achieved by increasing the toughness of the material or decreasing the radius of the grains, i.e., smaller grains. To avoid the sintering of the ceramic powders, the toughness is usually increased via the use of binders.

When dealing with ceramics, it should be always kept in mind in mind that the used materials should show good compatibility with the conditions and environment. For instance, one of the common ceramic materials, silicon nitride, has a tendency to react with water from ambient humidity. The surface of silicon nitride is softened by water, leading to chemo-mechanical polishing in the sliding contact. This reaction releases ammonia ($Si_3N_4 + H_2O \rightarrow SiO_2 + NH_3$). Upon further contact with water, silicon oxide is hydrated, which further softens the surface ($SiO_2 + 2H_2O \rightarrow Si(OH)_4$). The hydrated $Si(OH)_4$ layer exhibits a low shear strength, thus reducing friction and wear. This induces a general dependence on ambient humidity as well as velocity (an increase in velocity facilitates temperature increase and thus reduces the access of the water to the sliding interfaces). Notably, oxide-based ceramics usually demonstrate reverse dependence on humidity. An increasing humidity tends to facilitate stress-corrosion cracking in the materials as a result of the water attack on the oxygen bonds. The general wear map ceramic materials are given in Figure 1.43.

1.8.3 POLYMERS

Polymers are commonly used in various industrial and consumer products, including automotive and aerospace components, medical devices, and consumer electronics. Polymers generally consist of long molecule chains, mostly made of carbon in combination with oxygen, fluorine, hydrogen, nitrogen, etc. They have a lower melting point, low thermal conductivity, and low hardness/strength in comparison to metals and ceramics. Generally, polymers tend to have a low COF and their wear resistance may vary notably. Furthermore, the ability of polymers to lubricate is an important tribological property. Some polymers, such as polytetrafluoroethylene (PTFE), are known for their excellent self-lubricating properties and capability to form beneficial tribo-films. These polymers are often used as coatings or additives in lubricants to improve their effectiveness. The polymers can be divided into thermoplastics and thermosets, which differ in their structure, mechanical properties, and tribological performance. **Thermoplastics**, such as polymethyl methacrylate (PMMA) or polydimethylsiloxane (PDMS) start to flow and stick easily, showing a behavior rather similar to metals. The wear behavior of thermosets, such as epoxy, which are highly cross-linked polymers, is fairly different. Their lower ductility leads to the generation of smaller and dust-like wear debris, inducing finer brittle fracturing in the material, which makes their performance more comparable with ceramics.

Typically, when tribologically stressed, the COF of polymers firstly tends to increase with an increasing normal load, but will stay constant once the normal load exceeds a critical threshold from a certain level on. In contrast, increasing the velocity initially does not the resulting COF, while friction tends to decrease after passing a certain velocity. This behavior can be explained by the $P \cdot v$-limit, which is defined by the frictional heating of the polymer limited by the melting point (or glass transition) of the polymer. Consequently, the friction behavior of the polymers is thermally controlled, while the sliding systems adjust friction in response to the experienced conditions.

EXERCISE 1.12

Let's assume a polyethylene (PE; $T_m = 150°C$) structure in contact with a steel substrate (contact dimension is $2a$, whereby $a = 6$ mm at a contact pressure of 2.8 MPa), rotating at room temperature around the center of the symmetry on the steel substrate. The radius of the curvature is $\gg a$ (assume constant velocity in the contact), but low enough to equilibrate the system and consider reaching a uniform temperature profile. The coefficient of friction is 0.3 (typical for PE), the thermal conductivity for PE is 0.4 W/m·K, and the thermal conductivity for steel is 40 W/m·K.

- What is the $P \cdot v$-limit for the system?
- How would doubling the contact pressure or doubling the contact size affect the $P \cdot v$-limit?

Many applications require the pressure and velocities to exceed the $P \cdot v$-limits. This can be achieved by improving the tribological characteristics of the polymers (lowering the COF), increasing the safe temperature regime (increasing the melting or glass transition temperatures, or introducing additional cooling approaches, such as reduction in the operation temperature of effective heat dissipation), or improving the thermal conductivity of the polymers. These modifications became possible with the use of composites when polymeric matrices are reinforced with fillers.

1.8.4 Composites

Composites offer a viable approach for improving the friction and wear character-istics by combining different classes of materials. Let's assume a softer matrix with COF μ_m and wear rate k_m reinforced with a harder and more wear-resistant filler (μ_f, k_f) at a concentration of χ. Therefore, the concentration of the matrix is $1 - \chi$. To estimate the tribological characteristics of the composite, the apparent contact area A with a perfectly rigid and flat counter-body, we obtain the total area of matrix in contact $A \cdot (1 - \chi)$ as well as the area of filler in contact $A \cdot \chi$. If the pressure experi-enced by the matrix is P_m and the one experienced by the filler is P_f, we can calculate the normal force as

$$F_N = P_m \cdot A \cdot (1 - \chi) + P_f \cdot A \cdot \chi. \tag{1.121}$$

Consequently, the overall friction force becomes

$$F_f = \mu_m \cdot P_m \cdot A \cdot (1 - \chi) + \mu_f \cdot P_f \cdot A \cdot \chi, \tag{1.122}$$

and the COF is

$$\mu = \frac{F_f}{F_N} = \frac{\mu_m \cdot P_m \cdot A \cdot (1 - \chi) + \mu_f \cdot P_f \cdot A \cdot \chi}{P_m \cdot A \cdot (1 - \chi) + P_f \cdot A \cdot \chi}. \tag{1.123}$$

The wear volume of the composite after sliding the distance s can be estimated by

$$W_V = k_m \cdot P_m \cdot A \cdot (1 - \chi) \cdot s + k_f \cdot P_f \cdot A \cdot \chi \cdot s, \tag{1.124}$$

whereby

$$\Delta h_m = k_m \cdot P_m \cdot s \tag{1.125}$$

and

$$\Delta h_f = k_f \cdot P_f \cdot s \tag{1.126}$$

represent the height reduction of the matrix and the filler, respectively. Since the lat-ter must be equal (otherwise only the one with a smaller height change would support the load), we can derive

$$P_f = \frac{k_m}{k_f} P_m, \tag{1.127}$$

which indicates the load redistribution capability of composites with more wear-resistant component (filler) experiencing larger contact pressures than the less wear-resistant component (matrix). Using this, we can derive a general expression for the COF in dependency on the wear rates and the concentration:

$$\mu = \frac{\mu_m \cdot k_f \cdot (1-\chi) + \mu_f \cdot k_m \cdot \chi}{k_f \cdot (1-\chi) + k_m \cdot \chi}, \tag{1.128}$$

whereby $\mu = \mu_m$ for $\chi = 0$ and $\mu = \mu_f$ for $\chi = 1$. The respective wear volume of the composite

$$W_V = k_m \cdot P_m \cdot d \cdot A, \tag{1.129}$$

and the wear rate equals the one from the matrix. However, the filler allows that the pressure experienced by the matrix is lower since the filler contributes to the load support (load sharing). Therefore, $k = k_m$ for $\chi = 0$ and $k = k_f$ for $\chi = 1$.

Note that the expressions derived earlier are a very simplified formulation of the underlying processes (neglecting tribo-chemical processes and the formation of tribo-films, etc.) and assume uniform wear of materials undergoing the tribological process. However, it is crucial to recognize that real-world applications may exhibit diverse material behaviors. In practical scenarios, materials may not wear uniformly; instead, the matrix itself can become a focal point for stress propagation. In such cases, the entire filler particle or fiber might be extracted from the material, rather than experiencing uniform wear. Hence, the significance of binders and the arrangement of wear-resistant fillers cannot be overstated. The role of these elements becomes pivotal as they influence whether the wear point will be the matrix or the wear-resistant fillers. Also, the distribution of these fillers plays a crucial role in determining the overall performance of the material in a tribological setting. In a broader context, considering that asperities define the distribution of contacts, it is imperative that wear-resistant particles are larger than the asperities themselves. This ensures that the particles are not simply dislodged from the contact points, emphasizing the need for careful consideration of particle size and distribution to optimize material performance in tribological applications.

1.9   TRIBO-EFFECTS

Besides the observed friction and wear behavior, tribological contacts may result in a rise of interesting effects accompanying the processes. These effects can largely affect the reliability of the tribological systems and should be considered.

1.9.1   TRIBO-CORROSION

Tribo-corrosion is a complex phenomenon that occurs when two or more materials come into contact and experience both mechanical wear and corrosion simultaneously. The resulting wear and corrosion processes can interact with each other, leading to accelerated material degradation and failure. The severity of tribo-corrosion depends on a variety of factors (Figure 1.44), including the materials involved, the environment in which the materials operate, the nature of the mechanical stresses

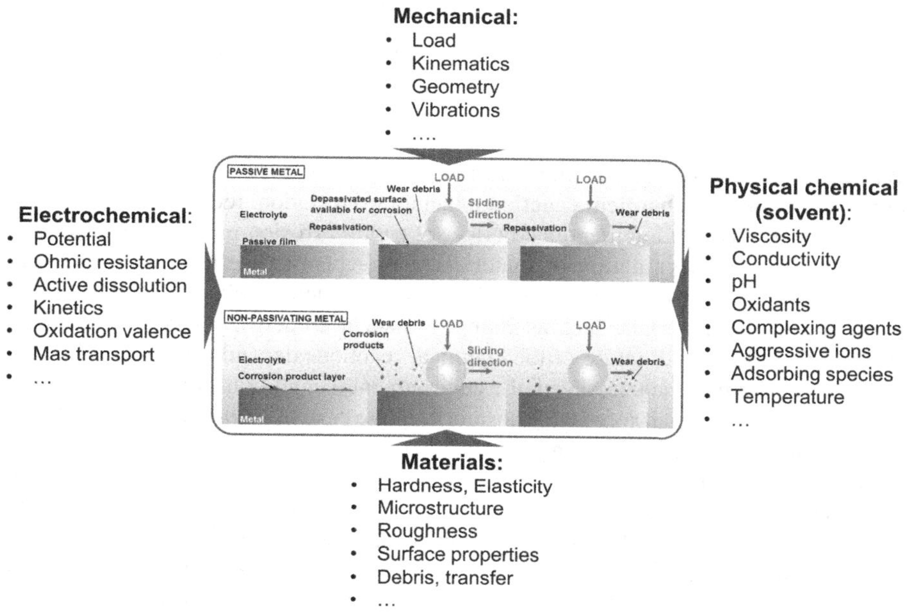
Mechanical:
• Load
• Kinematics
• Geometry
• Vibrations
• ....
Electrochemical:
• Potential
• Ohmic resistance
• Active dissolution
• Kinetics
• Oxidation valence
• Mas transport
• ...
PASSIVE METAL
LOAD
Wear debris
Depassivated surface available for corrosion
Electrolyte
Sliding direction
Repassivation
Passive film
Metal
NON-PASSIVATING METAL
Wear debris
Corrosion products
Electrolyte
Corrosion product layer
Physical chemical (solvent):
• Viscosity
• Conductivity
• pH
• Oxidants
• Complexing agents
• Aggressive ions
• Adsorbing species
• Temperature
• ...
Materials:
• Hardness, Elasticity
• Microstructure
• Roughness
• Surface properties
• Debris, transfer
• ...

FIGURE 1.44 Schematic illustration of potential tribo-corrosion mechanisms in passive and non-passivating (center) tribo-corrosion systems with the respective affecting variables. (Redrawn and modified from Refs. [56,57] with permission.)

placed on the materials, and the respective electrochemical activity. The tribo-corrosion processes occur in three main stages:

- Initiation
- Propagation
- Acceleration

During the **initiation stage**, the material surface is damaged due to mechanical wear, which exposes the underlying metal to the surrounding environment. This exposure can lead to the formation of localized corrosion sites, such as pits or cracks. During the **propagation stage**, the corrosion sites grow and deepen, leading to further mechanical wear and material loss. At this stage, the mechanical wear and corrosion processes can interact with each other, leading to a self-accelerating feedback loop that can result in rapid material failure. The **acceleration stage** occurs when the material has experienced significant wear and corrosion damage and approaches the end of its usefulness life. At this stage, the rate of material degradation increases rapidly, leading to catastrophic failure. The process is often accompanied by dramatic increase in friction and generation of large material wear particles and debris delaminated from the material surface. Tribo-corrosion can occur in a variety of settings, such as in machinery, biomedical implants, and automotive applications. Currently, the applications of the imminent interest include biomedical applications, mostly involving bioimplants such as artificial hip joints or dental implants. Since the body fluids are highly corrosive environments and most bioimplants rely on the use of metals, the tribo-corrosion processes pose the major concern of reliability and lifetime of materials.

1.9.2 Tribo-Chemistry

When two materials come into contact and are subjected to frictional forces, they can undergo a variety of **chemical reactions**, including oxidation, reduction, and surface activation. While the aforementioned effect of tribo-corrosion mostly focuses on the oxidation-induced degradation of materials, tribo-chemistry effects can positively affect friction and wear. These chemical reactions can lead to changes in the surface properties of the materials, such as changes in surface energy, roughness, and wettability. The extent and nature of the chemical reactions depend on a variety of factors, such as the materials involved, the environmental conditions, and the nature and intensity of the frictional forces. For instance, in the case of metal-on-metal contacts, the formation of metal oxides can occur due to the exposure of the metallic surface to ambient oxygen. Concerning lubricated systems, tribo-chemistry plays a critical role in the performance of the lubricant. The chemical reactions that occur at the interface between the lubricant and the material surfaces can lead to the formation of protective films or boundary layers that reduce wear and friction. The choice of lubricant and its composition can significantly affect the tribo-chemistry of the system. One interesting example of the tribo-chemistry is when the interface activity facilitated by the local heating and contact stresses created favorable conditions for reconstructions of solid lubricants toward forming onion-like carbon. As a result, friction may drop to negligible values ("**superlubricity**") and wear of the system can be eliminated.

1.9.3 Tribo-Catalysis

The basic principle of tribo-catalysis is that the mechanical forces generated during friction can create conditions that enhance **the rate of chemical reactions by lowering the activation energy** required for the reaction to occur. In contrast to the tribo-chemical activity, the material surfaces in contact can act as tribo-catalysts, providing active sites for the reaction to take place. Specifically, to initiate the growth of a tribo-film, the system should gain a certain activation energy to overcome the energy barrier. The **Arrhenius model** has been employed to understand the mechanochemistry of the sliding contacts and to estimate the growth rate Γ for the tribo-film formation as a function of temperature and stress conditions

$$\Gamma = \Gamma_0 \, e^{-\frac{\Delta U_{act} - \sigma \Delta V_{act}}{k_B T}}, \qquad (1.130)$$

where $\Gamma_0 = \nu V_m$ depends on the effective attempt frequency ν and the molar volume of the growth species V_m, ΔU_{act} is the internal activation energy (i.e., the energy barrier in the absence of stress), σ represents the mean value of the stress component affecting the activation barrier (assumed to be the compressive contact pressure, i.e., Hertzian contact pressure), ΔV_{act} stands for the activation volume, k_B reflects Boltzmann's constant, and T is the absolute temperature. As shown for the ZDDP film growth or growth of organic species from methyl thiolate, n-decane, or allyl alcohol vapor both on the nano- and the macro-scale, the activation energy is easier to provide by applying shear stresses rather than normal stresses. The effectiveness of tribo-catalysis depends on a variety of factors, such as the nature and morphology of the catalysts, the nature of the reactants, and the environmental conditions, such as temperature and pressure. The processes can involve various processes, such as mechanical activation, surface oxidation, and the formation of reactive species. Tribo-catalysis has a wide range of potential applications. For instance, it can be used in the development of self-lubricating materials, where the surface acts as a catalyst for the formation of a protective layer of lubricant during frictional contact. In energy conversion systems, tribo-catalysis can be employed to enhance the efficiency of energy conversion processes (tribo-catalytic fuel cells, where mechanical forces are used to catalyze the oxidation of fuel). Tribo-catalysis also has potential applications in environmental remediation. For instance, it can be utilized to enhance the degradation of pollutants in contaminated soils and water by catalyzing the breakdown of the pollutants.

1.9.4 TRIBO-CHARGING

Tribo-charging, also known as contact electrification, is the process by which two materials become electrically charged when brought into contact before being again separated. This phenomenon has been known since ancient times and was already observed by the Greek philosopher Thales, who noticed that rubbing amber against fur produced a static charge. The process of tribo-charging involves the transfer of electrons between both materials in contact due to the frictional forces generated when they are rubbed together. When the materials come into contact, electrons are transferred from one material to the other, resulting in an electrical charge separation. This creates an electric field that can build up a static electricity. The amount and polarity of the charge generated during tribo-charging depend on the materials involved, the humidity of the surrounding environment, and the speed and force of the rubbing. Materials that are more likely to donate electrons, such as those with a lower work function, will become positively charged, while materials that are more likely to accept electrons (higher work function) will become negatively charged.

Tribo-charging has a wide range of practical applications. For instance, it is used in electrostatic precipitators to remove particles from industrial exhaust gases, ink-jet printers to control the movement of ink droplets, and photocopiers to transfer toner particles to paper. Tribo-charging plays also a role in the adhesion and bonding of materials, such as in the manufacturing of composite materials. However, tribo-charging can also have negative effects, which can be seen in electrostatic discharge, which can occur when a static charge build-up is suddenly released, leading to potentially damaging sparks or shocks. Tribo-charging can also build up a static electricity in fuel tanks, which can create a fire hazard. To mitigate these negative effects, engineers and materials scientists can employ a variety of strategies, such as using antistatic materials or coatings, designing systems to minimize rubbing and friction, and ensuring proper grounding of conductive materials.

1.9.5 Tribo-Plasma

Tribo-plasma is a multifaceted phenomenon that manifests when two materials come into contact and experience extreme frictional forces, generating a **plasma state of matter**. The generation of tribo-plasma is a complex process that involves several stages, including **mechanical deformation, damage**, and the **creation of nucleation sites** on the material surfaces. During the frictional contact, surface molecules are vaporized, and their ionization ensues due to the high energy levels present leading to the formation of a plasma. The tribo-plasma generated can exert various effects on the material surfaces, including the formation of new chemical species and surface property modification such as adhesion, roughness, and wettability.

1.10 MECHANISMS OF FRICTION AND WEAR CONTROL

Controlling friction and wear is crucial in various engineering and industrial applications to improve the longevity and efficiency of mechanical systems. This implies the selection of the right materials for the respective applications that feature

- Chemical inertness → Low adhesion
- High hardness and elastic modulus → Good deformation behavior and resistance to abrasion
- Residual stresses → Resistance to surface spallation
- Resistance against oxygen, coatings, lubricants, the ambient media as well as the counter-body material → Limited tribo-chemical reactions

If the materials themselves cannot be freely chosen due to application-specific requirements, several mechanisms and strategies can be employed to manage and reduce friction and wear:

- Minimize the tribological **stress collective**, i.e., the applied load or pressure (via the geometry), velocity or kinematics, ambient temperature, humidity, pressure, and atmosphere. This resembles the most direct approach without the necessity to change the involved materials. However, it can only be influenced within certain limits as it is mainly driven by the application's primary function and requirements.
- **Liquid lubricants** are commonly used to reduce friction and wear between moving components in mechanical systems. The basic principle behind their functioning is the formation of a lubricating film between contacting surfaces, which separates them and minimizes direct material contact. This will be discussed in more detail in Chapter 3.
- The surfaces, i.e., their topography as well as properties, can be modified by **surface engineering**. Controlling the surface topography (i.e., roughness) by polishing or grinding can substantially impact friction and wear. Smoother surfaces tend to reduce the contact area and, therefore, friction, while controlled roughness can retain lubricants and promote hydrodynamic lubrication (Chapter 3).

Techniques like carburizing, nitriding, and heat treatment can improve the hardness of the surface, making it more wear-resistant. Coating the surfaces with materials such as polymers, ceramics, or thin films of lubricants can reduce friction and wear. These coatings act as a barrier between the contacting surfaces, preventing direct contact. Some applications benefit from the use of solid lubricants like graphite, molybdenum disulfide, or PTFE (Chapter 4). These materials reduce friction by providing low-friction tribo-films on the surface.

Some desirable characteristics when applying low-friction and low-wear surface modifications are

- Good adhesion of a boundary layer or coating to the base body
- Low friction against the counter-body
- No or lowest possible influence of the base body during treatment or coating, i.e., no dimensional changes and no drop in hardness, as well as easy to re-machine and finish
- Low costs (uncomplicated, fast, environmentally friendly)
- Decorative coloring

Combining liquid lubricants with surface engineering can further enhance the effectiveness of friction and wear control in various industrial applications. The choice of strategy depends on the specific requirements of the system and operating conditions.

1.11   CHECK YOURSELF

- What are the key components of a tribological system and how are they defined?
- How can you differentiate between open and closed systems in the context of tribological systems, and what are their characteristics?
- What is the system function in a tribological system, and how is it related to the interacting surface pair?
- Provide examples of energy-converting, information-converting, and material-converting tribological systems.
- How does the scale of tribological systems vary, and what are some examples of applications at different scales, from nano- and micro-level to macro-level?
- What are the four broad categories of solid engineering materials, and what are their unique properties and applications in engineering?
- What is Hooke's law, and how does it relate to the elastic behavior of materials? How is the yield strength defined for ductile materials?
- In the context of tribology, what additional material properties beyond classic elastic or plastic properties are important, and how do they influence surface interactions and performance in various applications?
- What are some of the structural alterations that can occur at the surface of a material, and how do they affect its properties?

- How is surface tension defined, and why is it important in various processes involving liquids and solid surfaces?
- What are the key parameters used to quantify surface roughness, and how do they influence the performance of components in engineering applications?
- What are the main types of contacts in tribology, and how do they differ in terms of forces and relative motions between contacting bodies?
- What is the difference between conformal and non-conformal contacts? Provide examples for each type.
- Describe the different kinematic types of contacts, including sliding, spinning, rolling, and rolling-sliding contacts. How do they vary in terms of relative motion between the contacting bodies?
- Explain the key parameters such as sum velocity, relative velocity, slide-to-roll ratio, and mean or effective velocity in the context of contact mechanics.
- What are the key assumptions and conditions that must be met for the Hertzian theory to be applicable in contact problems?
- How are the contact area dimensions, maximum pressure, and deformation calculated for different contact configurations, such as sphere on sphere, sphere on plane, cylinder on plane, and cylinder on cylinder?
- What are the main strength hypotheses used to determine equivalent uni-axial stress states for multi-axial stress conditions, and what are the applications and limitations of each of these hypotheses in the context of contact mechanics and material failure analysis?
- What are the key differences adhesive and non-adhesive contacts?
- How do tangential forces affect the distribution of stress in a contact zone, and what is the significance of the friction coefficient in this context?
- What are residual stresses in tribological contacts, and how can they impact the overall stress state of a material?
- How do rough surfaces differ from idealized smooth surfaces in contact mechanics, and what is the concept of the "real area of contact"?
- Describe the Archard and Greenwood–Williamson models for modeling rough surface contacts. How do they account for the complex nature of surface topography and interactions between asperities?
- What is the coefficient of friction and how is it defined?
- Explain the difference between static friction and dynamic kinetic friction.
- What is the stick-slip phenomenon, and how does it impact various fields, including machinery and geology?
- Describe the three phases of friction processes, namely energy introduction, energy conversion, and energy dissipation.
- What is the role of adhesion in friction, and how can it be quantified?
- Describe the deformative friction mechanism, particularly the role of surface asperities in generating abrasive friction, and explain how different material properties affect this mechanism.
- How does friction generate heat, and what factors influence the temperature increase in tribological contacts during friction? List and briefly

explain the different wear measures used to quantify the effects of wear on a material or component.

- How do abrasion, adhesive wear, and tribo-chemical reactions contribute to wear in sliding contacts, and what are the characteristic changes or wear phenomena associated with each mechanism?
- What is Archard's wear law, and how does it relate to the volumetric wear coefficient in tribological systems?
- Describe the key stages in the progression of wear in a tribological system, and explain how the nature of wear mechanisms can change throughout these stages.
- What is the primary difference between component wear and system wear in a tribological context, and why is it important to distinguish between the two?
- What factors influence the tribological properties of metals, and how do factors like chemical composition, microstructure, and mechanical properties affect their friction and wear behavior?
- Describe the common wear mechanisms experienced by metals, particularly at high pressures and under different conditions like delamination wear, corrosion wear, and melt wear.
- Why do ceramics generally have low coefficients of friction compared to metals, and what factors, such as grain size and surface energy, influence their wear resistance?
- Explain how ambient conditions, like humidity and sliding velocity, can impact the tribological behavior of ceramics, and why different ceramics may exhibit different dependencies on these factors.
- What are the primary characteristics and behaviors of polymers in tribological systems, and how do their properties differ from metals and ceramics in terms of friction, wear, and lubrication?
- What are the main stages of tribo-corrosion, and how do they contribute to material degradation in tribological systems?
- How does tribo-chemistry affect friction, wear, and the performance of lubricants, and what factors influence the extent of chemical reactions in tribological contacts?
- What is the principle of tribo-catalysis, and how can it be used to enhance the efficiency of energy conversion processes and environmental remediation?
- How does tribo-charging occur, and what practical applications and potential negative effects are associated with this phenomenon?

1.12   SOLUTIONS FOR THE EXERCISES

EXERCISE 1.1

Tribo-system	Base Body	Counter-Body	Intermediate Medium	Environmental Medium	System Type
Gear transmission	Gear wheel tooth I	Gear wheel tooth II	Gear oil	Oil–air mixture	Closed
Wheel/rail	Wheel	Rail	Humidity	Humidity, air	Open
Journal bearing	Bearing shell	Shaft	Bearing oil	Oil–air mixture	Closed
Excavator	Shovel	Dredged material	Particles	Air, dust	Open
Crushing plant	Grinding wheel	Crush-bakes	Ground material	Air, dust	Open

EXERCISE 1.2

- $R_t = R_p + R_v = z_{max} + |z_{min}| = A + |-A| = 2 \cdot A$

$$R_a = \frac{1}{L} \int_0^L |z(x)| \, dx = \frac{1}{L} \left[\frac{L \cdot A}{4} + \left| \frac{L \cdot (-A)}{4} \right| \right] = 0.5 \cdot A,$$

whereby the integral simply represented the area of the triangles. It is interesting to note that the resulting averaged roughness does not depend on the selected representative length as long as it accounts for the features seen on the surface. As such, selected $0.5 \cdot L$ as representative length, we would get the same R_a. For the sake of simplicity, we use $0.25 \cdot L$ for the representative length:

$$R_q = \sqrt{\frac{4}{L} \int_0^{\frac{L}{4}} z(x)^2 \, dx} = \sqrt{\frac{4}{L} \left(\frac{4A}{L} \right)^2 \int_0^{\frac{L}{4}} (x)^2 \, dx} = \sqrt{\frac{64 \cdot A^2}{3 \cdot L^3} [x^3]_0^{\frac{L}{4}}}$$

$$= \sqrt{\frac{64 \cdot A^2}{3 \cdot L^3} \left[\left(\frac{L}{4} \right)^3 - 0 \right]} = \sqrt{\frac{A^2}{3}} \approx 0.58 \cdot A$$

- $R_t = 2 \cdot A$

$$R_a = \frac{1}{L} \left(\int_0^{\frac{L}{2}} A \cdot \sin\left(\frac{2\pi}{L} x\right) dx - \int_{\frac{L}{2}}^{L} A \cdot \sin\left(\frac{2\pi}{L} x\right) dx \right)$$

$$= \frac{1}{L} \left(-\frac{AL}{2\pi} \left[\cos\frac{2\pi}{L} x \right]_0^{\frac{L}{2}} + \frac{AL}{2\pi} \left[\cos\frac{2\pi}{L} x \right]_{\frac{L}{2}}^{L} \right) = \frac{AL}{2\pi L}(-(-1-1)+(1+1))$$

$$= \frac{2A}{\pi} \approx 0.64 \cdot A$$

Again, this result is independent of the chosen representative length.

$$R_q = \sqrt{\frac{4}{L} \int_0^{\frac{L}{4}} A^2 \sin^2\left(\frac{2\pi}{L} x\right) dx} = \sqrt{\frac{1}{L} 4 \int_0^{\frac{L}{4}} A^2 \left[\frac{1}{2} - \frac{1}{2}\cos\left(\frac{4\pi}{L} x\right)\right] dx}$$

$$= \sqrt{\frac{2}{L} A^2 \left[x - \frac{L}{4\pi} \sin\left(\frac{4\pi}{L} x\right)\right]_0^{L/4}} = \frac{A}{\sqrt{2}} \approx 0.71 \cdot A$$

- $R_t = 2 \cdot A$

$$R_a = \frac{1}{4 \cdot A} \pi A^2 = \frac{\pi \cdot A}{4} \approx 0.79 \cdot A$$

We use the radius diameter of the circle as a representative length:

$$R_q = \sqrt{\frac{1}{4 \cdot A} 2 \int_0^{2A} \left(A^2 - (x-A)^2\right) dx} = \sqrt{\frac{1}{2A} \int_0^{2A} \left(2xA - A^2\right) dx}$$

$$= \sqrt{\frac{1}{2A} \left(\left[Ax^2\right]_0^{2A} - \left[\frac{x^3}{3}\right]_0^{2A} \right)}$$

$$= \sqrt{\frac{1}{2A} \left(4A^3 - \frac{8A^3}{3}\right)} = \sqrt{\frac{2}{3}} A \approx 0.82A$$

Based on these examples, we can observe that larger R_a/R_q ratios lead broader shapes for the peaks.

EXERCISE 1.3

- Sum velocity:

$$u_s = |u_1 + u_2| = |r \cdot \omega_1 + u_2| = \left| 10^{-2}\,\mathrm{m} \cdot 50\frac{1}{\mathrm{s}} + (-1)\frac{\mathrm{m}}{\mathrm{s}} \right| = 0.5\frac{\mathrm{m}}{\mathrm{s}}$$

- Relative velocity:

$$u_r = |u_1 - u_2| = \left| 0.5\frac{\mathrm{m}}{\mathrm{s}} - (-1)\frac{\mathrm{m}}{\mathrm{s}} \right| = 1.5\frac{\mathrm{m}}{\mathrm{s}}$$

- Slide-to-roll ratio:

$$SRR = 2\frac{|u_1 - u_2|}{u_1 + u_2} = -6$$

- Type of contact:

$$u_1 > 0 \wedge u_2 < 0 \wedge |u_1| < |u_2|$$

or

$$v_r > |u_2| \wedge v_s < u_2 \wedge -\infty < SRR < -2$$

→ Pure sliding (in opposite directions)
- Sum velocity:

$$u_s = 1\frac{\mathrm{m}}{\mathrm{s}}$$

- Relative velocity:

$$u_r = 1\frac{\mathrm{m}}{\mathrm{s}}$$

- Slide-to-roll ratio:

$$SRR = -2$$

- Type of contact:

$$u_1 = 0 \wedge u_2 < 0$$

or

$$v_r = |u_2| \wedge v_s = u_2 \wedge |SRR| = 2$$

→ Pure sliding

- Sum velocity:

$$u_s = 1.5 \frac{m}{s}$$

- Relative velocity:

$$u_r = 0.5 \frac{m}{s}$$

- Slide-to-roll ratio:

$$SRR = 0.67$$

- Type of contact:

$$u_1 > 0 \wedge u_2 > 0 \wedge |u_1| < |u_2|$$

or

$$v_r \langle |u_2| \wedge v_s \rangle u_2 \wedge 0 < SRR < 2$$

→ Rolling-sliding
- Sum velocity:

$$u_s = 2 \frac{m}{s}$$

- Relative velocity:

$$u_r = 0 \frac{m}{s}$$

- Slide-to-roll ratio:

$$SRR = 0$$

- Type of contact:

$$u_1 = u_2$$

or

$$v_r = 0 \wedge v_s = 2u_2 \wedge |SRR| = 0$$

→ Pure rolling

EXERCISE 1.4

- Isotropic, homogenous, linear-elastic material ✓
  (mostly given for not too high stress, i.e., pressures below 4 GPa)
- Dry contact ☒
  (usually not the case due to lubrication by oils; more accurate for higher pressures and lower velocities)
- Normal contact ☒
  (would only be the case without motion or for very low friction)
- Ideally smooth and frictionless surfaces ~(not exactly the case, but very small surface roughness and low friction due to lubricant film)
- No residual stresses ✓
- Flat contacts ~(more or less given; depends on the gear type and tooth profile)
- Small contact area compared to curvature ✓

EXERCISE 1.5

1. Sphere on sphere
2. Sphere on plane
3. Cylinder on plane
4. Cylinder on cylinder

EXERCISE 1.6

$$\rho^* = \frac{1}{0.5\,d_s} + \frac{1}{0.5\,d_s} + \frac{1}{0.5\,d_r} - \frac{1}{r_g} = \cdots \approx 0.26 \text{ mm}^{-1}$$

$$\cos\tau = \frac{\dfrac{1}{0.5\,d_s} - \dfrac{1}{0.5\,d_s} + \dfrac{1}{0.5\,d_r} + \dfrac{1}{r_g}}{\rho^*} \approx 0.70$$

$$\xi = 1.91, \eta = 0.607$$

$$E' = \frac{1-v^2}{E} = \cdots \approx 0.0043 \text{ GPa}^{-1}$$

$$a = \xi \cdot \sqrt[3]{\frac{3 \cdot F \cdot E'}{\rho^*}} = \cdots \approx 0.327 \text{ mm}$$

$$b = \eta \cdot \sqrt[3]{\frac{3 \cdot F \cdot E'}{\rho^*}} = \cdots \approx 0.104 \ \mu m$$

$$A = a \cdot b \cdot \pi = \cdots \approx 0.107 \ mm^2$$

$$p_{max} = \frac{3 \cdot F}{2 \cdot a \cdot b \cdot \pi} = \cdots \approx 1.4 \ GPa$$

- $\rho^* \approx 0.23 \ mm^{-1}$

$\cos \tau \approx 0.93$

$\xi = 3.59, \eta = 0.426$

$a \approx 0.642 \ mm$

$b \approx 0.076 \ \mu m$

$A \approx 0.153 \ mm^2$

$p_{max} \approx 1.0 \ GPa$

Note that by increasing the conformity of raceway with the ball, it is possible to substantially increase the contact area and decrease the maximum pressure.

EXERCISE 1.7

$$a = \sqrt[3]{\frac{3 \cdot F \cdot \frac{1-v^2}{E}}{4\left(\frac{1}{d_1} + \frac{1}{d_2}\right)}} = \sqrt[3]{\frac{3 \cdot 10 \ N \cdot \frac{1-0.3^2}{210000 \ N/mm^2}}{4\left(\frac{1}{20 \ mm} + \frac{1}{20 \ mm}\right)}} = 69 \ \mu m$$

$$p_{max} = \frac{1}{\pi} \sqrt[3]{\frac{6 \cdot F}{\left(\frac{1-v^2}{E}\right)^2} \left(\frac{1}{d_1} + \frac{1}{d_2}\right)^2}$$

$$= \frac{1}{\pi} \sqrt[3]{\frac{6 \cdot 10 \text{ N}}{\left(\frac{1-0.3^2}{210000 \text{ N/mm}^2}\right)^2} \left(\frac{1}{20 \text{ mm}} + \frac{1}{20 \text{ mm}}\right)^2} = 1.0 \text{ GPa}$$

Note that for the same materials and a 10 times higher normal load, we obtained the same contact pressure as in Exercise 1.6. This is due to highly concentrated contact here (sphere-on-sphere) compared to the conformity between sphere and raceway of Exercise 1.6.

- $a = \ldots = 55 \ \mu m$

$$p_{max} = \ldots = 1.6 \text{ GPa}$$

EXERCISE 1.8

- $a = l, b = \sqrt{\dfrac{4 \cdot F \cdot E^* \cdot d}{\pi \cdot l}} \ , p_{max} = \sqrt{\dfrac{F}{\pi \cdot l \cdot E^* \cdot d}}$

- $a = l, b = \sqrt{\dfrac{4 \cdot F \cdot E^*}{\pi \cdot l \left(\dfrac{1}{d_1} + \dfrac{1}{d_2}\right)}} \ , p_{max} = \sqrt{\dfrac{F}{\pi \cdot l \cdot E^*} \left(\dfrac{1}{d_1} + \dfrac{1}{d_2}\right)}$

EXERCISE 1.9

- Steel: $\Delta T = p \cdot \mu \cdot \dfrac{va}{k} = \dfrac{1 \cdot 10^6 \text{ Pa} \cdot 0.2 \cdot 1 \dfrac{m}{s} \cdot 5 \cdot 10^{-3} \text{ m}}{20 \dfrac{W}{m \cdot K}} = 50 \text{K}$

- Copper: $\Delta T = p \cdot \mu \cdot \dfrac{va}{k} = \dfrac{1 \cdot 10^6 \text{ Pa} \cdot 0.2 \cdot 1 \dfrac{m}{s} \cdot 5 \cdot 10^{-3} \text{ m}}{400 \dfrac{W}{m \cdot K}} = 2.5 \text{K}$

EXERCISE 1.10

- $\Delta T = p \cdot \mu \cdot \dfrac{v \cdot a}{2k} = \dfrac{1 \cdot 10^6\,\text{Pa} \cdot 0.2 \cdot 1\frac{\text{m}}{\text{s}} \cdot 5 \cdot 10^{-3}\,\text{m}}{2 \cdot 20\dfrac{\text{W}}{\text{m} \cdot \text{K}}} = 25\text{K}$

- In the case of 10 times sliding speed increase:

$$\Delta T = p \cdot \mu \cdot \frac{v \cdot a}{2k} = \frac{1 \cdot 10^6\,\text{Pa} \cdot 0.2 \cdot 10\frac{\text{m}}{\text{s}} \cdot 5 \cdot 10^{-3}\,\text{m}}{2 \cdot 20\dfrac{\text{W}}{\text{m} \cdot \text{K}}} = 250\text{K}$$

- For 100 times sliding speed increase:

$$\Delta T = p \cdot \mu \cdot \frac{v \cdot a}{2k} = \frac{1 \cdot 10^6\,\text{Pa} \cdot 0.2 \cdot 100\frac{\text{m}}{\text{s}} \cdot 5 \cdot 10^{-3}\,\text{m}}{2 \cdot 20\dfrac{\text{W}}{\text{m} \cdot \text{K}}} = 2500\text{K}$$

Such temperature increase seems to be unrealistically high.

EXERCISE 1.11

$$Pe = 100\frac{\text{m}}{\text{s}} \cdot 5 \cdot 10^{-3}\,\text{m} \cdot 0.2 \cdot 10^6 \frac{\text{s}}{\text{m}^2} = 10^5$$

$$\Delta T = 1.6\ Pe^{-\frac{1}{2}}\frac{\dot{q}_1 \cdot a}{k} = 1.6 \cdot \left(10^5\right)^{-\frac{1}{2}} \cdot \frac{1 \cdot 10^6\,\text{Pa} \cdot 0.2 \cdot 100\frac{\text{m}}{\text{s}} \cdot 5 \cdot 10^{-3}\,\text{m}}{20\dfrac{\text{W}}{\text{m} \cdot \text{K}}} = 25\text{K}$$

EXERCISE 1.12

- We assume that all generated heat is dissipated by the steel:

$$\Delta T = \frac{\dot{q} \cdot a}{k} = \frac{\mu \cdot P \cdot V}{k} a$$

$$PV = \frac{\Delta T \cdot k}{\mu \cdot a} = \frac{(150K - 25K) \cdot 40 \dfrac{W}{m \cdot K}}{0.3 \cdot 6 \cdot 10^{-3} m} = 2.8 \cdot 10^{6} \frac{N}{m \cdot s}$$

- This means that for $P = 2.8$ MPa, v is limited to 1 m/s.
- In the case of P being increased by a factor of 2 (up to ~5 MPa), v is limited to only 0.5 m/s. If a is increased by a factor of 2, the $P \cdot v$-limit decreases by a factor of 2, leading to lower applicable pressure and velocity.

Note: In the above solution, we considered a stationary contact. In the case of the moving contact, especially with high velocities not allowing to equilibrate the temperatures and reach stable heating in the contact, there is no single $P \cdot v$-limit.

REFERENCES

1. K. Holmberg, A. Erdemir, Influence of tribology on global energy consumption, costs and emissions, *Friction* 5 (2017) 263–284. https://doi.org/10.1007/s40544-017-0183-5.
2. H. P. Jost, *Lubrication: Tribology; Education and Research; Report on the Present Position and Industry's Needs.* Department of Education and Science, HM Stationery Office, 1966.
3. H. Jost, Tribology - Origin and future, *Wear* 136 (1990) 1–17. https://doi.org/10.1016/0043-1648(90)90068-L.
4. V. L. Popov, Is tribology approaching its golden age? Grand challenges in engineering education and tribological research, *Frontiers in Mechanical Engineering* 4 (2018) 11816. https://doi.org/10.3389/fmech.2018.00016.
5. D. Dowson, Men of Tribology: Leonardo da Vinci (1452–1519), *Journal of Lubrication Technology* 99 (1977) 382–386. https://doi.org/10.1115/1.3453230.
6. V. L. Popov, *Contact Mechanics and Friction: Physical Principles and Applications*, Springer, Berlin, Heidelberg, 2010.
7. D. Dowson, Men of Tribology: Guillaume Amontons (1663–1705) and John Theophilus Desaguliers (1683–1744), *Journal of Lubrication Technology* 100 (1978) 2–5. https://doi.org/10.1115/1.3453109.
8. H. van Leeuwen, Petrus van Musschenbroek (1692–1761), man of tribology, *Proceedings of the IMechE* 235 (2021) 2537–2551. https://doi.org/10.1177/13506501211042704.
9. D. Dowson, Men of Tribology: Charles Augustin Coulomb (1736–1806) and Arthur-Jules Morin (1795–1880), *Journal of Lubrication Technology* 100 (1978) 148–155. https://doi.org/10.1115/1.3453126.

10. D. Dowson, Men of Tribology: Heinrich Rudolph Hertz (1857–1894) and Richard Stribeck (1861–1950), *Journal of Lubrication Technology* 101 (1979) 115–119. https://doi.org/10.1115/1.3453287.

11. D. Dowson, Men of Tribology: John William Strutt (Lord Rayleigh) (1842-1919) and Beauchamp Tower (1845–1904), *Journal of Lubrication Technology* 101 (1979) 1–7. https://doi.org/10.1115/1.3453272.

12. D. Dowson, Men of Tribology: Robert Henry Thurston (1839–1903) and Osborne Reynolds (1842–1919), *Journal of Lubrication Technology* 100 (1978) 455–461. https://doi.org/10.1115/1.3453250.

13. M. Woydt, R. Wäsche, The history of the Stribeck curve and ball bearing steels: The role of Adolf Martens, *Wear* 268 (2010) 1542–1546. https://doi.org/10.1016/j.wear.2010.02.015.

14. D. Dowson, Men of Tribology: William Bate Hardy (1864–1934) and Arnold Sommerfeld (1868–1951), *Journal of Lubrication Technology* 101 (1979) 393–397. https://doi.org/10.1115/1.3453381.

15. H. Czichos, K.-H. Habig, *Tribologie-Handbuch*, Springer Fachmedien Wiesbaden, Wiesbaden, 2015.

16. A. I. Vakis, V.A. Yastrebov, J. Scheibert, L. Nicola, D. Dini, C. Minfray, A. Almqvist, M. Paggi, S. Lee, G. Limbert, J.F. Molinari, G. Anciaux, R. Aghababaei, S. Echeverri Restrepo, A. Papangelo, A. Cammarata, P. Nicolini, C. Putignano, G. Carbone, S. Stupkiewicz, J. Lengiewicz, G. Costagliola, F. Bosia, R. Guarino, N.M. Pugno, M.H. Müser, M. Ciavarella, Modeling and simulation in tribology across scales: An overview, Tribology International 125 (2018) 169–199. https://doi.org/10.1016/j.triboint.2018.02.005.

17. M. Marian, *Numerische Auslegung von Oberflächenmikrotexturen für geschmierte tribologische Kontakte*, FAU University Press, Erlangen, Germany.

18. P. Kügler, M. Marian, R. Dorsch, B. Schleich, S. Wartzack, A semantic annotation pipeline towards the generation of knowledge graphs in tribology, *Lubricants* 10 (2022) 18. https://doi.org/10.3390/lubricants10020018.

19. Q. J. Wang, D. Zhu, *Interfacial Mechanics: Theories and Methods for Contact and Lubrication*, First edition, CRC Press Taylor & Francis Group, Boca Raton, London, New York, 2020.

20. J. Peklenik, Industrie-Anzeiger 456–462.

21. H. Hertz, Über die Berührung fester elastischer Körper, *Journal für die reine und angewandte Mathematik* 1882 (1882) 156–171.
    Schaeffler Technologies AG & Co. KG, *Schaeffler Technical Pocket Guide*, Fourth edition, Herzogenaurach, Germany, 2014.

22. R. Grekoussis, T. Michailidis, Näherungsgleichungen zur Nach- und Entwurfsrechnung der Punktberührung nach Hertz, *Konstruktion* 33 (1981) 135–139.

23. F. Fritz, *Modellierung von Wälzlagern als generische Maschinenelemente einer Mehrkörpersimulation*. Zugl.: Karlsruhe, KIT, Diss., 2011, Print on demand, KIT Scientific Publishing, Karlsruhe, 2011.

24. G. Lundberg, Elastische Berührung zweier Halbräume, *Forsch Ingenieurwes* 10 (1939) 201–211. https://doi.org/10.1007/BF02584950.

25. H. Birkhofer, T. Kümmerle, *Feststoffgeschmierte Wälzlager: Einsatz, Grundlagen und Auslegung*, Springer, Berlin, Heidelberg, 2012.

26. K. L. Johnson, K. Kendall, A.D. Roberts, Surface energy and the contact of elastic solids, *Proceedings of the Royal Society of London. Series A* 324 (1971) 301–313. https://doi.org/10.1098/rspa.1971.0141.

27. B. Derjaguin, V. Muller, Y. Toporov, Effect of contact deformations on the adhesion of particles, *Journal of Colloid and Interface Science* 53 (1975) 314–326. https://doi.org/10.1016/0021-9797(75)90018-1.

28. E. Barthel, Adhesive elastic contacts: JKR and more, *Journal of Physics D: Applied Physics* 41 (2008) 163001. https://doi.org/10.1088/0022-3727/41/16/163001.
29. J. F. Archard, Elastic deformation and the laws of friction, *Proceedings of the Royal Society of London. Series A* 243 (1957) 190–205. https://doi.org/10.1098/rspa.1957.0214.
30. J. A. Greenwood, J.B.P. Williamson, Contact of nominally flat surfaces, *Proceedings of the Royal Society A: Mathematical, Physical and Engineering Sciences* 295 (1966) 300–319. https://doi.org/10.1098/rspa.1966.0242.
31. V. Kolli, A. Winkler, S. Wartzack, M. Marian, Micro-scale deterministic asperity contact FEM simulation, *Surface Topography: Metrology and Properties* 10 (2022) 44011. https://doi.org/10.1088/2051-672X/acac42.
32. J. A. Greenwood, C. Putignano, M. Ciavarella, A Greenwood & Williamson theory for line contact, *Wear* 270 (2011) 332–334. https://doi.org/10.1016/j.wear.2010.11.002.
33. J. A. Greenwood, J.H. Tripp, The contact of two nominally flat rough surfaces, *Proceedings of the Institution of Mechanical Engineers* 185 (1970) 625–633. https://doi.org/10.1243/PIME_PROC_1970_185_069_02.
34. A. W. Bush, R.D. Gibson, T.R. Thomas, The elastic contact of a rough surface, *Wear* 35 (1975) 87–111. https://doi.org/10.1016/0043-1648(75)90145-3.
35. J. Jamari, D.J. Schipper, An elastic-plastic contact model of ellipsoid bodies, *Tribology Letters* 21 (2006) 262–271. https://doi.org/10.1007/s11249-006-9038-3.
36. J. A. Greenwood, A simplified elliptic model of rough surface contact, *Wear* 261 (2006) 191–200. https://doi.org/10.1016/j.wear.2005.09.031.
37. T. Hisakado, Effect of surface roughness on contact between solid surfaces, *Wear* 28 (1974) 217–234. https://doi.org/10.1016/0043-1648(74)90163-X.
38. D. Whitehouse, J.F. Archard, The properties of random surfaces of significance in their contact, *Proceedings of the Royal Society of London. Series A* 316 (1970) 97–121. https://doi.org/10.1098/rspa.1970.0068.
39. M. Ciavarella, J.A. Greenwood, M. Paggi, Inclusion of "interaction" in the Greenwood and Williamson contact theory, *Wear* 265 (2008) 729–734. https://doi.org/10.1016/j.wear.2008.01.019.
40. B. Zhao, S. Zhang, Z. Qiu, Analytical asperity interaction model and numerical model of multi-asperity contact for power hardening materials, *Tribology International* 92 (2015) 57–66. https://doi.org/10.1016/j.triboint.2015.05.027.
41. N. Tayebi, A.A. Polycarpou, Modeling the effect of skewness and kurtosis on the static friction coefficient of rough surfaces, *Tribology International* 37 (2004) 491–505. https://doi.org/10.1016/j.triboint.2003.11.010.
42. N. Yu, A.A. Polycarpou, Contact of rough surfaces with asymmetric distribution of asperity heights, *Journal of Tribology-Transactions of the ASME* 124 (2002) 367–376. https://doi.org/10.1115/1.1403458.
43. N. Yu, A.A. Polycarpou, Combining and contacting of two rough surfaces with asymmetric distribution of asperity heights, *Journal of Tribology-Transactions of the ASME* 126 (2004) 225–232. https://doi.org/10.1115/1.1614822.
44. T. Tomota, R. Masuda, Y. Kondoh, T. Ohmori, K. Yagi, Modeling solid contact between rough surfaces with various roughness parameters, *Tribology Transactions* 64 (2021) 178–192. https://doi.org/10.1080/10402004.2020.1820123.
45. W. R. Chang, I. Etsion, D.B. Bogy, An elastic-plastic model for the contact of rough surfaces, *Journal of Tribology-Transactions of the ASME* 109 (1987) 257–263. https://doi.org/10.1115/1.3261348.
46. Y. Zhao, D.M. Maietta, L. Chang, An asperity microcontact model incorporating the transition from elastic deformation to fully plastic flow, *Journal of Tribology* 122 (2000) 86. https://doi.org/10.1115/1.555332.

47. J. Halling, K.A. Nuri, Elastic/plastic contact of surfaces considering ellipsoidal asperities of work-hardening multi-phase materials, *Tribology International* 24 (1991) 311–319. https://doi.org/10.1016/0301-679X(91)90033-6.

48. A. Majumdar, B. Bhushan, Fractal model of elastic-plastic contact between rough surfaces, *Journal of Tribology-Transactions of the ASME* 113 (1991) 1–11. https://doi.org/10.1115/1.2920588.

49. B. N. Persson, Elastoplastic contact between randomly rough surfaces, *Physical Review Letters* 87 (2001) 116101. https://doi.org/10.1103/PhysRevLett.87.116101.

50. B. N.J. Persson, F. Bucher, B. Chiaia, Elastic contact between randomly rough surfaces: Comparison of theory with numerical results, *Physical Review B* 65 (2002). https://doi.org/10.1103/PhysRevB.65.184106.

51. R. I. Taylor, Rough surface contact modelling-a review, *Lubricants* 10 (2022) 98. https://doi.org/10.3390/lubricants10050098.

52. F. P. Bowden, D. Tabor, *The Friction and Lubrication of Solids, [Nachdr.]*, Clarendon Press, Oxford, 2001.

53. K. -H. Zum Gahr, *Abrasiver Verschleiß metallischer Werkstoffe*, Verein Deutscher Ingenieure, 1981.

54. J. F. Archard, Contact and rubbing of flat surfaces, *Proceedings of the Royal Society A: Mathematical, Physical and Engineering Sciences* 24 (1953) 981–988. https://doi.org/10.1063/1.1721448.

55. A. López-Ortega, J.L. Arana, R. Bayón, On the comparison of the tribocorrosion behavior of passive and non-passivating materials and assessment of the influence of agitation, *Wear* 456–457 (2020) 203388. https://doi.org/10.1016/j.wear.2020.203388.

56. M. T. Mathew, P. Srinivasa Pai, R. Pourzal, A. Fischer, M.A. Wimmer, Significance of tribocorrosion in biomedical applications: Overview and current status, *Advances in Tribology* 2009 (2009) 1–12. https://doi.org/10.1155/2009/250986.

2 Tribological Testing Theory

Friction and wear are the two most important parameters studied in tribology. The underlying mechanisms and properties are important to elucidate and optimize the resulting friction and wear characteristics. Therefore, most testing approaches are directed toward three major aspects, which resemble the:

- quantification of the coefficient of friction (COF)
- quantification of the wear rate
- characterization of the worn surfaces

2.1 TRIBOMETRY

All efforts dedicated towards the characterization of tribological properties of materials and systems including COFs, wear volumes, wear rates, temperature distributions, vibrations, among others, can be summarized under **tribometry** (i.e., tribo-testing). However, the significance of various quantifiable parameters often depends on the underlying mechanisms, the measurement method, and the specific objectives. Since tribological testing always involves numerous overlapping influencing factors, it needs to be carefully designed and should include statistical experimental planning and evaluation. Based on a simplified system structure, stress conditions, and environmental conditions, it can be categorized into six categories as summarized in Figure 2.1. While in **field or operational tests**, original and complete systems are replicated and tested against real operating and environmental conditions (applied tribometry), tribo-testing is frequently realized in **bench tests** under laboratory conditions with only practical operating conditions. In **aggregate and component tests**, the focus is further reduced to the testing and examination of individual original

	Application or application-related testing (original structure and simplified stress collective)				Model structure and simplified stress collective	
Category	I	II	III	IV	V	VI
Measuring and testing technology	Machinery field test	Machinery bench test	System bench test	Component test	Specimen test	Model test
System, assembly, model	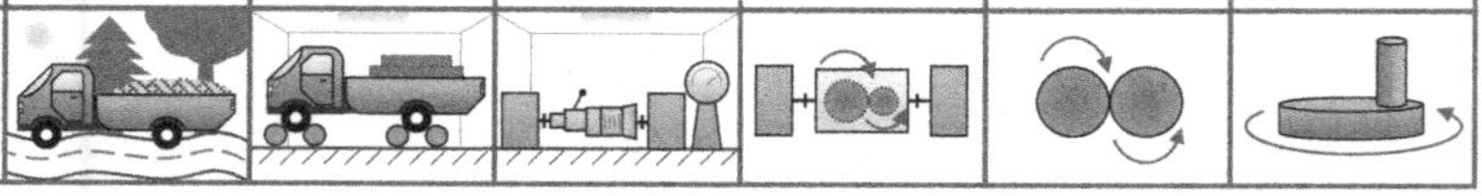					

FIGURE 2.1 Categories of tribological testing. (Redrawn from Ref. [1] with permission by CC BY 4.0.)

 DOI: 10.1201/9781003397519-2

aggregates or components. In **specimen tests**, tests with component-like specimen bodies are conducted under similar stress conditions. Finally, in **model tests**, fundamental investigations of friction and wear processes with simplified bodies under defined stresses are carried out (R&D tribometry). The advantages of the individual testing categories can be linked through an appropriate testing chain.

Tribometry also covers all typical **length scales** of tribology. At small scales, due to the relative size of the surface roughness or undulations, and the contacting bodies, the real or true surface area assumes a vastly important role. A good example and verification for this statement connects with the nanoscale contact zone, which is established in atomic force microscopy (AFM) measurements between a sample and a probing tip. In contrast, at bulk scales, roughness is not that significant, while the applied normal load is of greater impact, such as the load in a passenger vehicle relative to the smoothness of the paved road. Since microscopic phenomena govern the frictional behavior at small scales, intrinsic material properties (surface adhesion, shear, slip, among others), and surface modification (thin films, coatings) are used to control friction. At macroscopic scales, full-film oil-, or grease-based lubrication, and the overall component geometry are more essential to determine and modify the corresponding frictional performance. Therefore, scale-appropriate measurement techniques need to be identified and used to adequately quantify tribological properties.

Friction can generally be measured/quantified by several methods, which can be grouped based on the approach to measure the respective normal and tangential forces, whose origins date back to the days of nascent modern science. Despite being centuries old, the underlying fundamental principles are still used to develop modern-day measurement equipment. The basics of **friction measurement** connect with the

- **Weight Ratio**: Based on the studies by Leonardo da Vinci, measuring friction between a block and a table by changing the load hanging on a cord to quantify and detect the onset of sliding. In this case, the COF can be defined as m_{weight}/m_{block}.
- **Spring Balance**: Relies on pulling a spring connected to a block leading to slow increase in the spring force until the block starts to slide. The reading on the spring balance determines the friction force. In this case, the COF is given by $F_{spring}/F_{block} = k \cdot \Delta x / m_{block} \cdot g$.
- **Tilted Plane**: Uses a block placed on a tilted plane. The block begins to slide upon increasing the tilt angle θ. In this case, the COF can be expressed as $\tan(\theta)$.

Nowadays, friction and wear are typically analyzed using so-called **tribometers** or **tribo-testers**. A modern tribometer has various fundamental components. These are a sample stage, on which the sample is firmly affixed, a load cell to apply the load, a counter-face or -probe transferring the load onto the sample, sensors to detect and control the load applied as well as to measure the resulting frictional force, a motor to move the stage and/or the probe, and potentially a height sensor to estimate the linear wear loss. Ancillary parts include temperature- and environment-controlled lubricant baths or chambers as well as one or multiple forms of *in-situ*, in-line characterization tools. The sensitivity of the used sensors, the range of motion, and sample sizes notably vary and depend on the considered length scale (nano-, micro-, or macro-scale testing). It follows that the **steps in a generic tribometry setup** are

- Fix the sample on the sample stage and the counter-body into the holder.
- Bring the bodies into contact and apply the normal load.
- Move the sample and/or the counter-body, whereby **motion** can be rotational or reciprocating. **Rotational sliding** involves the movement of an object in a circular manner, which can occur in two directions (clockwise or counter-clockwise). It is often referred to as unidirectional sliding because the object moves in a single direction without changing its orientation. In contrast, **reciprocating sliding** involves back-and-forth linear motion, reversing its direction periodically.
- Measure normal and lateral forces (to compute the COF), which can be done using load cells, force transducers, displacement measurement (using capacitance change), or deflections.

The experimental challenge common to many tribometers is that the system's components may not be perfectly aligned at all times. More precisely, the used sample may not be perfectly normal to the counter-body or the load cell due to sample preparation and/or experimental misalignments. The consequences of this misalignment with an angle α for a sensor relative to the actual friction and normal forces are schematically illustrated in Figure 2.2.

While the actual COF during sliding in just one direction would be

$$\mu = \frac{F_F}{F_N},\qquad(2.1)$$

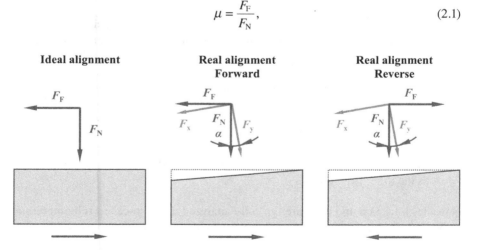

FIGURE 2.2 Perfect alignment (left) and real case with an exaggerated illustration of the misalignment in a reciprocating experimental set-up during forward (middle) and reverse motion (right). In the real case, please note that the faces of the loaded sample are not parallel due to polishing artifact as denoted by dotted line.

the measured COF is

$$\mu' = \frac{F_x}{F_y}, \tag{2.2}$$

where the measured F_x and F_y are the sum of the projections from friction and normal forces given by

$$F_x = F_F \cdot \cos\alpha - F_N \cdot \sin\alpha, \; F_y = F_N \cdot \cos\alpha + F_F \cdot \sin\alpha. \tag{2.3}$$

Consequently, the measured COF can be expressed as

$$\mu' = \frac{F_F \cdot \cos\alpha - F_N \cdot \sin\alpha}{F_N \cdot \cos\alpha + F_F \cdot \sin\alpha} = \frac{\mu \cdot F_N \cdot \cos\alpha - F_N \cdot \sin\alpha}{F_N \cdot \cos\alpha + \mu \cdot F_N \cdot \sin\alpha} = \frac{\mu \cdot \cos\alpha - \sin\alpha}{\cos\alpha + \mu \cdot \sin\alpha}. \tag{2.4}$$

Please note that μ' equals to μ for a perfect alignment, i.e., $\alpha = 0°$. For small misalignments, $\cos\alpha \to 1$ and $\sin\alpha \to \alpha$, for which the above equation simplifies to

$$\mu' = \frac{\mu - \alpha}{1 + \mu\alpha}. \tag{2.5}$$

If the angle of misalignment is just around 2° (best practice in lab-scale tribo-testing) and assuming a real COF of 0.5 or 0.1, we obtain the alignment-based experimental error of around ±0.043 (value 9%) or ±0.035 (about 35%), respectively. Hence, the error becomes quite significant when it comes to measuring very low COFs.[1]

To tackle this aspect, let us consider the reciprocating motion as shown in Figure 2.2. In this case, the forces F_x and F_y are measured during forward and reverse motion, while the COF can be determined as the ratio of the averaged frictional force (which is positive for forward motion and negative for reverse motion) and the average of the normal force

$$\mu' = \frac{\frac{1}{2}\left(F_{xf} - F_{xr}\right)}{\frac{1}{2}\left(F_{yf} + F_{yr}\right)}, \tag{2.6}$$

where

$$\begin{aligned} F_{xf} &= F_F \cdot \cos\alpha - F_N \cdot \sin\alpha = \mu \cdot F_N \cdot \cos\alpha - F_N \cdot \sin\alpha, \\ F_{yf} &= F_N \cdot \cos\alpha + F_F \cdot \sin\alpha = F_N \cdot \cos\alpha + \mu \cdot F_N \cdot \sin\alpha, \\ F_{xr} &= -F_F \cdot \cos\alpha - F_N \cdot \sin\alpha = -\mu \cdot F_N \cdot \cos\alpha - F_N \cdot \sin\alpha, \\ F_{yr} &= F_N \cdot \cos\alpha - F_F \cdot \sin\alpha = F_N \cdot \cos\alpha - \mu \cdot F_N \cdot \sin\alpha. \end{aligned} \tag{2.7}$$

---

[1] Making it extremely challenging to measure COFs down to superlubricity, i.e., COFs < 0.01.

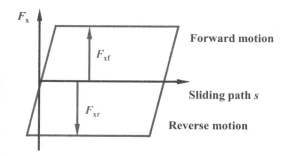

FIGURE 2.3   Friction loops in forward and reverse sliding motion.

By substitution, we obtain

$$\mu' = \frac{F_{xf} - F_{xr}}{F_{yf} + F_{yr}} = \frac{\mu \cdot F_N \cdot \cos\alpha - F_N \cdot \sin\alpha + \mu \cdot F_N \cdot \cos\alpha + F_N \cdot \sin\alpha}{F_N \cdot \cos\alpha + \mu \cdot F_N \cdot \sin\alpha + F_N \cdot \cos\alpha - \mu \cdot F_N \cdot \sin\alpha}$$

$$= \frac{2 \cdot \mu \cdot F_N \cdot \cos\alpha}{2 \cdot F_N \cdot \cos\alpha} = \mu.$$

(2.8)

This approach provides an accurate solution for measuring friction through continuous monitoring of forward and reverse values that generate the well-known "friction loops" as shown in Figure 2.3, for which the misalignment α just shifts the loop up or down. This concept became the fundamental base for many common approaches used to design modern tribometry set-ups, which will be detailly reviewed in subsequent sections.

2.1.1 Nanoscale Tribometry

Nanoscale tribometry focuses on studying friction, wear, and lubrication properties of materials at the nanometer range while applying loads on the nN level. These experiments provide a crucial, fundamental understanding of the tribological behavior of materials at small load and length scales. These tests in conjunction with conventional bulk experiments help to develop a holistic understanding of the phenomena happening on different scales such as surface interactions, adhesion, and molecular-level processes thus guiding the design of efficient tribo-systems.

The smallest contact evaluation can be done using a picoindenter probe (**picoindentation**) mounted onto the sample holder of a transmission electron microscope (TEM), see Figure 2.4. The test is highly versatile since the probe tip can have a radius of a few nm, and the loads applied are on the orders of pN. However, this approach remains rarely used and reserved for high-niche, fundamental studies, such as to quantify atomic plane interactions. There are practical challenges associated with the sample preparation and material stability under high vacuum, and the exposure to a highly energetic electron beam.

There are more robust picoindenter systems that can be used inside a scanning electron microscope (SEM) for *in-situ* testing and analysis. The application cases include phase-specific, nano-scale scratch/friction experiments, measuring hardness

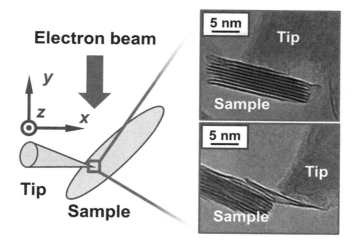

FIGURE 2.4   Schematic of picoindentation (left) and actual picoindenter probe at two different time steps (right). (Adapted from Ref. [2] with permission.)

and scratch resistance of small phases, or tribo-films, as well as studying material deformation of nanopillars.

Furthermore, **surface force microscopy** apparatus has been extensively used in the past decades to study nanoscale tribological phenomena. This has evolved into the development of **AFM**, which is nowadays a versatile technique that can be used for both imaging and tribological measurements. It involves scanning a sharp probe tip across the surface of a sample resulting in data with atomic-scale resolution. The approach implies that a sharp tip attached to a flexible cantilever is scanned over the sample of interest. As the tip moves across the surface, interactions between the tip and the sample cause the cantilever to deflect, which, in turn, can be translated into

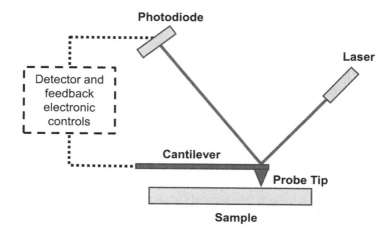

FIGURE 2.5   Schematic illustration of AFM showing the fundamental components and overall working principle.

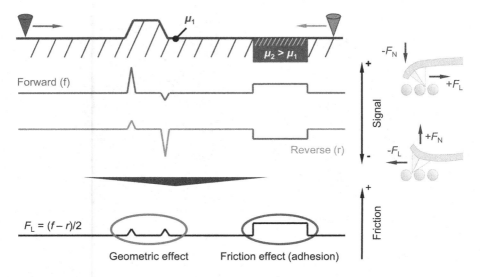

FIGURE 2.6   2D illustration of the resulting signal and lateral force detection in FFM during forward and reverse motion of a probe tip sliding over the sample to be characterized, thus allowing for the construction of friction maps.

the underlying roughness profile (Figure 2.5). Due to the precisely known geometry and mechanical properties of the cantilever and tip, these deflections can be translated into forces with great accuracy. AFM has further specialized in **lateral force microscopy (LFM)** and **friction force microscopy (FFM)**, which enable a highly precise and accurate assessment of friction and wear at the nanoscale with nearly atomic-scale resolution. The advancements in probe and cantilever materials manufacturing, high-sensitivity electronics, and ultra-high-precision motion control have enabled the design of sophisticated experiments including tribo-catalytic synthesis, in-situ corrosion measurements, phase evolution quantification, among others.

While traditional AFM focuses on imaging the topography of surfaces, FFM is fine-tuned to quantify and provide insights into the frictional forces experienced between the probe's tip and the sample's surface. While being generally like the AFM technique for imaging, the FFM probe can scan the sample in two main modes of deflection. It can record the normal bending mode, while the tip detects the normal forces and torsion of the cantilever due to the lateral and/or shear forces experienced by the tip (Figure 2.6). These are compiled and digitally reconstructed to produce 3D force maps that show variations in the frictional response across the surface.

2.1.2   Microscale Tribometry

To study tribology on the microscale, some set-ups make use of load/unload shuttle comb drives to bring a surface into normal contact with an oscillating counter-face. **Micro-electromechanical system (MEMS) tribometers**, as shown in Figure 2.7, consist of two comb drives, which can move the slider in two orthogonal directions, while an actuation voltage is used to bring the slider into contact with a counter-body. The normal force of the load is calculated from the excess of the threshold actuation

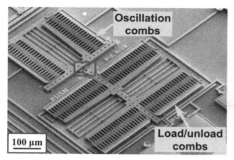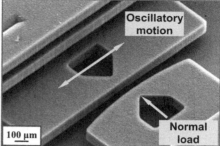

FIGURE 2.7 SEM image of the overall structure (left) as well as zoom into the contact area (right) of a MEMS sidewall tribometer. (Reprinted and adapted from Ref. [3] with permission.)

voltage and the known restoring spring constant of the normal load comb drive. The force output of the comb drives is calculated from electrostatics ($F = \alpha \cdot V^2$, where α is a constant determined by the actuator geometry and V is the potential difference between the opposing combs). The lateral force (i.e., friction force) can then be obtained from the balance between the electrostatic force generated by the push/pull lateral motion actuator, the restoring force of the suspension springs, and the friction force at the contact location, which depends on the lateral stiffness of the suspension springs. Other MEMS tribometers involve thermal actuations of the contact as a result of thermal expansion of actuation wires.

2.1.3 MACROSCALE TRIBOMETRY

Macroscale tribometry refers to techniques used to evaluate friction, wear, and lubrication on the length scale of several hundred micrometers. These approaches can be subdivided into ideal contact-condition experiments (i.e., laboratory) and application-mimicking component- and system-level tests.

2.1.3.1 Model-Level

Macroscopic lab experiments make use of tribometers, in which tribo-pairs mate in simple, specific, mathematically well-defined contacts. The commonly controlled variables in these experiments refer to the applied load, sliding speed, test environment, distance/time of test, diameter of ball or cylinder or other counter-body shape, radius of rotation, and kinematics. Thereby, one typically attempts to transfer the real application conditions as accurately as possible to the model level. The tests are then typically structured as parametric studies, for which one variable is changed while keeping the other constant to record the influence on friction and wear. The contact conditions commonly encountered in macroscopic lab experiments are plane-, point-, line-, and elliptical geometries (Figure 2.8). Thereby, third bodies can be introduced to mimic abrasive conditions.

Macroscale tribometry deals with contact pure sliding, rolling-sliding, or rolling conditions, which can be conformal (flat-on-flat) and concentrated in nature (Section 1.5.1). With respect to the resulting contact area/geometry, conformal contacts can be established by **flat-on-flat** geometries, such pin-on-disk/-block, block-on-block,

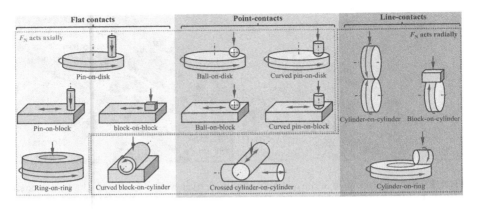

FIGURE 2.8 Contact conditions in common tribological test setup to achieve flat, point- and line-contacts with different kinematics.

curved block-on-cylinder, front face ring-on-ring, etc. **Point-contacts** result from ball- or head-curved pin-on-disk as well as cylinder-on-cylinders crossed at 90° setups, while **line-contacts** can be created from (parallel) cylinder-on-cylinder (see Videos 2.1 and 2.2 in ebook+), block-on-cylinder, or cylinder-on-ring configurations. By introducing radii on the pin or cylinder specimens, the contact geometry can also be modified to be in between point and line contact thus reflecting elliptical contact geometries. The underlying motion can be linear reciprocating or rotational. To achieve rolling proportions, both surfaces have to be curved and driven (or at least not restricted). This can be realized by a (parallel) cylinder-on-cylinder (also known as two-disk or twin-disk) setup, so-called four-ball testing, and a driven-ball-on-disk configuration. Although being fairly remote from most actual applications, pure sliding conditions with point contacts are most commonly found in tribometry, employing **ball- or pin-on-disk** geometries.

Irrespective of the approach used, these techniques render a great versability since they allow for the determination of tribological characteristics for a wide range of materials, applied loads, and sliding speeds. Frequently, the specimens are placed within a tempered lubricant bath to run tests under lubricated conditions. Furthermore, climate or vacuum chambers can be employed to allow for controlled temperatures and relative humidity conditions as well as an inert gas or vacuum atmosphere. Ball, pin, and block materials can be easily interchanged, which is another important benefit of these approaches. Due to their rather simple experimental set-up, online/*in-situ* measurement techniques may allow for condition monitoring thus characterizing the evolution of the underlying surface topography, surface chemistry, and compositional changes when connected to respective devices.

When targeting the assessment of tribological properties under rolling conditions, the **four-ball tester** is frequently used. In this test-rig, three stationary spheres placed in the plane support the fourth ball located on the top of them as schematically depicted in Figure 2.9. The spheres are typically enclosed in a cage or chamber, which can be filled with a lubricating oil. Upon the application of a normal load, the top ball is set into rotation. The COF is measured using a digital load cell mounted in a precise

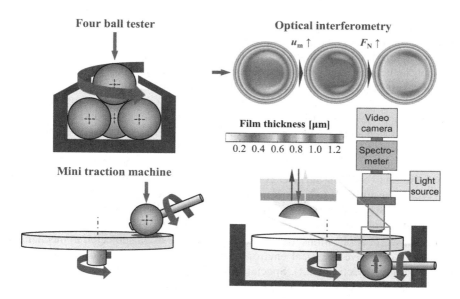

FIGURE 2.9   Four-ball tester (upper left) as well as mini traction machine (bottom left) set-ups and the overall principle of EHL film thickness measurement using optical interferometry (right).

location to obtain friction data with high resolution and accuracy. This test is particularly suited to assess the friction and wear characteristics of lubricants and solid materials (spheres) under high speeds (up to 2000 min^{-1}) and high contact pressures (up to 20 GPa), usually orders of magnitude larger than those seen in pin/ball-on-disk experiments. The setup typically allows for heating the lubricant present in the chamber up to 200°C, which is monitored and controlled by an integrated thermocouple/thermometer attachment. The normal load can be maintained constant or mimicking different step profiles with increasing or decreasing tendencies. The great advantage of the four-ball tester connects with its highly reproducible testing conditions, making it a valuable tool for quality control and research purposes. There are established and well-defined controlled test conditions and standardized procedures for consistent and repeatable measurements. Examples of these standards include ASTM D2266 or DIN 51350-1,2,4,6, to name a few. A four-ball tribometer can assess the load-carrying capacity of lubricants, providing insights into their ability to protect machine components from excessive wear and damage under various loads, loading rates, ranges, and with temperature. Furthermore, it is possible to extend the number of lower balls and/or insert smaller spacer balls in between, thus providing flexibility and the option to adapt. These features have made the four-ball equipment one of the highly adopted test-rigs in industry branches, spanning heavy machinery to automotive engines for both R&D and routine analysis.

Another example is the so-called **mini-traction machine (MTM)** used for the analysis of materials, mostly lubricants, on a small scale to evaluate their tribological characteristics under controlled sliding-to-rolling conditions (including opposed motion of the samples). MTMs can help to determine the load-bearing capacity of a lubricant by assessing how well it prevents wear and damage under different loads. The MTM often uses a ball-on-disk testing geometry (Figure 2.9, lower left). During testing, a small amount of the lubricant is applied to the contact interface, or the contact can be immersed into a lubricant bath, while friction is monitored under control of several test parameters, including the applied load (force), sliding, or rolling speed, temperature, and test duration. Systematic and controlled testing allows for the derivation of so-called friction maps, which provide a detailed visualization of friction in dependency of the varied parameters. MTM-like test set-ups employing a glass disk with a semi-reflective coating against a ball counter-body as well as precision optics enable the analysis of the interference patterns generated by the lubricant film between two sliding surfaces (Figure 2.9, right). Consequently, accurate calculations of the resulting **lubricant film thickness** at a specific point in the image can be achieved, which resembles a valuable tool for tribologists to validate numerical models of elastohydrodynamic contacts (Section 3.2.2).

In real-world applications, tribological systems are often subjected to mechanical stresses in corrosive chemical environments, which are referred to as tribo-corrosive conditions (Section 1.9.1). Specialized **tribo-corrosive tribometers** have been developed to properly study tribological systems exposed to corrosive media. A tribo-corrosive test-rig (Figure 2.10, left) consists of a standard macro-scale tribometer (unidirectional motion, linear reciprocating sliding or fretting conditions) with the sample stage submerged in a corrosive electrolyte. A source of potential and/or current such as a potentiostatic cell is connected to impose specified potentials (corrosion environments) and to evaluate the reaction kinetics of the rubbed metal thus *in-situ* study tribo-corrosion. Besides measuring the resulting normal and friction force to estimate the resulting COF, tribo-corrosion testing can be used to shed more light on the corrosive aspects. A generic tribo-corrosion cell consists of a counter electrode, usually platinum or other inert electrode, a reference electrode such as saturated calomel electrode or Ag/AgCl, and the sample connected as the working electrode. The contacts between the electrodes and the sample surface exposed to the electrolyte are insulated to prevent short-circuiting. This can be achieved by either water-proof concealment of the electrical contact using epoxy, or, by placing

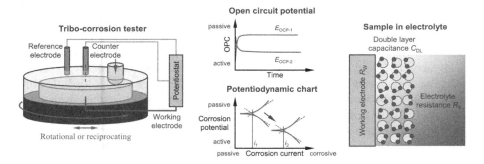

FIGURE 2.10   Schematic of tribo-corrosion tester (left), representative open circuit potential, and potentiodynamic charts obtained with and without application of tribological sliding (middle) as well as elements of a sample when immersed into an electrolyte.

the electrical contact on the other side of the electrolyte-facing surface as shown in Figure 2.10, left. The reference and counter electrodes are kept in proximity but with sufficient care to not touch the sample surface. After placing the electrodes in the appropriate locations and making a connection to the potentiostat, the electrolyte is filled into the sample cup, and the normal load is applied on the counter-face. The changes in the potential with no mechanical motion, and no external potential bias or current flow are recorded. The potential developed under these conditions is called open circuit potential (OCP). The OCP is allowed to stabilize with time, during which the surfaces attain thermodynamic equilibrium, and fluctuations in the potential become negligible. This is called "stabilized OCP" and the time to attain this state is recorded. After establishing an electrochemically stable set-up, the system is perturbed by introducing motion (reciprocating or unidirectional sliding), and fluctuations in the OCP are recorded. shows a representative OCP chart of the tribo-system under stationary conditions (E_{OCP}-1) and during sliding (E_{OCP}-2). Potentiodynamic polarization tests may be conducted under stabilized potential conditions to measure the corrosion current as shown in Figure 2.10, middle. The normalization of the current by the area of the sample exposed yields the corrosion current density (I_{corr}), which can be used to calculate the corrosion rate either as grams or mm per minute. The potentiodynamic graphs tend to be lower and shifted to the right-hand side, which suggests that the corrosion is more aggressive relative to a pristine sample. This behavior is attributed to the loss/rupture of the superficial oxide and thereby the electrochemical passivation from the tribological contacts. These experiments are performed with a knowledge and awareness that the underlying assumptions in the electrochemical theories such as the Butler-Volmer Equation and Stern-Geary Equation may not be satisfied, and the measurements are performed for a qualitative, broad understanding of materials performance. A simpler experiment connects with potentiostatic tests, for which the electrochemical response is recorded using a constant applied potential applied to a working electrode under loading and sliding while measuring the resulting current over time. The resulting time-ampereogram tends to demonstrate an increasing current with time, (i.e., wear volume loss). A comparative analysis helps to assess which materials are better suited in terms of tribo-corrosion.

A more sophisticated approach to assess tribo-corrosion relates to electrochemical impedance spectroscopy (EIS). EIS is a powerful technique that provides valuable information about the impedance (resistance and capacitance) of an electrochemical system as a function of frequency. A representative schematic of a sample when immersed into an electrolyte is shown in Figure 2.10, right. There are primarily three important components, namely, the working electrode, the resistance denoted with R_w, the solution resistance or electrolyte resistance R_s, and the double-layer charge capacitance, C_{DL}. A fourth component such as superficial oxide or tribo-film may be added on the surface of the working electrode. The resistance from the working electrode and solution can be assumed to be known. The capacitance and/or resistance offered by the C_{DL} originates from the thin interface between the surface and the fluid. At this interface, two layers of electric charge with opposing polarity tend to form (one at the surface of the electrolyte, and one on the electrode). These elements and the oxide capacitance are quantified by sweeping the sample across a frequency range at a fixed external potential bias. The resistance and capacitive elements are fitted in appropriate series and parallel arrangements in a representative electrical circuit (such as Randalls Cell) to quantify the corresponding circuit values. These circuit elements (fitted values) from EIS experiments provide valuable information about the tribological interface capacitance and resistance thus helping to assess the susceptibility of materials to corrosion and providing information about the corrosion mechanisms involved. This information is highly valuable to design tribo-corrosive systems with an overall aim to avoid the degradation of the involved materials thus extending the component lifetimes. In addition to submerged or wet-corrosive experiments, emerging lubrication systems based on solid lubrication can also benefit from EIS data. Impedance values can be used to investigate the formation and stability of protective tribo-films or passive layers on sliding surfaces. These passive films are of significance not only in the tribological context, but also in mechanical, optical, and electrical properties under dynamic sliding conditions.

2.1.3.2 Component- and System-Level Tribo-Testing

In contrast to the idealized lab experiments, applied, component-level and system-level testing approaches tribometry more from a holistic point of view thus focusing on the proper mechanical functioning of components and devices (bearings, gears, brakes, clutches, among others), which may include condition monitoring to detect failures or signatures preceding actual failure in early stages. This typically goes hand in hand with continuous quality control with the overall aim to generate data to predict and optimize the useful lifetime of mechanical components and systems.

Component-level tribo-testing aims at evaluating the friction, wear, and lubrication performance of individual mechanical components or systems rather than just materials or lubricants (see Videos 2.3 and 2.4 in ebook+). This type of testing is highly essential for different industries such as automotive, aerospace, manufacturing, and machinery, where the performance and longevity of components are critical. The most striking aspect of component- and system-level testing connects with the adaptation of realistic operating conditions and environments, in which the mechanical component to be studied functions. This is crucial to enable a more realistic assessment of the component's tribological behavior compared to standardized laboratory tests since the latter are often oversimplified and cannot properly replicate the interplay between the multiple parts of the component or system considered. The advantage of component- and system-level testing is that these interactions can be adequately captured, providing insights into how different elements within a system affect friction and wear. Approaches on component- and system-level are also highly beneficial for performance optimization and failure analysis. Thereby, test-rigs frequently feature tailored testing set-ups, specifically designed to replicate the unique operating conditions of the intended application, providing a platform for precise and targeted evaluations. An example of a customized component test-rig to study the tribological behavior of a cam/bucket tappet contact of the valve trains in combustion engines valve trains is revealed in Figure 2.11 (see also Videos 2.5 and 2.6 in ebook+). The test-rig consists of a base frame, a temperature-controllable lubricant unit, the drive, and a control unit (not shown here). To ensure realistic conditions and a good applicability, commercially manufactured components such as valves, springs, and cams (separated from the camshafts) are used. The lubricant

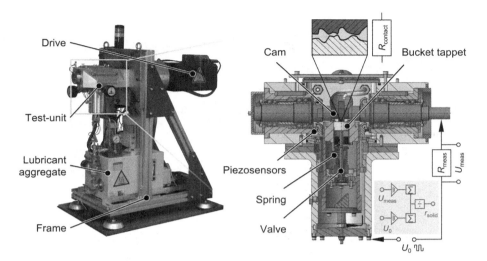

FIGURE 2.11 Cam-tappet friction test-rig (left) and schematic representation of the test unit in the sectional view with friction as well as resistive lubrication state measurement system (right). (Reprinted and modified from Ref. [4] with permission by CC BY 4.0.)

is continuously supplied from the tempered oil unit through throttle and lines in a minimal amount. The test-unit (Figure 2.11, right) is enclosed at the top by four piezoelectric sensors, arranged at 90° intervals and preloaded for high stiffness, and elastically supported at the bottom. This design allowed for the precise determination of the tangential friction force in the cam/tappet contact. Additionally, the test rig is equipped with a lubrication state measurement system, enabling its correlation with the resulting friction behavior.

Apart from such customized solution, there are numerous standardized or established test set-ups. For instance, the **FE8 bearing test rig** is a specialized, yet standardized (DIN 51819-1) platform to evaluate the performance of new materials and lubricants in bearing systems under controlled and reproducible testing environments (Figure 2.12). These rigs are designed to simulate real-world operating conditions, allowing to assess the durability, friction, and wear characteristics of different bearing configurations. FE8 test-rigs are particularly valuable for their ability to replicate a variety of load and speed conditions, providing a comprehensive understanding of how bearings behave under diverse operational scenarios. Despite user-definable testing programs, FE8 test-rigs are capable to conduct wear tests according to DIN 51819-2 for greases, according to DIN 51819-3 for oils, pitting tests according to VW PV 1483 and ZF 0000 702 232 for transmission oils as well as white etching crack (WEC) tests according to FVA 707 for oils. Another standardized example is the so-called **FZG back-to-back test-rig**, which is designed to simulate the complex interactions and dynamic conditions that gears experience in practical applications. By subjecting gears to controlled loads, speeds, and lubrication regimes, FZG back-to-back test-rigs allow researchers to systematically study wear, fatigue, and efficiency under realistic operating conditions. FZG back-to-back test-rigs play

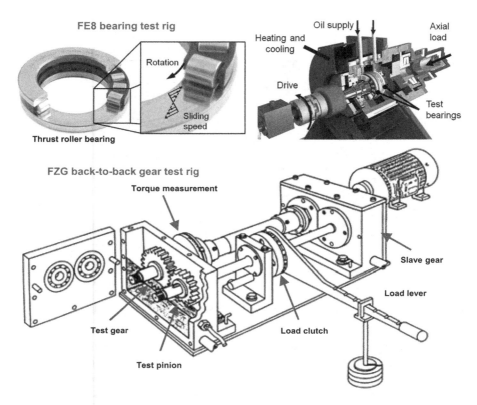

FIGURE 2.12 FE8 bearing test rig (top) and FZG back-to-back gear test rig (bottom). (Reprinted and adapted from [5,6] with permission by CC BY 4.0.)

a crucial role in advancing gear technology, ensuring that gears meet the stringent demands of modern machinery and contribute to overall system efficiency and longevity. It should be noted that these test-rigs are just some chosen examples from a broad range of developed and available component or system test-rigs for manifold applications.

2.1.4 Wear Assessment in Tribometry

The wear analysis can be either conducted as a direct quantification of the mass loss or a post-testing assessment of the induced changes related to surface morphology, surface topography, and contact geometry (Section 1.7). The former is simple, easy to quantify, and a physical measure, but may experience problems with precision and accuracy. The latter involves mapping the topography (by a contact or non-contact technique), thus estimating the respective wear volumes and/or wear rates. Assuming a known density, the wear volume can be also transformed into a mass loss. The latter approach resembles some more benefits since the involved techniques and analysis of the surface topography already capture many features that help to understand the wear mechanism. In this regard, the wear volume is initially quantified at a coarser scale followed by an evaluation at a finer scale for accuracy and precision. The coarse-scale topography of the worn surfaces can be obtained by light microscopy (LM). This technique is suitable to detect/record the overall wear mechanisms resulting from macroscale tribological experiments. This rudimentary analysis allows to gain a basic understanding of the degree of the material damage induced by the tribological tests. This paves the way for subsequent, more sophisticated surface studies in pre-defined areas of interest by LM.

Subsequently, techniques to assess the surface roughness (Section 2.2.1) can be used for the analysis of the worn surface. In this regard, common techniques are contact and optical profilometry, extensively used for the visualization of the wear tracks and/or wear marks. Therefore, let us consider the example of the wear produced during ball-on-disk macroscale testing depicted in Figure 2.13. The initial condition connects with a perfectly elastic contact without any plastic deformation on either of the two bodies in contact. The corresponding image of the ball (also called as counter-face) shows a pristine, round surface with a reflection of the light source indicating smoothness (Figure 2.13, left). After tribo-testing, two scenarios are possible. On the one hand, the ball may remain its original shape and geometry without any significant damage or deformation, while the surface of the flat substrate demonstrates a pronounced wear loss (Figure 2.13, middle). The top view of the wear track shows a distinct, well-defined material loss region, whereas no quantifiable wear loss was detected on the ball. Some material dislodged from the surface of the flat sample was observed to adhere to the ball, known as "transfer-layer." This resembles the classic case of the ball having a greater wear resistance than the flat sample. On the other hand, if the wear resistance of the ball is lower than that of the substrate, major damage is introduced to the ball, which results in the formation of a plane called ball flat or counter-body cap, or wear cap (Figure 2.13, right). The extent of the materials

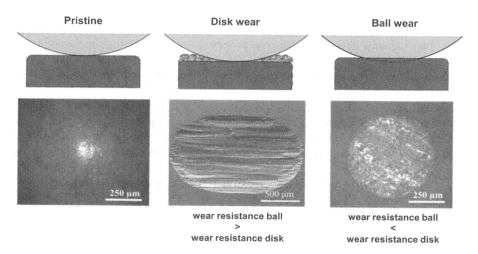

FIGURE 2.13   Schematics for the wear induced during ball-on-flat macroscale tribological testing. (Reprinted and adapted from Ref. [7] with permission from Elsevier.)

loss is shown with the dotted line, and an example of such a worn-out ball surface is shown in the corresponding photograph. Since, under pure sliding conditions, the ball experiences permanent contact under these conditions, this model is commonly used for the wear evaluation of self-mated materials.

The wear volume V in the self-mated case can be calculated using

$$V = \left(\frac{\pi h}{6}\right)\left(\frac{3d^2}{4} + h^2\right), \tag{2.9}$$

where h is the height of the counter-body cap, and d stands for the diameter of the counter-body scar. In this regard, h can be estimated as

$$h = r - \sqrt{r^2 - \frac{d^2}{4}}, \tag{2.10}$$

with r being the radius of the counter body. Based on these equations, it follows that the diameter of the wear-cap needs to be determined experimentally for a known radius of the counter-body. This measurement can be done using a common LM. In conclusion, the wear volume sliding can be completed using a simple set of equations and measuring the diameter of the scar using LM.

After the initial coarse-scale characterization of the worn surface by LM, a higher resolution quantification of the respective wear volumes and wear rates can be carried out by white light interferometry (WLI) or laser scanning confocal microscopy (LSCM). If wear is observed on both surfaces, the quantification of the respective losses shall be carried out for both rubbing surfaces. In addition to the absolute wear volumes, it is important to estimate the wear rates based on Archard's law (Section 1.7.1). Both WLI and LSCM being high-resolution surface topography mapping techniques help to provide valuable insights into the wear analysis, allowing researchers to visualize and identify surface changes at the micro- and sub-micrometer scales. These techniques play a crucial role in understanding wear mechanisms, optimizing material properties, and improving the durability of components and devices.

It should be noted that the accuracy of the friction and wear analysis as well as the post-characterization (reviewed in detail in Section 2.2) highly depends on the steps undertaken during sample preparation, removal, and post-processing, such as cleaning. The cleaning process aims at removing contaminants, wear debris, and any residues that may have accumulated during tribo-testing. Consequently, the adopted procedure varies depending on the nature of the tribological test and the materials involved. It can be very gentle like dry-cleaning, using dry wipes or compressed air to remove loose particles and debris from the sample surface, or more thorough like wet-cleaning, using appropriate solvents, detergents, or surfactants wash, with or without sonication, to dissolve organic residues or contaminants. In the case of wet-cleaning, the selection of appropriate cleaning agents depends on the reactivity of the studied materials. Once the samples are cleaned, they should be additionally rinsed if needed with other solvents or water as well as dried using air or an inert gas, sometimes also with the use of heat to evaporate any residues. In the case of post-characterization, access to the cross-section of the wear track is required. In this regard, the cross-sections can be prepared using simple approaches like electrical discharge machining (EDM) or wire cutting, followed by polishing or etching, or more elaborated procedures using focused ion beam (FIB) milling, which is reviewed in Section 2.2.2.

2.2 CHARACTERIZATION OF THE UNDERLYING FRICTION- AND WEAR-MECHANISMS

As we learned in the previous module, tribological characteristics are highly sensitive to and dependent on the materials' properties. This implies that material characteristics and properties must be holistically studied to understand the resulting tribological behavior. In this regard, the physical characteristics of materials, such as hardness, elastic modulus, surface roughness, density, coefficient of thermal expansion, surface energy, among others, have to be studied since they may directly affect the resulting contact area and contact stresses as well as processes occurring in the tribological contact zone. Moreover, chemical characteristics including surface composition as well as their chemical or molecular structure that define their chemical compatibility, nature and type of bonding, free energy, chemical affinity, and oxidation tendency, must be precisely assessed.

Therefore, the overall tribological assessment is only not limited to friction and wear measurements but also requires an in-depth analysis of the materials involved in sliding, both prior to and after tribological testing, and in some cases even *in-situ*. Thereby, *in-situ* characterization refers to the analysis of a material or system during operation, which implies that the measurements are taken while the tribological experiment is carried out. In contrast, *ex-situ* characterization involves analyzing a sample after tribological testing and removing it from the tribo-tester. In addition to physical and chemical properties, the surfaces need to be analyzed for their surface topography, surface, and sub-subsurface microstructure, all of which influence the measured frictional and wear behavior. A list of commonly measured properties and the techniques available for their assessment and quantification is presented in Table 2.1.

2.2.1 PHYSICAL CHARACTERIZATION OF TRIBOLOGICALLY-ACTIVE SURFACES

2.2.1.1 Roughness

Surface roughness (Section 1.4.3) is essential in tribometry since it invariably influences many aspects of surface phenomena including friction, adhesion, wear resistance, and lubricant retention. There are several ways surface roughness and/or topography can be measured and expressed. The following discussion is focused on aspects relevant to applied tribology with the overall aim to guide students and/or practitioners.

On the one hand, **contact methods** involve physically touching the surface with a measuring stylus or probe. The stylus follows the surface contours, and its vertical movement is converted into data points that represent the surface profile. Some common contact methods include:

TABLE 2.1

Commonly Measured Properties and the Techniques Used for Their Quantification

Properties	Techniques
Surface visualization	Optical Microscopy, Scanning Electron Microscopy, Transmission Electron Microscopy, Surface Probe Microscopy
Topography	White Light Interferometry, Laser Scanning Confocal Microscopy, Atomic Force Microscopy, Scanning Tunneling Microscopy
Structure	Time-of-Flight Secondary Ion Mass Spectroscopy, X-Ray Photoelectron Spectroscopy, Atom Probe Tomography, Auger Spectroscopy, Transmission Electron Microscopy
Composition	X-ray Photoelectron Spectroscopy, Auger Spectroscopy, Time-of-Flight Secondary Ion Mass Spectroscopy, Fourier Transformed Infrared Spectroscopy, Raman Spectroscopy
Crystallography	X-ray diffraction, Transmission Electron Microscopy, Electron Back Scattered Diffraction
Spatial distribution	Time-of-Flight Secondary Ion Mass Spectroscopy, Atom Probe Tomography, Atomic Force Microscopy, Raman Spectroscopy, Transmission Electron Microscopy
Thickness	Atomic Force Microscopy, White Light Interferometry, Ellipsometry, Raman Spectroscopy
Energetics	Contact angle measurement, Scanning Electrochemical Potential Microscopy, Atomic Force Microscopy

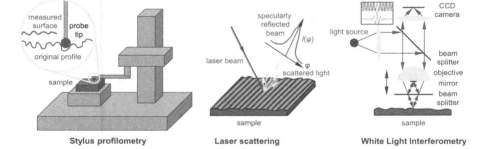

FIGURE 2.14 Principles of contact- and non-contact-based methods to measure the corresponding surface roughness.

- **Stylus Profilometry**: In this method, a stylus with a sharp tip is moved across the surface and the vertical displacement of the stylus is recorded (Figure 2.14, left).
- **Scanning Probe Microscopy (SPM)**: Similar to the stylus method, techniques like AFM and Scanning Tunneling Microscopy (STM) use a sharp probe to scan the surface at the nanoscale and provide high-resolution roughness data.

Contact-based methods are well-established and applicable to a wide range of materials, including metals, ceramics, polymers, and composites. Moreover, they can not only quantify roughness but also form deviations larger than waviness, allowing for a more comprehensive assessment. However, since these methods involve point-to-point physical contact, they are not overly fast when it comes to large measurement areas and the stylus may cause indentation on soft materials or not be able to follow fine geometric features (e.g., when the stylus tip is larger than the feature), leading to measurement artifacts and errors. Moreover, there is a potential risk of damaging delicate surfaces, especially if the stylus is not carefully chosen. In contrast, extremely rough and hard surfaces might cause excessive wear on the stylus, potentially affecting its accuracy over time.

In contrast, **non-contact methods** do not require physical contact with the surface to be studied. These techniques use light, lasers, or other forms of energy to measure surface characteristics. Some common non-contact methods include:

- **Laser Scanning Profilometry**: This technique uses a laser beam to scan the surface. The reflected light is analyzed to determine surface height variations, allowing for rapid measurements over larger areas.
- **Confocal Microscopy**: Confocal microscopy utilizes a focused laser beam to create optical sections of the surface. By analyzing the intensity of scattered light (Figure 2.14, middle), it generates a three-dimensional image of the surface topography.
- **Interferometry**: Techniques like **White Light Interferometry** (WLI; Figure 2.14, right) and Phase-Shifting Interferometry make use of interference patterns to measure surface height variations with nanometer-level precision.
- **Optical Coherence Tomography (OCT)**: Similar to medical imaging, OCT uses light waves to measure surface profiles and is useful for transparent materials as well as layered structures.
- **Airborne Ultrasonic Profilometry**: Ultrasonic waves are directed towards the surface, and the reflected waves are analyzed to determine surface roughness.

Non-contact methods eliminate the risk of damaging or altering delicate surfaces during measurement, which makes them suitable for soft or easily deformable materials that could be adversely affected by contact-based methods. Some methods, such as laser scanning profilometry and confocal microscopy, can rapidly scan surfaces and provide quick measurements, making them ideal for applications that require high throughput. Furthermore, non-contact methods excel in covering large surface areas efficiently, allowing for a comprehensive view of surface characteristics in a shorter amount of time without operator-induced variability, also resulting in more consistent measurements. However, non-contact methods are limited by the material and surface types they can accurately measure. Transparent or highly reflective surfaces, for instance, can pose challenges due to the way these techniques interact with light. Certain techniques have specific measurable ranges and may struggle with surfaces that have extreme slopes or certain types of textures.

Environmental factors such as vibrations, temperature changes, or air currents can interfere with the accuracy of non-contact measurements. Furthermore, the higher precision and technology of non-contact methods often come with a higher cost compared to simpler contact methods.

Generally, the derivable surface roughness parameters (Section 1.4.3) largely depend on the vertical and lateral resolutions of the employed instruments. Consistency and comparability are at least provided when measuring with devices of the same or similar resolutions, while different statistical parameters for the same surface can be obtained from instruments of differing accuracy.

2.2.1.2 Hardness and Modulus

To allow for the full interpretation of frictional data acquired by various tribometry techniques, it is often necessary to determine nano-scale mechanical properties. It usually holds true that nano-scale mechanical properties can be significantly different compared to the mechanical bulk characteristics. Moreover, when having multiple phases in a material, bulk hardness measurements only provide an "average" information of the entire ensemble, whereas **nanoindentation** can generate phase-specific hardness values. The third and common reason to use nanoindentation connects with the usage of thin films – either tribo-films that form due to rubbing, or engineered/deposited coatings that need analysis, isolated from the substrate.

The overall working principle involves applying a controlled load to a very sharp intender tip. The resulting depth of penetration, slope of the unloading curve and shape of the graphs can be mathematically fitted to calculate the hardness and reduced elastic modulus of the material. The probe tip is usually made of highly polished diamond having a cube corner, Berkovich, or pyramidal geometry. The technique is called nanoindentation since the applied load is on the order of nN, producing nanometer-size indentations. A schematic of an indenter contacting the sample substrate and the resulting indentation as well as a representative graph generated from nanoindentation are shown in Figure 2.15.

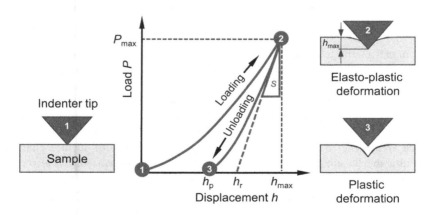

FIGURE 2.15 Schematic illustration of nanoindentation using a sharp tip to obtain the respective load versus displacement graphs.

There are three steps in each nanoindentation experiment, which connect with loading the sample, holding the sample at a predefined maximum load, and finally unloading the sample. The displacement of the tip while applying and removing load is recorded using a sensitive load and displacement sensor, which results in three load-displacement curves (loading, holding, and unloading curves). These simple graphs can be used to calculate hardness, Young's modulus, stress-strain response, creep, fracture toughness, and plastic and elastic energies of the material studied. The maximum depth of penetration of the indenter is on the order of tens to hundreds of nanometers and is denoted as h_{max}, while the corresponding maximum load is P_{max}. Upon loading the tip, there is some permanent plastic deformation set into the material, whereas some elastic recovery is commonly observed upon unloading. The slope of the unloading curve is calculated as

$$s = \frac{dp}{dh} \tag{2.11}$$

and represents a measure of the contact stiffness. The indentation hardness calculation follows the fundamental relationship

$$H = \frac{P}{A}, \tag{2.12}$$

where A represents the area of the indentation. For the hardness evaluation based on nanoindentation, there are a few more steps to be considered. The contact depth h_c is the difference of the maximum depth of penetration (h_{max}), and elastic strain multiplied by the plasticity:

$$h_c = h_{max} - \varepsilon \frac{P_{max}}{s}, \tag{2.13}$$

with the strain component ε, a constant dependent on the indenter shape. The contact area can be calculated as

$$A_c = 24.5 \cdot h_c^2, \tag{2.14}$$

for a Vickers or Berkovich indenter or

$$A_c = 2 \cdot \pi \cdot R \cdot h_c^2, \tag{2.15}$$

for a spherical indenter, respectively.

Similarly, the reduced Young's modulus can be calculated as

$$E_r = \frac{\sqrt{\pi}}{2\beta} \frac{S}{\sqrt{A_c \cdot h_c}}, \tag{2.16}$$

where β depends on the indenter shape and is equal to 1.034 for a Berkovich indenter. These equations are mathematical expressions used to quantify the materials' properties purely based on the indenter's geometry, known constants, and the shape (slope) of the load-displacement curve.

SEM is usually used to characterize and quantify the shape of the indentations. Exemplary micrographs of several indentations made on a two-phase material are depicted in Figure 2.16.

It is a general practice that nanoindentation experiments on bulk samples are performed in an array, matrix, or serpentine shape. The resulting indentations typically resemble three geometrical shapes, see Figure 2.16, bottom, while specialized cases of each type may also occur. Thereby, it is common to observe an almost perfect reproduction of the tip's geometry into the materials' surface, while the material can also be pushed out. Consequently, excessive strain causes the material to "pile up" on the edges of the residual impressions, indicating that the material is ductile, and deforms easily. The third case connects with material "sink-ins," which imply that the surface experienced purely elastic contact. The shapes of the residual indentations need to be carefully determined to enable an adequate calculation and correction of the involved area. This is important to avoid an over- or underestimation of the resulting hardness and modulus values. Moreover, the fracture toughness can be also estimated based on nanoindentation experiments. The residual indentation may exhibit radial cracks as commonly observed for hard and/or brittle materials. The length of the crack tip is measured and used to calculate the corresponding fracture

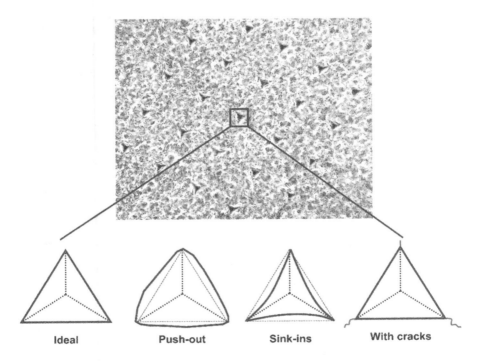

FIGURE 2.16 Optical micrograph with several nanoindentations (top) as well as different indentation shapes with potential cracks caused by brittle materials (bottom).

TABLE 2.2

Physical Meaning of Various Hardness and Modulus Ratio Expressions

Parameter	Physical Meaning
$\dfrac{H}{E}$	Deformation relative to yielding
$\left(\dfrac{H}{E}\right)^2$	Transition from elastic to plastic deformation on mechanical contact
$\left(\dfrac{H^2}{2E}\right)$	Modulus of resilience
$\left(\dfrac{H}{E_r^{\,2}}\right)$	Resistance to the plastic indentation

toughness. Examples of cracks on Berkovich and Vickers indenters are indicated in Figure 2.16, bottom.

The measured hardness and modulus using nanoindentation play an important role in the assessment of wear and for predictive purposes. The physical meaning of hardness is understood as resistance to indentation deformation under load, while the elastic modulus represents the resistance to the material's ability to resist elastic deformation upon an applied stress. Based on these measures, four more meaningful parameters can be defined as summarized in Table 2.2.

In addition to pure indentation measurements, most nanoindenters also offer a scratch mode, wherein the frictional response can also be quantified. The working principle is similar to the indentation experiments, with a difference that the loading is along a line, versus a singular point, and the forces measured are expressed as friction.

2.2.2 Chemical Characterization of Tribologically-Active Surfaces

Under tribological testing, oxidation and/or tribo-chemical changes can occur, which may notably change the overall surface chemistry of one or even both rubbing surfaces thus greatly affecting the resulting friction and wear response. After the careful characterization of the initial surfaces and appropriate tribo-testing, a detailed wear analysis must be conducted to understand how materials degrade and interact with their environment over time due to mechanical, chemical, or other forms of stress. In this regard, it is essential to consider both rubbing surfaces to fully derive the underlying mechanisms. The analysis of only one involved surface may lead to misleading or even wrong interpretation of the acting phenomena.

It is recommendable to pursue a multi-step approach to characterize the underlying surface chemistry prior to and after tribo-testing. For this purpose, we can utilize a number of established methods with different lateral and depth resolutions, a selection of which is shown in Figure 2.17. For conductive, macro-scale samples, SEM coupled with energy-dispersive X-ray spectroscopy (EDX) is an excellent way to generate mappings of the elemental distribution. It should be noted that this

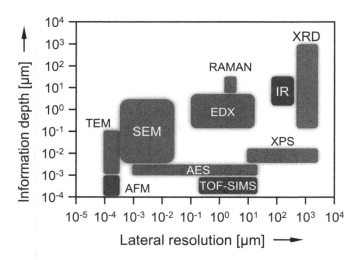

FIGURE 2.17 Lateral and depth resolution of various characterization tools to assess the surface chemistry.

method does not assess the superficial chemistry since the respective information is generated in a depth of a few microns. However, for homogeneous, bulk samples, this method can provide valuable information about the overall chemistry and the corresponding changes. Subsequently, more advanced and sophisticated techniques should be selected to investigate the surface chemistry such as Raman spectroscopy, Fourier-transform infrared spectroscopy, and X-ray photoelectron spectroscopy (XPS). As surfaces experience wear, there can be alterations in the chemical bonds and molecular structure of the materials. The mentioned spectroscopic methods can detect changes in functional groups and chemical bonds, providing insights into the wear mechanisms and the extent of wear on a material's surface. For tribo-systems with the possibility to form thin tribo-films, which resemble thin layers formed during rubbing, spectroscopic techniques can be used to analyze the composition and structure of these tribo-films, helping researchers understand their formation and effectiveness in reducing friction and wear.

2.2.2.1 SEM/EDX

Among the most common techniques for surface characterization is the elemental mapping of the surfaces since it provides a quick assessment of the changes in the surface elements due to tribo-testing. For instance, monitoring the changes in the elemental mapping of oxygen across the wear track helps to identify the corrosion processes during sliding. Similarly, carbon is often indicative of the underlying tribo-film formation owing to mechanical and tribo-catalytic processes. This analysis is often performed by SEM, which operates by directing a focused electron beam onto the sample surface and detecting the interaction signals, such as secondary electrons, providing information about the sample's topography, and backscattered electrons, revealing contrasts in the atomic number (compositional variations). These signals are collected by specific detectors and used to create

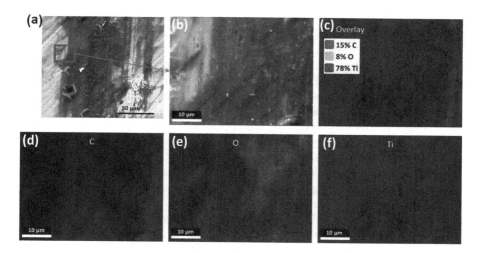

FIGURE 2.18 SEM-EDX map of the crated (a, b) wear track during tribological testing as well as (c) overlay map and detailed (d) carbon, (e) oxygen, and (f) titanium maps, demonstrating the uniformity of the titanium and oxygen concentration inside and outside wear track, while the contribution of carbon was predominant inside the wear track. (Reprinted and adapted from Ref. [8] with permission from Elsevier.)

images thus analyzing the sample's composition and morphology. To obtain a good, nano-scale resolution in SEM, the elimination of any charge build-up events on the surfaces is required, which is attainable by coating the samples with a thin layer of a conductive material (gold or carbon), or attaching it with copper or carbon tapes. Elemental mapping in SEM is performed by EDX detectors to identify and quantify the elements present by measuring the characteristic X-rays emitted when the incident electron beam interacts with the sample (Figure 2.18). In this specific case, carbon mapping was performed to support the concept of the embedment of nanodiamonds during sliding as a mechanism for surface reinforcement and friction reduction of titanium interfaces. This revealed the potential of carbon nanodiamonds to be used in artificial joints to improve their performance and prolong the operation lifetime [8,9].

2.2.2.2 TEM

Transmission electron microscopy (TEM) allows to gain high-resolution insights into structural changes and transformations of tribologically active nanomaterials or the formation process and structure of beneficial tribo-films. After the high-resolution preparation of the TEM specimen, a focused electron beam is generated, which passes through the specimen and interacts with the atoms in the material. These interactions include elastic and inelastic scattering. The interactions between the electron beam and the specimen led to the creation of a variety of signals, including transmitted electrons, diffracted electrons, and scattered electrons. These signals are captured using detectors and converted into an image that represents the specimen's internal structure. In general, TEM images are black-and-white and provide information about the specimen's internal structures as well as crystal lattice arrangements, and defects.

In this regard, TEM becomes accessible by two approaches. On the one hand, generated wear debris can be directly collected from the tribological interface (Figure 2.19). After redispersing it into a suitable solvent (ethanol or isopropanol), it can be drop-casted in low concentration on the respective TEM grid used as a sample holder. This approach relates to the most straightforward way to conduct TEM studies in tribology. Following this approach, TEM analysis can provide valuable analysis of the ongoing processes in different stages of sliding for tribo-catalytic reactions.

On the other hand, areas of interest can be defined based on the obtained information by the performed complementary materials characterization to extract highly local TEM lamellae by focused ion beam microscopy (FIB). Using the "lift-out" technique in FIB (Figure 2.20), a small sample can be extracted for further analysis. In this regard, the lift-out process begins with preparing the sample of interest, which is covered by a protective layer like platinum. Afterwards, FIB is used to mill or etch away material around the region of interest. The goal is to create a thin "stem" that connects the protective layer to the rest of the sample. Once the sample is sufficiently thin and still stiff enough, the FIB beam is used to sever the stem, allowing the protected sample to be lifted out using a micromanipulator. The lifted-out

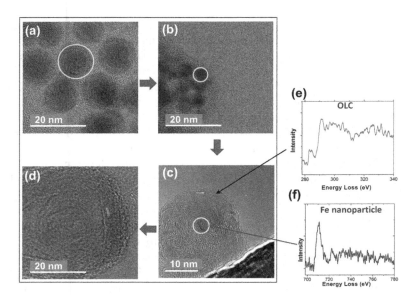

FIGURE 2.19 (a–f) TEM evaluation of wear debris collected from the wear track at different time intervals during the tribological test performed using iron nanoparticles+graphene solid lubricants. It was demonstrated that high contact pressures at the sliding interface promote tribo-chemical reaction between Fe NPs and graphene leading to the formation of onion-like-carbon nanostructures (OLCs) that lead to the superlubricity, or near-zero friction, sliding behavior. (Reprinted and adapted from Ref. [10] with permission from Wiley.)

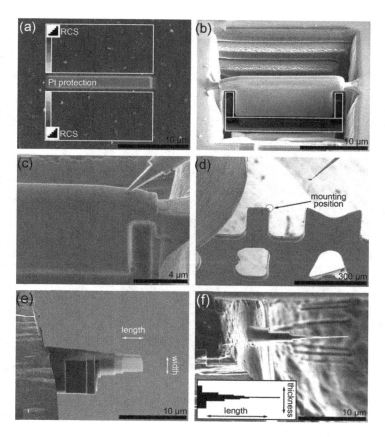

FIGURE 2.20 (a–f) Approach to prepare a plane-view TEM specimen from a specific site based on the FIB lift-out technique. (Reprinted and adapted from Ref. [11] with permission from Elsevier.)

sample is transferred to a suitable substrate, such as a TEM grid for further analysis. In this regard, TEM can provide local, highly resolved information about micro-structural changes underneath the wear-track, which helps to fully characterize the sub-surface microstructure. Based on this analysis, the formation of twins or deformation bands can be made visible, which helps to determine the mechanically affected/deformed zones. This can give valuable insights into the friction-induced energy input, which provides the respective energy to change the sub-surface micro-structure thus inducing microstructural defects.

Similar to SEM, TEM can be coupled with EDX, thus enabling elemental map-ping with an extremely high resolution, which can shed more light on the involved nano-scale chemical changes and processes (Figure 2.21). In systems with the capability to form tribo-films based on tribo-chemistry, the local thickness and homogeneity as well as the internal morphology of the tribo-film can be assessed by TEM and TEM-EDX. Following this approach, the morphological constituents

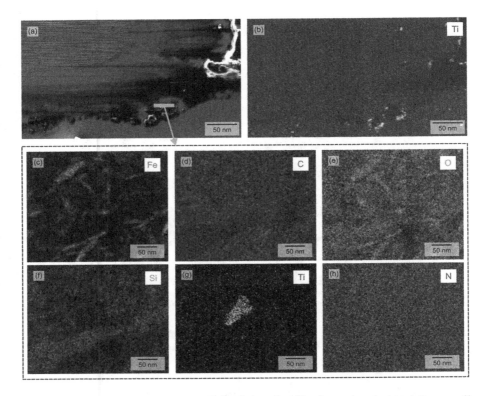

FIGURE 2.21 (a–h) SEM-EDX analysis of the tribo-film formation to detect the overall wear features and accumulations of Ti in MXene tribo-films. TEM-preparation in the zone marked in (a) to perform TEM-EDX mappings to shed more light on the involved chemical changes in zones of interest in the tribo-film. (Reprinted and adapted from Ref. [12] with permission from ACS.)

can be characterized with the possibility to distinguish between amorphous and (nano)-crystalline phases and arrangements. Moreover, TEM can provide insights into the composition, morphology, and size distribution of wear debris generated during tribological processes. This information is crucial for understanding the wear mechanisms and potential effects on the surrounding environment.

2.2.2.3   Raman Spectroscopy and Fourier-Transform Infrared Spectroscopy

The basic principle of Raman spectroscopy involves shining monochromatic light (usually from a laser) onto a sample and collecting the scattered light. The scattered light can have a different energy (frequency) than the incident light due to interactions with the molecular vibrations of the sample (Stokes and Anti-Stokes scattering). The frequency shifts in the scattered light are indicative of the vibrational modes of the molecules present in the sample. Consequently, Raman spectroscopy can be used to characterize the chemical composition and structure of the surfaces of materials

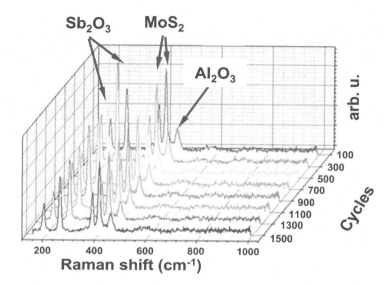

FIGURE 2.22 Evolution of the in-situ Raman spectra acquired during tribo-tests performed for PEO-chameleon coatings. The Raman peaks confirmed no oxidation signs inside the wear track suggesting that consistent low friction performance is the result of continuous 2D material lubrication effect. (Reprinted and adapted from Ref. [13] with permission from Elsevier.)

involved in tribological processes. It can help identify surface contaminants, oxides, and other chemical species that can affect friction and wear behavior, which is schematically illustrated in the cycle-dependent Raman measurements for chameleon coatings in Figure 2.22.

Due to the good special resolution of Raman spectroscopy, areas of interest can be identified by SEM in the worn surface, and, subsequently, Raman spectroscopy can be used to differentiate the chemical species and occurred changes in different parts of the wear track. Figure 2.23 exemplarily shows the conducted Raman analysis for pure $Ti_3C_2T_x$ and MoS_2 as well as hybrid composite coatings, which can provide important information about the formed tribo-films.

Besides Raman spectroscopy, Fourier-transform infrared spectroscopy (FTIR) can be used to analyze the vibrations of molecules. It provides information about functional groups, chemical bonds, and the overall molecular structure of a substance. The principle of IR spectroscopy is based on the interaction between infrared radiation and molecular vibrations. Similar to Raman spectroscopy, IR spectroscopy can provide insights into the chemical changes that occur on surfaces due to friction, wear, and lubrication processes.

Considering the overall principles of Raman and infrared spectroscopy, both techniques appear to be interesting since they deliver complementary information due to their susceptibility to generate the respective signals. Some chemical molecules are Raman active and inactive for IR and vice versa, which depends on the possibility to generate dipole moments (IR spectroscopy) and the respective polarizability of the chemical molecule (Raman spectroscopy).

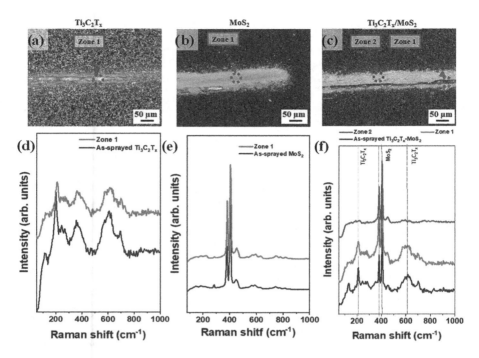

FIGURE 2.23 (a–c) Overall assessment of the formed wear tracks by light microscopy and (d–f) the corresponding Raman signals inside and outside the wear track for pure MXene, MoS₂, and their hybrid coatings. (Reprinted and adapted from Ref. [14] with permission from Elsevier.)

2.2.2.4 XPS Analysis

For a more sophisticated analysis of the initial surface chemistry, XPS is recommended to fully characterize the surface chemistry of the involved surfaces (depth resolution of about 10 nm) [15]. In this regard, XPS, which is useful to characterize the elemental composition, chemical state, and electronic structure of materials, operates based on the photoelectric effect and is sensitive toward the outermost few nanometers of a material's surface. The principle of XPS involves the interaction of X-rays with a sample's surface atoms and the measurement of the emitted photoelectrons (Figure 2.24). XPS typically requires that the obtained raw data are post-treated including background subtraction and physically correct peak fitting, since a wrong post-treatment can lead to a completely incorrect data interpretation.

2.2.2.5 Auger Electron Spectroscopy

Auger electron spectroscopy (AES) uses an electron beam to interact with the sample's surfaces (in comparison to XPS, where an X-ray beam is used to eject an electron). In AES, the collection depth is limited to 1–5 nm due to the small escape depth of electrons, which permits the analysis of the first 2–10 atomic layers with a spot size being around 10 nm (lateral resolution). Similar to XPS, AES measures the

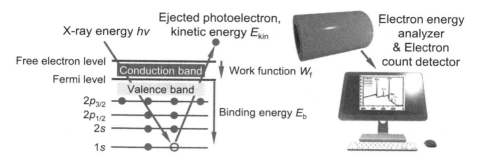

FIGURE 2.24 Schematic illustration of the photoionization process for the XPS characterization. (Reprinted and adapted from Ref. [16] with permission from Elsevier.)

kinetic energy of an electron to determine its binding energy, which depends on the element and the electronic environment of the nucleus, thus allowing to distinguish elements and their oxidation states.

2.2.2.6 X-Ray Diffraction

In the case of metallic samples, the analysis of the cross-sectional sub-surface microstructure (grain sizes, grain orientation, texture information, defect density, among others) is crucial, since the induced tribo-energy can induce grain refinement or increase the number of defects present across the tribological contact zone. In this context, coarse-scale X-ray diffraction (XRD) provides valuable information about the crystallographic structure of crystalline materials, which helps to determine the arrangement of atoms within a solid, the lattice parameters, and the orientation of crystal planes. XRD is based on the principle of wave interference and Bragg's law. In short, when X-rays strike the regular arrangement of atoms in a crystal lattice, they undergo constructive and destructive interference as they are scattered by the atoms. The resulting interference patterns are generated satisfying Bragg's law, which relates the angles of incidence and the spacing between crystal planes to the wavelength of the X-rays and the observed diffraction angles. It can be expressed as

$$n\lambda = 2d \sin(\theta), \tag{2.17}$$

where n is the order of diffraction, λ relates to the used X-ray wavelength, d refers to the spacing between crystal planes, and θ stands for the detected angle of diffraction. The positions and intensities of the diffraction peaks provide information about the crystal lattice parameters (unit cell dimensions), the symmetry of the crystal, and the arrangement of atoms within the lattice. In this regard, XRD can provide valuable information on the existing phases (quantitative and qualitative) as well as grain sizes and grain orientation (macroscopic texture), see Figure 2.25. More sophisticated measuring techniques based on XRD also allow for the detection of macroscopic residual stresses in the samples after tribo-testing. However, when considering results based on XRD, it should be also considered that this information stems from the bulk of the sample and represents a volume-average information due to the notable penetration depth of the used X-ray radiation.

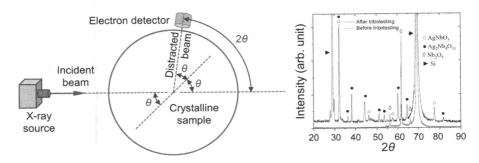

FIGURE 2.25 Schematic illustration of the measuring principle of XRD. Analysis is performed on $Nb_2O_5/Ag_2O/Nb_2O_5$ composite coatings demonstrating phase-transformation into ternary oxides $AgNbO_3$ and $Ag_2Nb_4O_{11}$ during sliding. This phase transformation is responsible for friction reduction. (Reprinted and adapted from Ref. [17] with permission.)

2.2.2.7 Electron Backscatter Diffraction

For a more advanced and refined analysis of the sub-surface micro-structure, electron backscatter diffraction (EBSD) helps to analyze the crystallographic properties of materials at a micro- and nanoscale level. The measurements of EBSD are realized in a scanning electron microscope using an additional EBSD detector, which analyzes the interaction between an electron beam and a sample's crystal lattice (Figure 2.26). When an electron beam strikes a crystalline material, the electrons interact with the lattice atoms, leading to elastic and inelastic scattering. Backscattered electrons with

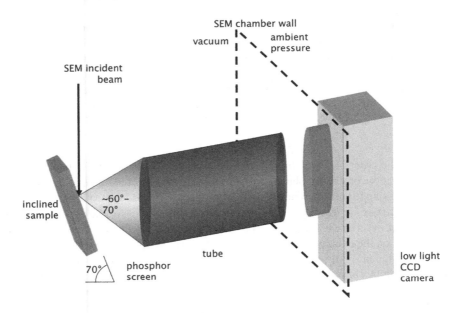

FIGURE 2.26 Schematics of the measuring principle of EBSD. (Reprinted from Ref. [18] with permission by CC BY-NC-ND 3.0.)

minimal energy loss can create an additional source of electrons within the material near the material's surface. This secondary electron source emits electrons, which are diffracted by adjacent lattice planes under the condition of satisfying the Bragg condition. The intersection of the diffracted electrons with a fluorescent phosphoscreen creates a characteristic pattern, which is known as the Kikuchi pattern. Kikuchi patterns contain information about the crystallographic orientation of the sample and are used to determine the orientation of individual crystal grains. Consequently, EBSD permits to analyze the local phase situation, grain sizes, grain orientation, defect situation, among others, which makes it a powerful tool to characterize the local micro-structure with nanoscopic resolution.

A comparison of the situation prior to and after tribological testing can generate essential information about the internal stress states (bulk) as well as changed grain sizes and orientation and defect situation (nano-scale). In this regard, particularly, EBSD is a very powerful tool since it can track and quantify changes in the micro-structure of materials as they undergo wear, which include grain refinement, grain growth, phase changes, and the formation of surface layers. It can help to identify wear mechanisms by analyzing the microstructural features that develop on worn surfaces. For example, different wear mechanisms like adhesive, abrasive, or fatigue wear may result in distinct patterns of grain deformation, fracturing, and orientation changes. EBSD is capable to assess changes in crystallographic texture (preferred orientation) of materials before and after wear, which provides valuable information about how wear processes affect material anisotropy and mechanical properties. Regarding the defect situation, friction and wear often lead to the formation of new grain boundaries (2D microstructural defects), grain boundary migration, and changes in grain boundary character. EBSD can quantify grain boundary misorientations and distribution changes on worn surfaces, which are important for understanding material behavior under wear. The obtained data may be correlated with wear rate measurements to better understand the relationship between microstructural changes and wear behavior.

2.2.2.8 Atom Probe Tomography and Time-of-Flight Secondary Ion Mass Spectrometry

In case of having the necessity to observe the respective elemental distribution with an atomic resolution, the characterization of choice is atom probe tomography (APT). In this regard, APT operates by field evaporating individual atoms from the surface of a needle-shaped specimen, which is typically prepared using a combination of FIB milling and electropolishing to create a fine tip to be attached to a microtip holder (Figure 2.27). The evaporated atoms are then detected and analyzed based on their time of flight to a position-sensitive detector. By collecting information about the atoms' positions and their time of flight, APT reconstructs a three-dimensional map of the atomic positions and compositions within the specimen.

In a tribological context, APT can provide detailed insights into the composition, distribution, and segregation of elements at surfaces and interfaces of materials undergoing tribological interactions. Moreover, APT can analyze wear tracks and debris generated during tribological tests. By examining the chemical composition of wear debris and mapping the subsurface microstructure, important information

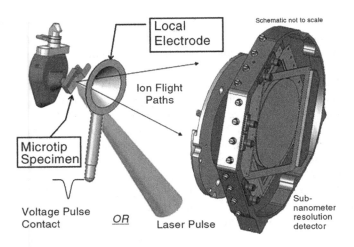

FIGURE 2.27 Schematic illustration of the APT with the main components. (Reprinted from Ref. [19] with permission by CC BY 4.0.)

about the underlying friction and wear mechanisms can be derived. This technique can be also used to characterize the atomic structure and composition of these coatings, offering insights into their performance and durability.

Time-of-flight secondary ion mass spectrometry (TOF-SIMS) is similar to the overall APT concept. Although the resolution of this technique is limited and it is not as sensitive as APT, it does not require extensive sample preparation and can be formed directly on the samples after tribological testing. TOF-SIMS begins with a primary ion beam, typically composed of noble gas ions like cesium (Cs^+) or bismuth (Bi^+). These ions are accelerated and directed toward the sample's surface. Upon impact, they start to create secondary ions (sputter process). The impact of the primary ion beam causes the ejection of these secondary ions, including atomic and molecular species, from the sample's surface, which carry valuable information about the sample's composition, including the elemental and molecular composition. The ejected ions are then analyzed using time-of-flight mass spectroscopy, which measures the time it takes for the secondary ions to travel a known distance from the sample to the detector. This time, along with the known flight path, is used to determine the mass-to-charge ratio (m/z) of each ion. The resulting mass spectrum provides information about the presence and relative abundance of elements and molecules on the sample's surface. TOF-SIMS provides certain benefits for tribological analysis since it can be used for depth profiling to unravel the changes induced inside the wear track. Depth profiling in TOF-SIMS involves removing material layer-by-layer and analyzing the composition at different depths.

2.2.2.9 Surface Energy and Wettability

Since the tribological contact is highly sensitive to surface energetics, enabling some basic assessment of the surface energy is valuable to predict the reactivity of the involved surfaces. This can be done by performing contact angle measurements. These measurements are essential for both dry and lubricated contacts since they

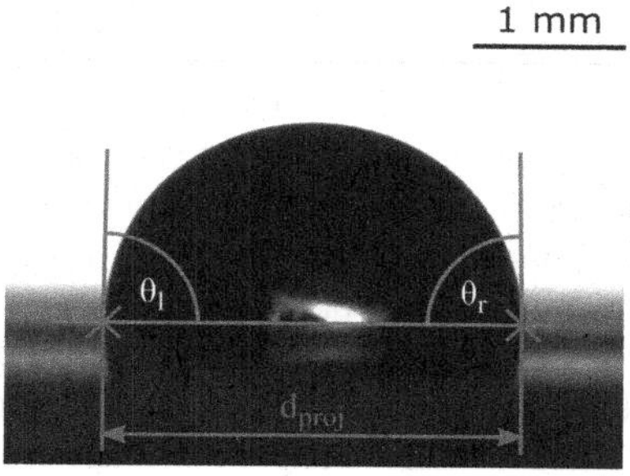

FIGURE 2.28 Representative shadowgraph for apparent contact angle measurements of a water droplet on a solid material. (Reprinted from Ref. [20] with permission by CC BY 4.0.)

provide basic information about the interaction of materials with lubricants as well as with humidity and vapors present in the environment. The contact angle is the angle formed at the three-phase contact line where a solid surface, a liquid droplet, and a gas (usually air) meet (Section 1.4.2). Contact angle measurements are typically performed using contact angle goniometers, which position a droplet of the test liquid on the solid surface and capture images of the contact angle (Figure 2.28) and thus assessing the wettability of a solid surface. By comparing the measured contact angles for a series of liquids with known surface tensions (polar and apolar liquids), the Lifshitz–van der Waals and Lewis acid–base components of the surface energy can be determined using various models like the Owens–Wendt–Rabel–Kaelble (work) method. This method employs Young's equation, delineating the equilibrium of forces at the three-phase contact line where air, liquid, and solid interfaces converge (Section 1.4.2).

2.2.3 CHARACTERIZATION OF LUBRICANTS

In the case of lubricants, their characterization after tribo-testing can provide valuable insights regarding the underlying lubrication mechanism. In this regard, **FTIR** is a powerful approach widely employed in various industrial applications. This technique provides insights into the chemical composition and quality of lubricants, aiding in maintenance, quality control, and troubleshooting processes. FTIR allows for the identification of individual lubricant components. Since lubricants contain a variety of additives, base oils, and contaminants (Chapter 3) with distinct chemical IR signatures and lubricants are exposed to high temperatures and mechanical stress during operation (oxidation and degradation over time), the comparison of FTIR spectra of a lubricant sample with reference spectra or libraries can pinpoint specific compounds present, helping to ensure that the lubricant meets the required specifications. FTIR can detect changes in the lubricant's chemical composition, such as the appearance of oxidation products and the depletion of antioxidants. Monitoring these changes enables the prediction of the lubricant's remaining useful life and facilitates timely maintenance or replacement. Furthermore, FTIR is highly sensitive to the presence of contamination, including water, dust, metal particles, and other impurities.

FTIR analysis can quickly identify these contaminants and their concentration, allowing for targeted corrective actions.

In recent years, **quartz crystal microbalance (QCM)** became another popular approach for probing the stability of the lubricants and their lubrication characteristics through the interactions at solid/liquid interfaces. QCM allows for an *in-situ* monitoring of the surface changes in a non-destructive way. QCMs are made from AT-cut quartz crystals (usually 10 mm or 1 inch in diameter) oscillating in a shear mode upon applying alternating current to the involved top and bottom electrodes. Specifically, when a voltage is applied to the quartz crystal, it vibrates at its natural resonant frequency, often in MHz range. As mass is added or removed from the surface of the quartz crystal (due to the adsorption or desorption of molecules), the resonant frequency of the crystal changes. When mass is added, the frequency decreases, while mass removal leads to the opposite trend. This change in resonance frequency is directly proportional to the mass change and is highly sensitive to even sub-monolayer changes on the surface. Upon immersion of the QCM in a fluid, the resonant frequency of the QCM also decreases, while this change depends on the viscosity and density of the fluid (Section 3.1). QCM has been successfully utilized for a wide range of interfaces, starting from analyzing slip initiated at solid/solid interfaces formed between adsorbed monolayers and QCM electrode surface materials with the overall aim to evaluate the tribo-film formation and lubrication efficiency for solid/liquid interfaces.

In the case of tribological relevance, QCM has been used to quantitatively evaluate corrosion onsets [21], the contribution of tribo-heating [22,23], the activation energy necessary for tribo-film formation from oil additives [24], and the analysis of oil degradation [25,26]. For instance, as demonstrated in Figure 2.29 [26], QCM allowed to *in-situ* monitor the degradation of oil mixed with different amounts of water upon thermal cycling. This analysis demonstrated that the oil + water mixtures exhibited a decrease in the observed change in the frequency during heating. This difference was associated with surface modifications leading to sludge formation during the heating of the oil + water mixture, which affects the lubrication efficiency.

While QCM can provide sensitive indications about the lubricant stability, it requires a specially designed experimental set-up [27]. Meanwhile, other analytical techniques, such as **Differential Scanning Calorimetry (DSC)** and **Thermogravimetric Analysis (TGA)**, are more commonly employed to assess the oxidation stability of lubricants. DSC measures the heat flow associated with thermal transitions during sample heating, which allows to detect exothermic reactions associated with the breakdown of antioxidants or the initiation of lubricant oxidation. DSC is usually combined with TGA to monitor the weight change during thermal degradation, which helps to identify the temperature of decomposition and evaporation as well as the corresponding mass losses. The analysis can be also complemented by information extracted from **Mass Spectrometry (MS)** to identify compounds based on their mass-to-charge ratio (*m/z*). MS does not only provide detailed information about the composition of complex mixtures but can also detect traces of contaminants and changes in the additives as a result of lubricant use.

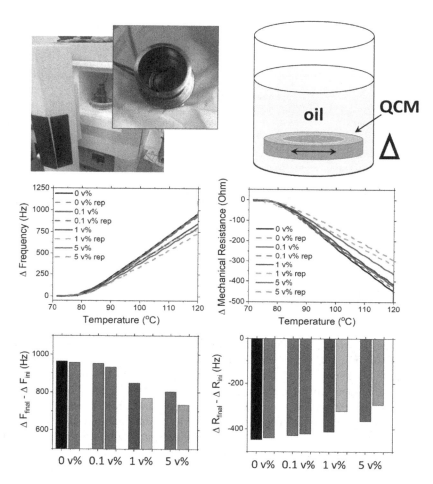

FIGURE 2.29 High-temperature QCM apparatus and schematic illustration of its operating principle as well as changes in the resonant frequency and mechanical resistance of the oscillations of the QCM immersed in lubricants during heating. Summary of the corresponding changes in the resonant frequency and mechanical resistance of QCM oscillations observed during heating when immersed in oils with different water contents. (Reproduced with permission from Ref. [26].)

2.3 DATA-GENERATION, -MANAGEMENT, AND -SCIENCE IN TRIBOLOGY

The generation of experimental data and knowledge in tribology is typically divided into three steps:

- Experimental planning
- Experimentation
- Analysis and interpretation

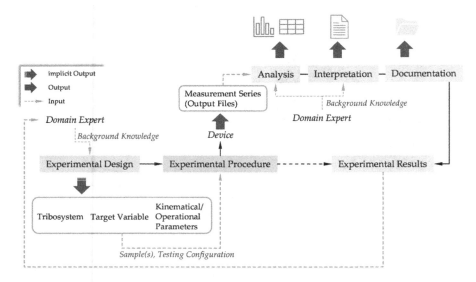

FIGURE 2.30   Process of producing experimental knowledge. (Reprinted from Ref. [1] with permission by CC BY 4.0.)

This process requires a substantial amount of background knowledge from experts (Figure 2.30). Initially, the tribologist designs an experimental set-up, including the fabrication of test specimens suitable for the studied tribo-system, defining the target variables, and specifying kinematic and operational parameters. The second step, the experimental procedure, is often conducted using standardized testing equipment described in Section 2.1 that generates output files containing data series. As tribological statements rely on comparing experimental results with those of a reference system, the experimental process involves various data series with a variation of the target variable. Furthermore, advanced characterization methods as introduced in Section 2.2 to reveal the underlying friction- and wear-mechanisms are carried out. The final step, experimental results, is achieved through the analysis of data series, interpretation of results, and documentation by the expert (typically in scientific publications or laboratory reports). Interpreting the results requires profound background knowledge of experimental standards and methodology in the field of tribology. The insights gained from these experiments serve as input for designing new experiments or selecting additional characterization methods. Without formalization of experimental knowledge, this process involves a significant amount of manual work for the tribologist.

Data serve as the foundational asset that imparts value to scientific investigations, with a growing emphasis on high-quality data that can seamlessly traverse research groups and infrastructures. This impetus has given rise to guiding principles known as the **FAIR** principles [28], emphasizing data's

- findability,
- accessibility,

- interoperability, and
- reusability.

Adhering to these principles has led to the development of detailed metrics evaluating the compliance of shared digital objects with these standards, ensuring value for data users in the future. The advantages of adopting FAIR principles extend beyond facilitating communication, encompassing enhanced trustworthiness of data. This, in turn, streamlines the transformation of data into knowledge and opens avenues for artificial intelligence (AI) and machine learning (ML) algorithms (Section 2.4). In the realm of tribology, generating FAIR research data presents unique challenges due to the interdisciplinary nature of the field. Complex tribological problems demand a comprehensive understanding of processes and mechanisms between contacting surfaces, complicating the establishment of discipline-specific data infrastructures. The absence of standards is partly attributed to tribologists interpreting results through diverse scientific backgrounds spanning various physical science and engineering fields. In tribology, in which the sequence of events and subtle external influences significantly impact experiment outcomes, data provenance is crucial for knowledge generation. For tribological experiments to be FAIR, they necessitate machine-actionable information sets encompassing all processes and equipment preceding the test. Addressing this challenge involves storing and potentially sharing detailed data to unravel friction and wear mechanisms. The pursuit of FAIR data in tribology hinges on synchronizing two essential efforts:

- establishing a schema of categories that generalizes tribological processes and objects, agreed upon by a critical mass of scientists, and
- developing a lab framework and digital infrastructure that allows flexible workflows and encourages recording detailed scientific processes in a machine-actionable manner.

Efforts to define common terms describing tribological specimens, equipment, and data manipulation aim to enhance communication, assist electronic lab notebook developers in designing user interfaces, and provide a foundation for computational algorithms. However, the tribological community lacks curated metadata repositories or strategies for controlled vocabularies, hindering the digitalization effort. Overcoming this challenge requires multilateral agreement on semantic definitions describing tribological experiments, underscoring the need for suitable tools facilitating collaboration in such endeavors.

In the past, generated data archives have typically been stored in traditional databases, such as **relational databases**. The latter are highly purpose-oriented, implying that they depend on data formats, organizational requirements, and possible on-top applications. This results in isolated and heterogeneous data sources, making it challenging to break down data silos through integration into a central data warehouse and hindering standardization of result treatment and data processing. To overcome these heterogeneous and ambiguous data sources and facilitate data exchange between isolated systems, semantic and linked data technologies are nowadays increasingly employed. A primary distinction between semantic technology and, for instance, relational databases, lies in the nature of data processing. While relational databases deal with the structure of data and thus rely on highly structured, clean input data, **semantic databases** are concerned with the meaning of data by utilizing **ontologies**, which provide an "explicit specification of a conceptualization." Ontologies offer a scalable framework for formalizing knowledge in tribology, enabling domain experts to encode their expertise in both human- and machine-readable formats. Generally speaking, ontologies define classes of entities (such as a tribometer or a human), their attributes (such as a descriptive text of a tribometer), individual instances (such as a tribometer with a serial number), and relationships between classes (such as a human operating a tribometer), see Figure 2.31. Using the Resource Description Framework (RDF), ontologies present facts in a structured manner, and the Web Ontology Language (OWL) facilitates the encoding of these RDF-based ontologies, creating graphs that not only serve as clear knowledge representations potentially applicable to AI/ML but also align with several FAIR principles. These principles include the "F" aspect of attaching rich metadata and assigning globally unique and persistent identifiers using Internationalized Resource Identifiers (IRIs), the "A" component

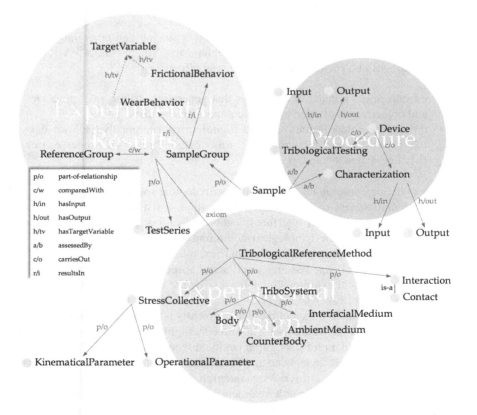

FIGURE 2.31   Core concepts (circles) and relationships (arrows) of the tribAIn ontology. (Reprinted from Ref. [1] with permission by CC BY 4.0.)

of offering a standardized communication protocol through the RDF query language (SPARQL), and the "I" element of ensuring data and metadata interoperability through traceable knowledge representation. Developing ontologies for tribology possesses a significant challenge, not only due to the interdisciplinary nature of research aspects but also because of the unique and often lab-specific experimental procedures and custom-built tribometers. An effective starting point involves leveraging existing higher-level ontologies, enhancing reasoning capabilities within established schemas shared by other ontologies [29].

The so-called tribAIn ontology was the first approach to provide a formal and explicit specification of knowledge in tribology to enable semantic annotation and the search of experimental setups and results [1]. For generalization, it was linked to the intermediate-level ontology EXPO (ontology of scientific experiments), supplemented with subject-specific concepts meeting the needs of the domain of tribology, thus being able to adopt heterogeneous data sources containing natural language texts and tabular data. TribAIn was later extended by a semantic annotation pipeline using state-of-the-art natural language processing (NLP techniques to semi-automatically adopt information from published literature) [30]. However, this only captures the information presented by the original authors, a practice that often falls short of meeting the FAIR data requirements. Therefore, the human intermediator between conducted experiments and recorded meta-data should be eliminated as far as possible by digitalization of an experimental environment, which requires a software infrastructure between controlled vocabularies (organized in an ontology) and the experimental data, while accounting for as many details as possible. Garabedian et al. provided a blueprint for FAIR data publication in experimental tribology based on electronic lab notebooks (ELNs) and the Karlsruhe Data Infrastructure for Materials Science (Kadi4Mat) [29]. This allows for the direct integration with custom tribometers (for at-source data collection) and export of data in both machine- and user-readable formats It should be noted that creating FAIR (meta-) data isn't a standalone addition to the functions of an experimental laboratory, but rather involves an integrated amalgamation of scientific, software, and administrative solutions. Figure 2.32 illustrates numerous groups and sequential pathways, exemplifying how diverse actors (depicted at the bottom) contribute to the process of digitalization, ultimately streamlining the workflow for laboratory scientists. The effectiveness of the back-end digital tools in communicating with users relies on the provision of accurate knowledge representation [29].

Thereby, it is crucial that these frameworks minimally disrupt ongoing research practices. Despite this challenge, the implementation of a lab-wide system provides an opportunity to enhance overall operational efficiency. For tribometers controlled by a Laboratory Virtual Instrument Engineering Workbench (LabVIEW; a system-design platform and development environment for a visual programming language), which are predominant in many labs, straightforward codes can be added, allowing seamless packaging and uploading data to the database. Conversely, analog processes in experimental tribology, such as specimen milling and cleaning, lack digital outputs and have traditionally been recorded in paper lab notebooks without formalized vocabulary or a system. To address this, guided user interfaces (GUI) run on tablet computers offer an intuitive means of collecting requisite details in a formalized manner (Figure 2.33) [29].

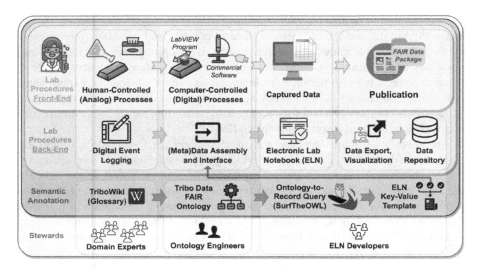

FIGURE 2.32   Component diagram for FAIR data production. (Reprinted from Ref. [29] with permission by CC BY 4.0.)

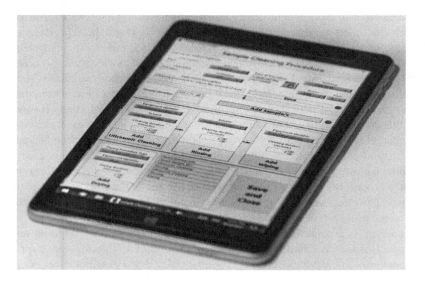

FIGURE 2.33   A GUI to collect pre-defined key-value descriptions of the event sequence comprising a typical specimen cleaning procedure – a digital interface for an inherently analog process. (Reprinted from Ref. [29] with permission by CC BY 4.0.)

From a managerial perspective, the ELN manages data storage, user interactions, and timestamps, essentially creating a comprehensive map of who did what and when. Equally crucial is the content of Garabedian's FAIR data package, originating from tribological experiments. The central information are the records, which are metadata (author, creation/revision time, persistent ID, license, tags, ontology class) and details about the represented entity. A record may include externally generated data, such as tables, text, images, videos, links to other records, or hierarchies of records within a collection, uniting all entities in a project. For sharing, a record has two forms: a human-readable PDF and a structured JSON file; if files were uploaded, a zipped archive can be included. Also, records can be uploaded to data repositories (such as Zenodo) and automatically receive a DOI. Notably, records can be anonymized before export to preserve researchers' privacy. The Zenodo hosts visual summaries, providing a links-based ontology-derived graph (similar to Figure 2.31) and a time-based workflow.

2.4 COMPUTATIONAL TRIBOLOGY

Historically, fundamental and applied research in tribology is mostly related to experimental trial-and-error approaches, which are time-consuming and costly. Moreover, a number of experiments, while keeping the experimental conditions constant, must be conducted to ensure the reproducibility of the obtained results. This aspect becomes further complicated when considering the friction and wear are system- and material-dependent, which implies that experiments conducted in a laboratory with a certain test-rig are not comparable to the work conducted by another laboratory. On the one hand, this urgently asks for some normalization strategies to make tribological research comparable (Section 2.3), while, on the other hand, it underlines the necessity for appropriate input from theory and numerical simulations. In this regard, computational methods are useful to back up experimental trends and to shed more light on the involved mechanisms and phenomena from a theoretical point of view. Besides the pure cross-correlation with experimental data, numerical methods are also expected to make smart predictions related to the expected performance of, for instance, new material systems, which would greatly reduce experimental effort and help to guide the experimentalists toward the right directions.

In this regard, computational methods, and mathematical algorithms with the overall aim to understand, elucidate, and predict friction, wear, and lubrication in complex tribological systems are resembled under the general term "**computational tribology**." In this regard, this complexity connects with the involved time and length scales in tribological systems. While macroscopic tribological systems can have characteristic length scales on the order of centimeters to meters, the rubbing surfaces typically touch and contact each other on the microscopic asperity level, which creates local contact areas on the order to micro- and nanometers. Consequently, adequately capturing the involved characteristics length scales is the first challenge, which must be solved in computational modeling. Apart from that, the mentioned asperity contact lasts for short period of times, since tribological systems are dynamic systems under a certain relative motion. This contact can induce rather high temperatures and pressures for short times, which may change the respective surface chemistry, microstructure, and overall topography due to potentially involved wear. Consequently, besides the complexity due to the involved time and length scales, potential interactions happening in the tribological contact zones further complicate the overall problem. Tribological processes can induce many potential effects in one or even both rubbing surfaces, which range from classical abrasive and adhesive wear to induce phase transformations and tribo-chemical reactions thus notably changing the sub-surface microstructure and surface chemistry. In many cases, these phenomena are superimposed, which even further complicates the search for appropriate simulation approaches, which do not oversimplify the encountered situation, but enable the calculation of physically-related data in reasonable times and computational effort. Therefore, numerical simulations cannot account for all involved features and phenomena. Consequently, the typical approach in computational tribology is to break down the complex systems into simpler ones, thus reducing the time and length scales to be considered. After having reduced the involved scales, more specialized simulation approaches can be adopted as illustrated in Figure 2.34.

Computational modeling, therefore, provides a great variety of numerical methods, which are particularly suited to cover certain time and length scales. When considering more macroscopic systems, numerical methods of choice typically relate to simulations on a system level as well as continuum mechanics approaches. In this regard, the fundamental base of these simulations generally connects to contact mechanics. Thereby, a complex interplay of various phenomena has to be considered and captured as illustrated in Figure 2.35.

Elastic Half-Space Approach: The basic contact mechanical approaches going back to the Hertzian theory (Section 1.5.1) can be considered as somewhat outdated and stress calculations in contact problems are usually performed with the assistance of computer-based methods. The semi-analytical **elastic half-space approach** has been originally developed to calculate wheel-rail contacts and has found wide applications. The theory determines the normal and tangential contact forces, as well as the dimensions of the contact zone, by minimizing the complementary

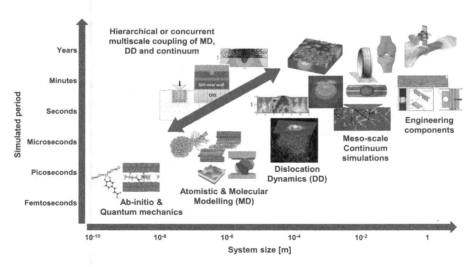

FIGURE 2.34 A time-vs. length-scales map of models developed in tribology highlighting the intrinsic link between multiscale/physics that needs to be captured to provide predictive tools for engineering applications. (Reprinted and adapted from Ref. [31] with permission from Elsevier.)

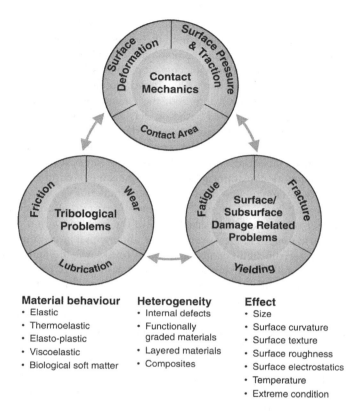

FIGURE 2.35 Relations between contact mechanics and tribological and damage-related problems. (Reprinted from Ref. [32] with permission from Elsevier.)

strain energy. Following a two-dimensional discretization of the contact zone, surface displacements for each discrete element can be calculated using the Boussinesq and Cerruti equations. These calculated displacements are then combined using the superposition principle to obtain the overall displacement. The computation of stresses inside the bodies can be also determined analytically from the contact forces on the surface of each discrete element, utilizing superposition. A significant advantage of the semi-analytical half-space model is that, in many cases, discretizing the two-dimensional contact area is sufficient, which leads to comparably low computational costs.

Boundary Element Method (BEM): Similarly, the **BEM** roots in the equations of the elasticity theory, often encapsulated in Hooke's law. These equations are reformulated in an integral form, allowing to establish a direct link between unknown values, such as stresses or displacements, and known boundary conditions and external loads. Through these equations, BEM transforms intricate partial differential equations into solvable forms on the discretized grid. A crucial aspect of applying BEM is the generation of an accurate boundary mesh aligning with real-world geometry as well as the treatment of singularities that may arise at sharp corners or edges.

The integral equations are often solved using either the collocation method or the Galerkin method. In the collocation method, the integral equations are evaluated at specific points of the boundary, while the boundary is approximated by a set of trial functions in the Galerkin method. Elastic contact problems are inherently non-linear due to the contact conditions. Therefore, an iterative approach is frequently employed to linearize the problem. At each iteration, the contact pressure distribution and boundary displacements are updated based on the solution from the previous iteration. This process continues until a convergence criterion is met.

Another emerging technique is the **conjugate gradient method (CGM)**. Originally devised for unconstrained problems, CGM can be extended to address problems with inequality constraints, such as those encountered in elastic contacts. The attractiveness of employing CGM in contact simulations is attributed to several key advantages. Firstly, CGM exhibits a super-linear rate of convergence. As a result, the number of iterations necessary to solve the problem with a desired level of accuracy tends to remain modest even when dealing with a large number of nodal points. This characteristic significantly contributes to lowering computational costs. Secondly, there is a well-established proof of CGM convergence for quadratic optimization problems with inequality constraints. This feature allows for the versatile application of CGM in solving contact problems. Polonsky and Keer specifically developed a CGM-based single-loop iterative procedure to simulate rough contacts. Further refinements have been introduced to enhance computational efficiency, to determine the contact area during contact pressure iteration, to enforce the required constraints and the force balance in each CGM iteration step, thus ensuring that the interfacial pressure and gap consistently satisfy compatibility requirements and eliminating the need for an additional loop to adjust the rigid body approach. These improvements allow for the simultaneous solution of contact pressure, real contact area, and rigid body approach with a streamlined, single-level iteration process. The deformation and pressure distribution for the contact of a plane with a rough sphere is exemplarily plotted in Figure 2.36. For more information on continuums approaches to model interfacial mechanic problems, the interested reader is referred to Ref. [33].

Finite Element Method (FEM)/Finite Element Analysis (FEA): The **Finite Element Method (FEM)**, also called finite element analysis (FEA), can be used to solve complex partial differential equations that govern the contact mechanical behavior by dividing a complex geometry into smaller, simpler sub-domains, which are called finite elements. These elements (triangles, quadrilaterals, tetrahedra, and hexahedra depending on the problem) are connected at specific points called nodes, forming a mesh that approximates the geometry of the system under investigation. Within each finite element, the governing equations that describe the physical behavior of the system are formulated. To represent the variation of these quantities within an element, mathematical functions known as shape functions are used. Once the governing equations are established within each element, they are assembled into a global system of linear algebraic equations. This captures the interactions and relationships between the elements in the entire structure. Solving this system numerically yields the desired solutions, such as displacements, stresses, and contact forces. All these aspects are essential to understand how forces are transmitted and distributed during sliding or rolling contact. Contact conditions and boundary conditions

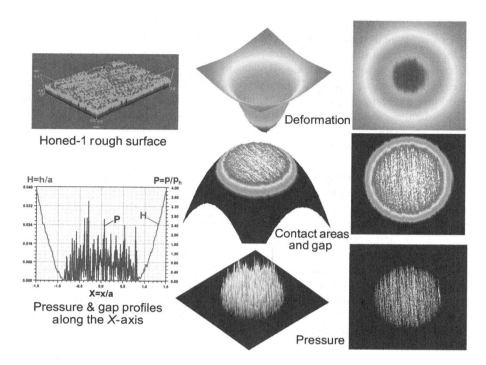

FIGURE 2.36 Numerical solution for a ball in contact with a flat, considering the effect of machined surface roughness. (Reprinted from Ref. [33] with permission from CRC Press.)

are imposed at the contact interface to govern how the bodies interact when they touch or separate (see Video 2.7 in ebook+ for the example of an artificial knee joint). FEM offers a high degree of flexibility since it can handle diverse geometries and materials, making it suitable for a wide range of contact mechanical problems. Another advantage is the variety of commercially available programs. However, calculations frequently are more expensive compared to the elastic half-space approach or BEM since the complete three-dimensional component(s) must be discretized. FEM can also help to model frictional interfaces under the assumption to surface roughness and locally varying material properties. With respect to advanced surface engineering techniques (Chapter 4), FEM can provide important insights into the design of surface textures and coatings and how they perform under specific tribological contact conditions.

Crystal Plasticity: Regarding the plastic deformation of materials under tribological stresses, which is an aspect highly likely to occur, **crystal plasticity** is the relevant constitutive framework under the assumption of rough surfaces and when assuming that the established contact spot is on the order to the grain size of the material. Therefore, crystal plasticity modeling is used to simulate the deformation behavior of crystalline materials at the microstructural level. It focuses on understanding how individual crystals or grains within a material deform, interact, and contribute to the overall mechanical response. To appropriately describe the involved microstructure of the material, crystal plasticity considers the crystallographic structure of individual grains, which is obtained by EBSD (Section 2.1.2). This aspect is highly essential to enable the estimation and calculation of the real deformation behavior of the involved materials. Besides the local grain size distribution, EBSD is capable to measure the local orientation of each grain. This input is also highly important to capture the anisotropic mechanical behavior. Based on the generated EBSD data, each grain within the material is treated as an individual entity with its own orientation and properties thus simulating the deformation of these grains, accounting for slip, twinning, and other deformation mechanisms. In this regard, the most common deformation mechanism in metallic materials connects with slip, which involves the movement of dislocations along specific crystallographic planes and directions. Depending on the underlying crystal structure, these models can also consider the activity of multiple slip systems to accurately capture the plastic deformation. These calculations are based on constitutive equations to relate the stress, strain, and strain rate of a grain, which depend on the crystallographic orientation, slip system activities, and hardening mechanisms. In the case of polycrystalline materials, the interactions between adjacent grains are important and cannot be neglected. Crystal plasticity can incorporate these interactions to adequately model the evolution of microstructure during deformation and tribological stressing. A snapshot of a polycrystalline CuNi alloy model after sliding under severe loading conditions is depicted in Figure 2.37.

Discretization Schemes: In the case of lubricated systems, FEM is useful to predict operation interval in terms of tribological conditions to establish full-film hydrodynamic, elastohydrodynamic, mixed, and boundary lubrication (Section 3.2). Related to full-film and elastohydrodynamic lubrication, FEM is capable to simulate the flow of lubricants, pressure distributions in lubricated contacts, and the formation of lubricant films, which are crucial to design efficient lubrication systems with the overall goal to minimize wear and friction.

In addition to FEM, numerical solutions for lubricated systems can be effectively obtained through alternative discretization schemes. The **Finite Difference Method (FDM)**, for instance, uses equations in their differential form, with support points strategically positioned at the corners of the volume elements. Meanwhile, the **Finite Volume Method (FVM)** adopts an approach that involves establishing solutions for the desired variables by integrating the differential equations at the central points of control volumes. These numerical methods provide valuable alternatives to model, for instance, (elasto-)hydrodynamic contacts.

Computational Fluid Dynamics (CFD) modeling is generally used to simulate and analyze fluid flow, heat transfer, and related physical phenomena. CFD is based on the Navier-Stokes equations, which account for conservation of mass, momentum, and energy, thus establishing the foundation for fluid flow simulations. From a tribological perspective, CFD can model the flow of lubricants in different types of tribological contacts, such as journal bearings, thrust bearings, and gearboxes (Figure 2.38). It helps to provide insights into the pressure distribution, lubricant film thickness, and fluid velocities under different lubricated conditions including full-film, elastohydrodynamic, mixed as well as boundary lubrication. In systems with pronounced tendency for cavitation (e.g., surface textures), CFD can

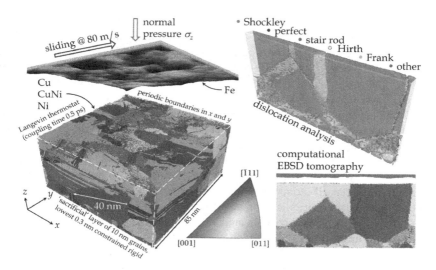

FIGURE 2.37 3D depiction of a CuNi and Fe tribo-system after a sliding event, presenting a separated view of the interacting bodies to provide clarity regarding the surface. The CuNi base body is color-coded to distinguish various crystallographic orientations, while the Fe counter-body is represented in gray scales to depict its topography. In the top right corner, dislocation analysis of a section of the initial system is presented before the sliding commences, with distinct colors indicating different dislocation types. The bottom right corner features a corresponding computational electron backscatter diffraction (EBSD) tomograph. (Reprinted from Ref. [34] with permission by CC BY 4.0.)

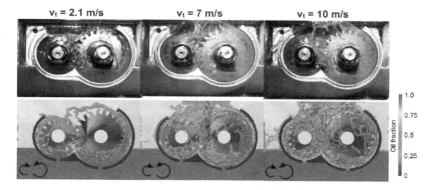

FIGURE 2.38 Comparison of measured (top) and CFD-simulated (bottom) oil distribution in a gear stage with guide plate. (Reprinted from Ref. [35] with permission by CC BY 4.0.)

adequately model cavitation and vaporization phenomena, which occur when local pressure drops lead to the formation of vapor bubbles or cavities within the lubricant (Section 3.2.1). Moreover, it can help to predict turbulent flows or the heat transfer within tribological contacts due to frictional heating.

Molecular Dynamics (MD): Apart from rather macroscale analysis, the involved phenomena in tribological processes also require an adequate consideration of the nanoscopic level considering short periods of time and length scales. In this regard, the method of choice relates to **MD simulations**, which model the behavior of atoms and molecules as a function of time, allowing for the study of tribological phenomena at the atomic scale. The typical scale of MD is tens of nanometers and tens of nanoseconds, which might be too small for unraveling the full tribological phenomena dynamics but is still informative to predict and explain the behavior of tribological interfaces that are usually not accessible for *in-situ* observations.

In general, MD simulations start with a detailed description of the initial conditions of the system including positions and velocities, as well as interactions (potentials) between all atoms or molecules in the system (Figure 2.39). They are often performed using periodic boundary conditions, where the simulation box (usually ~10 nm in dimensions) is replicated infinitely in all directions. This prevents edge artifacts and allows the simulation of bulk systems. The underlying interactions are described using a force field, which includes parameters for the bond lengths, angles, dihedral angles, and non-bonded interactions (such as van der Waals forces and electrostatic interactions). The force fields required for successful MD analysis are often provided by density functional theory (DFT) modeling. During simulations, the positions, and velocities of the particles in the system are updated for each time step using the equations of motion (Newton's second law). These equations take into account the forces acting on each particle, which are calculated based on the force field parameters and the current positions of the particles. The selection of the time step plays a critical role to ensure the success of the simulations as too small time steps could be incapable to capture the phenomena of interest, while too large time steps may result in numerical instability and inaccuracy.

Since tribological processes occur at the atomic and molecular level, MD simulations are particularly suitable for gaining insights into the mechanisms and behavior of materials in contact (Figure 2.40). From a fundamental point of view,

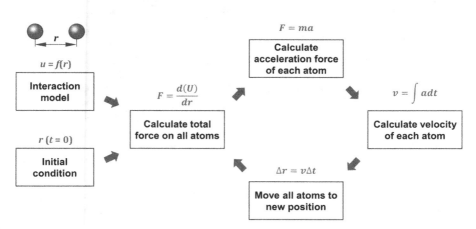

FIGURE 2.39 Schematic of the computational approach used for setting up the MD simulations.

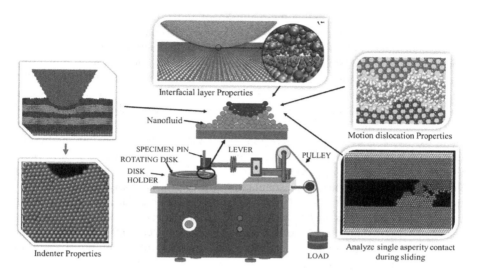

FIGURE 2.40 MD simulations to study indentation experiments, the formation of interfacial layers and the movement of dislocations and single-asperity sliding contacts. (Reprinted from Ref. [36] with permission by CC BY 4.0.)

MD simulations can elucidate the atomic-level processes responsible for friction and wear. By modeling the interactions between contacting surfaces and simulating their motion, it can be studied how atoms rearrange, form adhesion, and lead to wear over time. Concerning lubricated conditions, MD simulations can provide insights into the adsorption, spreading, and shear behavior of lubricant molecules, which resembles valuable information related to the design of more effective lubricants and understanding boundary lubrication. For systems, for which the tribological performance is predominated by tribo-chemical reactions, MD simulations shed more light on the involved chemical reactions occurring across the tribological interface, which may happen due to extreme conditions of pressure and temperature. In this regard, experimental tribological and characterization can be coupled and cross-correlated with the performed MD simulations. A similar cross-correlation of experimental and numerical data can be achieved for nano-tribological experiments using FFM.

There are certain limitations of MD simulations when correlating these results to experimental observations. The small time scales make it impossible to reproduce low sliding speeds and shear rates, while the low spatial scale limits the resolvable material features such as roughness and size of the defects. Another challenge is the limited library of available force fields and potentials, which restricts and narrows the number of materials to be modeled.

Density Functional Theory (DFT): DFT simulations are widely used to study the electronic structure and properties of atoms, molecules, and solids. In this context, DFT focuses on the electron density, which is used to predict a great variety of properties of the material system, including molecular structure, vibrational frequencies, ionization energies, electric and magnetic properties, reaction path, and others.

The central equation in DFT is the Kohn-Sham equation, which involves a set of non-interacting electrons moving in an effective potential. This effective potential includes an external potential (nuclei-electron interaction) and an exchange-correlation potential (resulting from electron-electron interactions). Regarding the latter, approximations are generally used to relate the electron density to the exchange-correlation energy. With regard to tribology, DFT simulations can provide insights into the fundamental processes that govern friction and wear on an atomic and molecular scale. This approach can be used to study the adhesion forces between surfaces, which is crucial in understanding friction and wear. DFT can predict the chemical reactions (reaction pathways, energy barriers, and the stability of reaction products) that occur at sliding surfaces, which are often responsible for wear and material degradation. These simulations are helpful to analyze the interactions between lubricant molecules and metal surfaces thus gaining a deeper understanding how lubricant molecules adsorb onto surfaces and form protective layers. In a similar regard, DFT simulations can shed light on the formation, composition, and stability of tribo-films. This approach also appears suitable to capture the combined effects of mechanical wear and corrosion, known as tribo-corrosion. Simulations can provide insights into the electrochemical processes that occur at sliding interfaces, impacting the degradation of materials.

Artificial Intelligence (AI)/Machine Learning (ML): In the last couple of years, there is a new trend in computational tribology, which relates to predictive modeling using **AI and ML approaches**. The latter are grounded in logic, probability theory, algorithm theory, and computing principles. In its initial stages, ML entails the design of computing systems tailored for specific tasks, capable of learning from training data over time. These systems develop and refine experience-based models to predict outcomes, enabling them to respond to queries within a particular domain. Therefore, these approaches are powerful tools to make predictions about future events or outcomes based on a sufficiently big set of data (experimental or numerical). The success of these approaches largely depends on the quality of the used data sets (see Section 2.3). This implies that the generation and post-treatment of the data are of utmost importance. In a given data set, special attention must be paid to choose relevant variables that are likely to have a significant impact on the prediction as well as to potentially create new features to better represent the underlying patterns in the data. Thereby, the data is typically split up and a significant part of the data is used to train the selected algorithm/model, while a reduced amount of data is needed for verification and further predictions. The performance of the selected approach is generally checked using appropriate metrics, such as mean absolute error (MAE), mean squared error (MSE), root mean squared error (RMSE), or classification metrics like accuracy, precision, among others. Various algorithms can be employed in ML, with their suitability strongly dependent on the specific task. Generally, algorithms are categorized as either "supervised learning" or "unsupervised learning." In supervised learning, algorithms learn relationships from a provided set of input and output data vectors. Throughout the learning process, a "teacher" (such as an expert) supplies correct inputs and outputs. In contrast, unsupervised learning involves the algorithm generating a statistical model describing a given dataset without external evaluation by a "teacher." Reinforcement learning, although occasionally classified as supervised learning, exhibits distinct characteristics. Instead of inducing knowledge from pre-classified examples, an "agent" engages in experimental interactions with the system. The system responds to these experiments with either reward or punishment. Consequently, the agent optimizes behavior with the objective of maximizing reward and minimizing punishment. While the classification of the three mentioned learning types is widely accepted, there is no unanimous agreement on which algorithms should be assigned to specific categories. The list of algorithms and models includes linear regression, decision trees, random forest analysis, support vector machines, among others. Artificial Neural Networks (ANNs), see Figure 2.41, are certainly the most prominent type and are fundamentally designed based on the architecture of natural brains. They constitute a computing system comprising simple yet highly interconnected processing elements, known as neurons, which process information through dynamic responses to external inputs. The transfer function calculates the network input of a neuron based on the weighted inputs. The output value is determined by the activation function, which considers a threshold value. In a training process, the weightings and thresholds for each neuron can be adjusted. The arrangement of neurons and interconnections, specifying the number of neurons in a layer and how they are arranged in parallel, is referred to as topology or architecture. The final layer, known as the output layer, may have several hidden layers positioned between the input and output layers, forming a multilayer ANN. Although single-layer networks also feature feedback loops, multilayer ANNs are commonly represented in graph theory notation, with nodes representing neurons and edges representing their interconnections.

In a tribological context, AI and ML modeling have significant potential to predict different tribological characteristics, which may lead to improved efficiency, reduced maintenance costs, and increased component lifespan when properly used, see Figure 2.42. These approaches can predict the wear rate of materials and components based on various parameters like load, speed, temperature, and surface roughness, which helps in optimizing maintenance schedules and material selection.

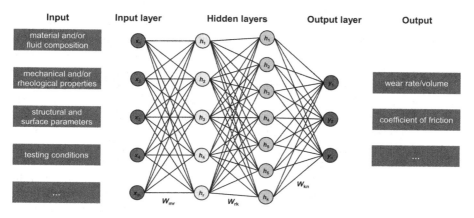

FIGURE 2.41 Correlation between material properties and testing conditions using artificial neural networks. (Reprinted and adapted from Ref. [37] with permission by CC BY 4.0.)

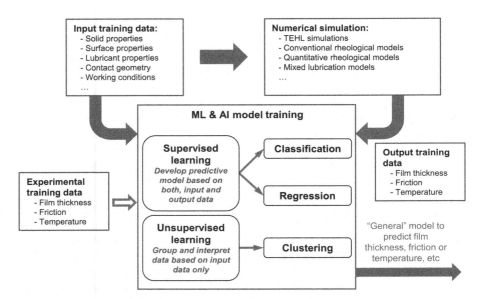

FIGURE 2.42 Main steps and parameters associated with the training process of ML/AI models to predict the lubrication performance from experimental and simulation data sets. (Reprinted and adapted from Ref. [37] with permission by CC BY 4.0.)

Moreover, ML algorithms can analyze wear debris data to detect early signs of component degradation or faults, allowing for timely maintenance interventions. With respect to the formulation of lubricants, ML algorithms can help in designing and optimizing lubricant formulations by predicting the performance of different oil and additive combinations, while AI may be helpful to recommend the appropriate lubricant or additive based on specific operating conditions and material combinations. Another advancing field for the implementation of AI and ML algorithms in tribology connects with online monitoring, for which AI-based predictive maintenance systems can monitor equipment in real time, using data from sensors and historical performance data to predict when components need maintenance or replacement. ML models can analyze vibration data, temperature data, and other sensor inputs to detect abnormal operating conditions that might lead to excessive wear. After the detection of a failure, AI can analyze failure modes and patterns to identify the root causes of wear-related failures, helping in designing more robust components and systems. To avoid the occurrence of failures, AI can assist in designing new materials with improved tribological properties by simulating material behavior under different conditions and optimizing material compositions. Although holding great potential for predicting tribological characteristics with impressive precision in the near future, special care needs to be taken with regard to the quality and completeness of the data since the entire quality and accuracy of the predictions depend on these. For more information on data-driven AI and ML in the context of tribology, the interested reader is referred to Refs. [37,38].

Another emerging trend in computational tribology connects with **physics-informed machine learning (PIML)**. The integration of physics-based knowledge into ML presents a powerful tool for comprehending and enhancing phenomena associated with friction, wear, and lubrication. Conventional ML approaches often rely exclusively on data-driven techniques, lacking the incorporation of fundamental physics. In contrast, PIML methodologies, including Physics-Informed Neural Networks (PINNs), utilize established physical laws and equations to guide the learning process. This results in models that are more accurate, interpretable, and transferable. PIML has been effectively applied to various tribological tasks, such as predicting lubrication conditions in hydrodynamic contacts or forecasting wear and damages in tribo-technical systems [39].

2.5 CHECK YOURSELF

- What are the categories into which tribological testing can be categorized based on system structure, stress conditions, and environmental conditions?
- How does the significance of various quantifiable parameters in tribological testing depend on factors such as underlying mechanisms, measurement methods, and specific objectives?
- Explain the importance of considering length scales in tribology and how microscopic and macroscopic phenomena influence frictional behavior.
- What are the fundamental principles and historical methods behind friction measurement?
- In the context of experimental challenges in tribometers, how does misalignment impact the measurement of the coefficient of friction, and what is the formula used to account for misalignment in the measured coefficient of friction?
- What is the focus of nanoscale tribometry, and why is it essential in understanding the tribological behavior of materials?
- What are the challenges associated with picoindentation experiments using a transmission electron microscope, and how does the technique contribute to fundamental studies?
- Describe the working principle of atomic force microscopy and its evolution into lateral force microscopy and friction force microscopy for tribological measurements at the nanoscale.
- How does friction force microscopy differ from traditional atomic force microscopy in terms of its ability to assess frictional forces, and what are the advancements that have enabled sophisticated experiments in tribology?
- In microscale tribometry, how do MEMS tribometers function, and what are the key components involved in calculating normal and lateral forces in these setups?
- What are the key variables controlled in macroscopic lab experiments in tribometry, and how are parametric studies structured?
- Explain the different contact conditions commonly encountered in macroscopic tribological test setups.

- How does a four-ball tester work, and what are its advantages in assessing friction and wear characteristics of lubricants and solid materials?
- Describe the purpose and functionality of a mini-traction machine in evaluating the tribological characteristics of materials, especially lubricants.
- What is the significance of tribo-corrosive conditions in real-world applications, and how do specialized tribo-corrosive tribometers study such conditions?
- Discuss the importance of component- and system-level tribo-testing in evaluating the mechanical functioning of components and devices, providing examples of standardized test-rigs.
- What are the physical and chemical characteristics of materials that must be studied to understand their tribological behavior?
- Explain the difference between *in-situ* and *ex-situ* characterization in the context of tribological assessment.
- Describe common contact-based methods for measuring surface roughness in tribometry.
- What are the advantages and limitations of non-contact methods for measuring surface characteristics in tribological studies?
- How is nanoindentation used to determine the mechanical properties of materials, and what are the key parameters derived from nanoindentation experiments?
- What are the potential effects of oxidation and tribo-chemical changes on the overall surface chemistry of rubbing surfaces during tribological testing?
- Why is it essential to consider both rubbing surfaces when conducting wear analysis to understand the underlying mechanisms?
- What is the role of SEM coupled with EDX in the characterization of tribologically active surfaces?
- How can Raman spectroscopy and FTIR contribute to understanding the wear mechanisms and surface changes during tribological processes?
- What information can be obtained from TEM in the context of tribology, and how is it applicable to wear analysis?
- Explain the basic principle of XPS and its significance in the characterization of surface chemistry.
- How does EBSD provide insights into the crystallographic properties of materials at a micro- and nanoscale level?
- What role does APT play in characterizing the atomic structure and composition of coatings and wear debris in tribological studies?
- How can QCM be utilized in tribology for in-situ monitoring of surface changes and lubrication efficiency?
- What analytical techniques, other than QCM, are commonly employed to assess the oxidation stability of lubricants, and what information do they provide?
- What are the three main steps involved in the generation of experimental data and knowledge in tribology, and why is background knowledge from experts crucial in this process?

- How do the FAIR principles contribute to the value of scientific investigations in tribology, and what challenges are unique to implementing these principles in the interdisciplinary field?
- What is the main challenge in experimental tribology that computational methods, specifically computational tribology, aim to address?
- What are some of the numerical methods used in computational tribology for modeling macroscopic systems, and what are their advantages and limitations?
- How does crystal plasticity modeling contribute to understanding the plastic deformation of materials under tribological stresses?
- What role does FEM play in predicting the operation interval in lubricated systems, and what are its advantages and disadvantages?
- How does CFD contribute to understanding fluid flow in tribological contacts, and what types of lubrication conditions can it simulate?
- What is the typical scale of MD simulations, and what insights can they provide into tribological phenomena at the atomic scale?
- How does DFT contribute to understanding friction and wear on an atomic and molecular scale?
- What is the role of AI and ML in computational tribology, and how can they be applied to predict tribological characteristics?

REFERENCES

1. P. Kügler, M. Marian, B. Schleich, S. Tremmel, S. Wartzack, tribAIn-towards an explicit specification of shared tribological understanding, *Applied Sciences* 10 (2020) 4421.
2. J.P. Oviedo, S. Kc, N. Lu, J. Wang, K. Cho, R.M. Wallace, et al., In situ TEM characterization of shear-stress-induced interlayer sliding in the cross section view of molybdenum disulfide, *ACS Nano* 9 (2015) 1543–1551.
3. D.A. Hook, S.J. Timpe, M.T. Dugger, J. Krim, Tribological degradation of fluorocarbon coated silicon microdevice surfaces in normal and sliding contact, *Journal of Applied Physics* 104 034303 (2008).
4. M. Marian, *Numerische Auslegung von Oberflächenmikrotexturen für geschmierte tribologische Kontakte*, FAU University Press, Erlangen, Germany, 2020.
5. C. Gachot, C. Hsu, S. Suárez, P. Grützmacher, A. Rosenkranz, A. Stratmann, et al., Microstructural and chemical characterization of the tribolayer formation in highly loaded cylindrical roller thrust bearings, *Lubricants* 4 (2016) 19.
6. T. Tobie, F. Hippenstiel, H. Mohrbacher, Optimizing gear performance by alloy modification of carburizing steels, *Metals* 7 (2017) 415.
7. A. Ayyagari, T.W. Scharf, S. Mukherjee, Dry reciprocating sliding wear behavior and mechanisms of bulk metallic glass composites, *Wear* 350–351 (2016) 56–62.
8. A. Shirani, N. Nunn, O. Shenderova, E. Osawa, D. Berman, Nanodiamonds for improving lubrication of titanium surfaces in simulated body fluid, *Carbon* 143 (2019) 890–896.
9. A. Shirani, Q. Hu, Y. Su, T. Joy, D. Zhu, D. Berman, Combined tribological and bactericidal effect of nanodiamonds as potential lubricant for artificial joints, *ACS Applied Materials & Interfaces* 11 (2019) 43500–43508.
10. D. Berman, K.C. Mutyala, S. Srinivasan, S.K.R.S. Sankaranarayanan, A. Erdemir, E.V. Shevchenko, et al., Iron-nanoparticle driven tribochemistry leading to superlubric sliding interfaces, *Advanced Materials Interfaces* 6(23) (2019) 1901416.

11. M. Schaffer, B. Schaffer, Q. Ramasse, Sample preparation for atomic-resolution STEM at low voltages by FIB, *Ultramicroscopy* 114 (2012) 62–71.

12. P.G. Grutzmacher, S. Suarez, A. Tolosa, C. Gachot, G. Song, B. Wang, et al., Superior wear-resistance of Ti3C2Tx multilayer coatings, *ACS Nano* 15 (2021) 8216–8224.

13. A. Shirani, T. Joy, A. Rogov, M. Lin, A. Yerokhin, J.-E. Mogonye, et al., PEO-Chameleon as a potential protective coating on cast aluminum alloys for high-temperature applications, *Surface and Coatings Technology* 397 (2020) 126016.

14. D.F. Zambrano-Mera, M.I. Broens, R. Villarroel, R. Espinoza-Gonzalez, J.Y. Aguilar-Hurtado, B. Wang, et al., Solid lubrication performance of sandwich Ti3C2Tx-MoS2 composite coatings, *Applied Surface Science* 640 (2023) 158295.

15. D.N.G. Krishna, J. Philip, Review on surface-characterization applications of X-ray photoelectron spectroscopy (XPS): Recent developments and challenges, *Applied Surface Science Advances* 12 (2022) 100332.

16. R. Ghobeira, P.S. Esbah Tabaei, R. Morent, N. De Geyter, Chemical characterization of plasma-activated polymeric surfaces via XPS analyses: A review, *Surfaces and Interfaces* 31 (2022) 102087.

17. A. Shirani, J. Gu, B. Wei, J. Lee, S.M. Aouadi, D. Berman, Tribologically enhanced self-healing of niobium oxide surfaces, *Surface and Coatings Technology* 364 (2019) 273–278.

18. A.J. Wilkinson, T.B. Britton, Strains, planes, and EBSD in materials science, *Materials Today* 15 (2012) 366–376.

19. M.K. Miller, et al., The future of atom probe tomography, *Materials Today* 15(4) (2012) 158–165.

20. P. Foltyn, F. Restle, M. Wissmann, S. Hengsbach, B. Weigand, The effect of patterned micro-structure on the apparent contact angle and three-dimensional contact line, *Fluids* 6(2) (2021) 92.

21. J. Lee, D. Berman, Inhibitor or promoter: Insights on the corrosion evolution in a graphene protected surface, *Carbon* 126 (2018) 225–231.

22. J. Krim, Atomic-scale origins of friction, *Langmuir* 12 (1996) 4564–4566.

23. A. Dayo, W. Alnasrallah, J. Krim, Superconductivity-dependent sliding friction, *Physical Review Letters* 80 (1998) 1690–1693.

24. A. Shirani, S. Berkebile, D. Berman, Promoted high-temperature lubrication and surface activity of polyolester lubricant with added phosphonium ionic liquid, *Tribology International* 180 (2023) 108287.

25. A. Shirani, T. Joy, I. Lager, J.L. Yilmaz, H.-L. Wang, S. Jeppson, et al., Lubrication characteristics of wax esters from oils produced by a genetically-enhanced oilseed crop, *Tribology International* 146 (2020) 106234.

26. K. Jacques, T. Joy, A. Shirani, D. Berman, Effect of water incorporation on the lubrication characteristics of synthetic oils, *Tribology Letters* 67 (2019) 105.

27. B. Acharya, M.A. Sidheswaran, R. Yungk, J. Krim, Quartz crystal microbalance apparatus for study of viscous liquids at high temperatures, *Review of Scientific Instruments* 88(2) (2017).

28. M.D. Wilkinson, M. Dumontier, I.J. Aalbersberg, G. Appleton, M. Axton, A. Baak, et al., The FAIR guiding principles for scientific data management and stewardship, *Scientific Data* 3 (2016) 160018.

29. N.T. Garabedian, P.J. Schreiber, N. Brandt, P. Zschumme, I.L. Blatter, A. Dollmann, et al., Generating FAIR research data in experimental tribology, *Scientific Data* 9(1) (2022) 315.

30. P. Kügler, M. Marian, R. Dorsch, B. Schleich, S. Wartzack, A semantic annotation pipeline towards the generation of knowledge graphs in tribology, *Lubricants* 10(2) (2022) 18.

31. A.I. Vakis, V.A. Yastrebov, J. Scheibert, L. Nicola, D. Dini, C. Minfray, et al., Modeling and simulation in tribology across scales: An overview, *Tribology International* 125 (2018) 169–199.

32. B. Li, P. Li, R. Zhou, X.-Q. Feng, K. Zhou, Contact mechanics in tribological and contact damage-related problems: A review, *Tribology International* 171 (2022) 107534.

33. Q. Wang, Zhu, D., *Interfacial Mechanics, Theories and Methods for Contact and Lubrication*, CRC Press, Boca Raton, FL, 2019.

34. S.J. Eder, P.G. Grutzmacher, M. Rodriguez Ripoll, D. Dini, C. Gachot, Effect of temperature on the deformation behavior of copper nickel alloys under sliding, *Materials (Basel)* 14(1) (2020) 60.

35. L. Hildebrand, F. Dangl, M. Sedlmair, T. Lohner, K. Stahl, CFD analysis on the oil flow of a gear stage with guide plate, *Forschung im Ingenieurwesen* 86 (2021) 395–408.

36. I. Srivastava, A. Kotia, S.K. Ghosh, M.K.A. Ali, Recent advances of molecular dynamics simulations in nanotribology, *Journal of Molecular Liquids* 335 (2021).

37. A. Rosenkranz, M. Marian, F.J. Profito, N. Aragon, R. Shah, The use of artificial intelligence in tribology-a perspective, *Lubricants* 9(1) (2020) 2.

38. M. Marian, S. Tremmel, Current trends and applications of machine learning in tribology-a review, *Lubricants* 9(9) (2021) 86.

39. M. Marian, S. Tremmel, Physics-informed machine learning-an emerging trend in tribology, *Lubricants* 11(11) (2023) 463.

3 Liquid Lubrication

One of the most common approaches to reduce friction and wear as well as address challenges in tribological systems involves the use of lubricants in the form of liquid or semi-liquid mixtures (oils and greases). **Liquid lubricants** are an integral part of many industrial operations since they provide a range of tribological benefits that make them invaluable in a wide variety of settings. The main functions of lubricants are:

- **Surface Separation**: By providing a lubricating film between the surfaces of tribological contacts, lubricants minimize direct contact and the associated frictional forces. Lubricants also help to distribute loads more evenly across the contacting surfaces, thus reducing stress concentrations. Consequently, lubricants reduce abrasive or adhesive wear and extend the operational life of machinery and components.
- **Heat Dissipation**: Lubricants help to dissipate heat generated during frictional contact. This is especially important in high-speed and high-load applications, in which excess heat can lead to component failure or reduced efficiency.
- **Contact Cleaning and Sealing**: Lubricants can help to maintain a clean and controlled environment within machinery and also form seals to keep out contaminants such as dust, dirt, and moisture, which can otherwise accelerate wear and corrosion.
- **Vibration and Noise Damping**: Effective lubrication can reduce vibrations and noise generated by friction between surfaces. This is essential for a smooth and quiet operation of machinery.
- **Corrosion Prevention**: Lubricants can create a protective barrier against corrosion by preventing moisture and oxygen from reaching metallic surfaces.

Therefore, lubricants must be **compatible** with the materials they come into contact with. They should not damage the surfaces they lubricate, ensuring the longevity of components, while at the same time do not adversely affect their environment. In the following sections, we summarize the wide selection of liquid lubrication solutions. However, before reviewing in detail the possible choices, the basic mechanisms and fundamentals of liquid lubrication must be understood.

3.1 LIQUID LUBRICANT PROPERTIES

A detailed understanding of the performance of lubricated tribo-contacts directly correlates with the properties of the lubricants as well as their rheology, which unifies the study of the flow and deformation of molecules they are made of. Some of the most relevant lubricant properties for tribological contacts are

DOI: 10.1201/9781003397519-3

- **density**
- **viscosity**
- **thermal conductivity**
- **specific heat capacity**
- **flash point**
- **pour point**
- **oxidative stability**
- **vapor pressure**

These properties are of relevance when comparing results from different experiments as well as performing theoretical and numerical calculations. However, it should be noted that their values are not constant when the lubricant is subjected to varying temperatures, pressures, or high shear gradients. Consequently, the consideration of realistic operational conditions is essential for a reliable design of lubricated contacts.

The **density** of liquid lubricants refers to the mass per unit volume of the lubricant and is usually expressed in kilograms per cubic meter (kg/m³) or grams per cubic centimeter (g/cm³). The density can be determined by various methods, including hydrometers, pycnometers, digital density meters, and gas comparison pycnometers. These approaches involve either weighing a known volume of the liquid or comparing it to a reference substance. Please note that the density of the lubricants in operation is not a constant value. As temperature increases, fluids tend to expand, and as pressure increases, they contract, resulting in a change in density, while maintaining a constant mass. In this regard, density decreases with increasing volume and vice versa. The reduction in volume under pressure, which is more pronounced in gases compared to liquids, is referred to as compressibility. The density depends on temperature T as

$$\rho(\vartheta) = \left[1 - (T - T_0) \cdot \beta\right] \cdot \rho_0, \tag{3.1}$$

using the reference density ρ_0 at temperature T_0, the applied pressure $P = 0$ and the thermal expansion coefficient given as

$$\beta = \frac{\partial V / \partial T}{V_0} = -\frac{\partial \rho / \partial T}{\rho_0}. \tag{3.2}$$

The densities of mineral oils often exhibit a linear temperature dependence, decreasing by approximately 5%–15% when the temperature increases by 100°C (at atmospheric pressure). This results in thermal volume expansion coefficients ranging between 0.7 and $1.2 \cdot 10^{-3}$ K^{-1} [1].

The relationship between pressure P and density can be expressed by

$$\rho(P) = \left(1 + \frac{A_{\rho 1} \cdot P}{1 + A_{\rho 2} \cdot P}\right) \cdot \rho(T). \tag{3.3}$$

As established by Dowson and Higginson [2] based on quasi-static experiments involving a single mineral oil and pressures up to 0.4 GPa, which has gained widespread recognition in literature. In that work, the coefficients $A_{\rho 1}$ and $A_{\rho 2}$ were pre-determined for the specific oil ($A_{\rho 1} = 0.6 \cdot 10^{-3}$, $A_{\rho 2} = 0.6 \cdot 10^{-3}$) indicating that an increase of pressure to 1 or 4 GPa yields density rises of 22% and 31%, respectively. By combining the previously derived isobaric and isothermal equations, we can obtain an equation that describes the dependence on temperature and pressure:

$$\rho(T,P) = \left(1 + \frac{A_{\rho 1} \cdot P}{1 + A_{\rho 2} \cdot P}\right) \cdot \left[1 - (T - T_0) \cdot \beta\right] \cdot \rho_0. \tag{3.4}$$

Another approach was suggested by Bode [3] and is given by

$$\rho(T,P) = \frac{(1 - \alpha_s \cdot T) \cdot \rho_s}{1 - D_{\rho 0} \cdot \left(\dfrac{D_{\rho 1} + D_{\rho 2} \cdot T + P}{D_{\rho 1} + D_{\rho 2} \cdot T}\right)}, \tag{3.5}$$

where the coefficients D_ρ, the density at absolute zero ρ_s (maximum packing density), and the thermal expansion coefficient of the solidified solid phase α_S are to be determined for each lubricant based on experimental data [1].

The **specific gravity** of a lubricant is a measure of the density of the lubricant relative to the density of water at a reference temperature. It is a dimensionless quantity, typically expressed as a ratio. The specific gravity of water is defined as 1.0, while most liquid lubricants have values larger than 1.0, which indicates that they are denser than water.

The **viscosity** of a liquid lubricant refers to the specific measure of the lubricant's resistance to flow or its internal friction when it is used to reduce friction between moving surfaces. It indicates how easily the lubricant can flow and deform under the influence of shear forces, which are generated by the relative motion of the lubricated components. A liquid lubricant with a high viscosity will be thicker and more resistant to flow, while one with low viscosity will be thinner and flows more easily. Viscosity is a critical property in lubrication because it affects the lubricant's ability to provide a protective film between contacting surfaces and reduce friction and wear. In this context, a distinction is made between

- the **dynamic viscosity** η with the unit Pa·s (or P; 1 P = 0.1 Pa·s) and
- the **kinematic viscosity** v with the unit m²/s (or St; 1 St = 10^{-4} m²/s)

where, the dynamic viscosity is the "physically correct" viscosity of a fluid, while the kinematic viscosity represents the ratio between the dynamic viscosity and the density:

$$v = \frac{\eta}{\rho}. \tag{3.6}$$

The **ISO VG** (Viscosity Grade) **oil classification** system is a standardized method for categorizing lubricating oils based on their viscosity, simplifying their selection and specification of lubricants and making it easier to choose the right oil for the specific application. The system assigns a numerical value to each oil, which reflects its kinematic viscosity characteristics at 40°C, used as a standard reference temperature.

It includes a range of viscosity grades, starting from ISO VG 2 (ultra-low viscosity) for light-duty applications, all the way up to ISO VG 1500 (extremely high viscosity) for heavy-duty and high-temperature applications.

Measuring the viscosity of liquid lubricants involves using viscometers, which are specialized instruments designed for this purpose. These viscometers typically employ rotational or capillary flow principles to assess the resistance of the lubricant to flow. The viscosity is determined by measuring the time it takes for the lubricant to flow through the instrument under controlled conditions.

There are different analytical approaches to describe the temperature-dependent viscosity using polynomial or exponential equations, among them is the approach established by Vogel [4]:

$$\eta(T) = A_{\eta 1} \cdot e^{\frac{A_{\eta 1}}{A_{\eta 1} + T}} \tag{3.7}$$

with the oil-specific coefficient, A_η, has been widely used due to its high accuracy.

A scheme for classifying the viscosity of engine oil stems from the Society of Automotive Engineers (**SAE**), using a two-part code, e.g., SAE 5W-30. The W[1]-rating indicates how the oil performs at lower temperatures, specifically, the oil's flowability at low temperatures, which is crucial for easy engine starting and protection under cold weather conditions. The lower 'W' numbers indicate a better low-temperature performance. The second number indicates the oil's viscosity at higher operating temperatures. A higher second number represents a thicker, more viscous oil. Another classification scheme is the so-called **viscosity index (VI)**, which is a measure used to describe how the kinematic viscosity of an oil changes with temperature and is specified in ASTM D2270. Specifically, it quantifies how the oil's viscosity varies as it experiences temperature fluctuations. A higher VI indicates that the oil's viscosity is less affected by temperature changes.

The pressure dependency (neglecting the temperature effect) can be expressed by an approach from Barus:

$$\eta(P) = e^{\alpha_p \cdot P} \cdot \eta_0, \tag{3.8}$$

with the reference viscosity η_0 and the pressure-viscosity coefficient

$$\alpha_p = \frac{\partial \eta / \partial P}{\eta_0} \tag{3.9}$$

which indicates the extent of pressure-induced changes in viscosity [5]. Substantial effects from pressure on the viscosity become apparent in the case of very high pressures, such as those found in Hertzian contacts (see Section 1.5.2). Since the Barus equation tends to yield excessively high viscosities at higher pressures, it has been extended through power or polynomial approaches. A widely adopted approach goes back to Roelands:

---

[1] The "w" stands for "winter."

$$\eta(P) = e^{\left\{ [\ln(\eta_0)+9.67] \left[-1 + \left(1 + \dfrac{P}{1.98 \cdot 10^9 \text{ Pa}} \right)^{\frac{\alpha_p \cdot 1.98 \cdot 10^9 \text{ Pa}}{\ln(\eta_0)+9.67}} \right] \right\}} \cdot \eta_0, \tag{3.10}$$

By combining Eq. 3.10 with the equation from Vogel (Eq. 3.7), we obtain an approach to express the pressure and temperature dependency:

$$\eta(T,P) = e^{\left\{ [\ln(\eta(P))+9.67] \left[-1 + \left(1 + \dfrac{P}{1.98 \cdot 10^9 \text{ Pa}} \right)^{\frac{\alpha_p \cdot 1.98 \cdot 10^9 \text{ Pa}}{\ln(\eta_0)+9.67}} \left(\dfrac{T-138}{T_0-138} \right)^{\frac{\beta_\eta \cdot (T_0-138)}{\ln(\eta_0)+9.67}} \right] \right\}} \cdot \eta_0. \tag{3.11}$$

The equation only requires knowledge about the coefficients α_p and β_η though, the accuracy of the equation declines as pressure rises [1]. Generally, it should be noted that measuring viscosity at very high pressures (> 1 GPa) poses challenges due to specialized equipment requirements (high-pressure viscometers or diamond anvil cells), precise temperature control, and sample containment. Therefore, available data is frequently extrapolated and should be treated carefully.

Apart from temperature and pressure, the viscosity of liquid lubricants exhibits a significant dependency on the shear rate, a phenomenon known as shear-rate dependency or shear-thinning behavior due to the alignment and deformation of the lubricant's molecular structure in response to the applied shear force. This can be illustrated by a liquid being placed between two plates with area A and distance ∂h (Figure 3.1). When the liquid adheres to both plates (Stokes' adhesion condition) and the upper plate is moved at velocity v_2 while the lower plate remains stationary ($v_1 = 0$), the fluid layers close to the plates have the same velocity as the plates and the layers in between move at different velocities relative to each other, resulting in laminar shear flow. The ratio of the velocity difference between the plates to the plate separation is referred to as the shear gradient or shear rate

$$\dot{\gamma} = \frac{\partial v}{\partial h}. \tag{3.12}$$

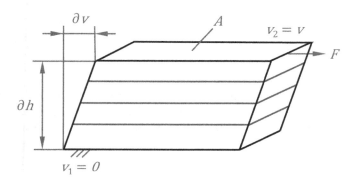

FIGURE 3.1   Shearing of a fluid between two parallel plates.

The ratio of the force F required for plate displacement (due to the internal friction of the fluid) to the plate's area A yields the frictional shear stress

$$\tau = \frac{F}{A}. \tag{3.13}$$

This can be interpreted as equivalent shear rate and stress composed of the shear rates and stresses from the two spatial directions:

$$\dot{\gamma} = \sqrt{\dot{\gamma}_{zx}^2 + \dot{\gamma}_{zy}^2} \tag{3.14}$$

and

$$\tau = \sqrt{\tau_{zx}^2 + \tau_{zy}^2}. \tag{3.15}$$

For fluids without shear thinning (Newtonian fluids), the dynamic viscosity connects with the frictional shear stress and the shear gradient in the following way:

$$\eta = \frac{\tau}{\dot{\gamma}} = \text{const.} \tag{3.16}$$

Fluids that do not conform to the equation mentioned above belong to the group of non-Newtonian fluids (Figure 3.2). **Newtonian fluids** (e.g., water) do not demonstrate a dependence on the viscosity of the shear rate, which implies that their viscosity remains constant regardless of the applied shear force. **Bingham fluids** also maintain a constant viscosity. However, they require a minimum yield stress to initiate flow. These fluids stay solid until a certain force is applied, after which they flow like a liquid. An example of this behavior is honey. In the case of **dilatant fluids**, the viscosity increases with the shear rate. This can, for instance, be observed for corn starch in water. Finally, for **structural viscous/pseudoplastic fluids**, the viscosity decreases as the shear rate rises (shear thinning). We know this behavior from, for example, ketchup. When ketchup is at rest in the bottle, it has a relatively high viscosity, meaning it is thick and doesn't flow easily. This is why it often appears to be stuck in the bottle when you first try to pour it. When you apply shear force to ketchup

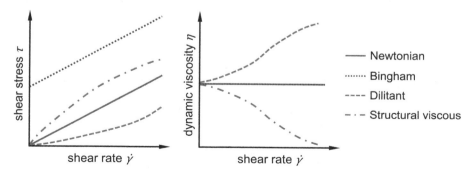

FIGURE 3.2 Flow (left) and viscosity curves (right) of different fluids with Newtonian, Bingham, dilatant, and structural viscous behavior.

by squeezing the bottle or tapping it, the ketchup experiences an increase in shear rate. In response to this increased shear rate, the ketchup's viscosity decreases significantly. As a result, it becomes more fluid and flows more readily. After the shear force is removed, ketchup gradually returns to its higher viscosity state, thickening up again over time. Similarly, many liquid lubricants show a shear-thinning behavior.

Various models have been derived to analytically describe the flow behavior of these fluids. A first structural viscous model has been introduced by Ostwald and De Waele, for which the shear rate

$$\dot{\gamma} = k \cdot \tau^n \tag{3.17}$$

is expressed as a function of the shear stress, the consistency coefficient k, which quantifies the lubricant's resistance to flow, and the flow behavior index n. The latter determines whether the fluid exhibits a shear-thinning behavior ($n < 1$) or shear-thickening behavior ($n > 1$). A value of $n = 1$ represents the Newtonian behavior.

In the Eyring [6] model, which can be mathematically expressed as

$$\dot{\gamma} = \frac{\tau_0}{\eta} \cdot \sinh\left(\frac{\tau}{\tau_0}\right), \tag{3.18}$$

the transition from Newtonian to non-Newtonian flow behavior occurs upon surpassing the shear stress τ_0. After this point, the shear stress τ continues to rise with an increasing shear gradient, albeit significantly less steeply than in the Newtonian scenario. Using this approach, Bair and Winer considered that the shear rate equals [7]

$$\dot{\gamma} = -\frac{\tau_{\lim}}{\eta} \cdot \ln\left(1 - \frac{\tau}{\tau_{\lim}}\right) \tag{3.19}$$

assuming a maximum achievable limit shear stress $\tau_{\lim}$, which cannot be surpassed. Rheological models for viscoelastic fluids are built upon the foundation of a **Maxwell fluid**, which exhibits both elastic (spring-like or storage) and viscous (damping or dissipative) characteristics. Therefore, the Bair and Winer model can be extended by a spring-term including the shear modulus G [1]

$$\dot{\gamma} = \overbrace{\frac{\partial \tau / \partial t}{G}}^{\text{spring}} - \overbrace{\frac{\tau_{\lim}}{\eta} \cdot \ln\left(1 - \frac{\tau}{\tau_{\lim}}\right)}^{\text{damper}}. \tag{3.20}$$

Some liquid lubricants, especially when involving long-chained polymers as frequently employed as additives (see Section 3.3), feature shear-thinning behavior in between two pronounced Newtonian plateaus (η_0 and η_∞), as shown in Figure 3.3, which can be described by the models from Cross [8]

$$\dot{\gamma} = \frac{\left(\dfrac{\eta_0 - \eta_\infty}{\eta - \eta_\infty} - 1\right)^{\frac{1}{n}}}{\lambda} \tag{3.21}$$

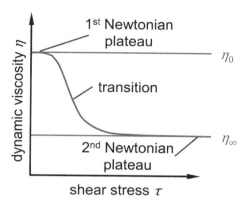

FIGURE 3.3 Typical viscosity shear-thinning response with two Newtonian plateaus.

or Carreau [9] $(a = 2)$ and Yasuda et al. [10]

$$\dot{\gamma} = \frac{\left[\left(\left(\frac{\eta - \eta_\infty}{\eta_0 - \eta_\infty} - 1\right)^{\frac{a}{n-1}} - 1\right)^{\frac{1}{a}}\right]}{\lambda}, \tag{3.22}$$

whereby n and λ are lubricant-specific fitting parameters. The elimination of the second Newtonian region is achieved by setting the viscosity that emerges at higher shear rates η_∞ to zero [1].

In some instances, the thermal properties of liquid lubricants, particularly their thermal conductivity and specific heat capacity, should be taken into consideration. **Thermal conductivity** measures a substance's ability to conduct heat and can be determined using a thermal conductivity meter, also known as a heat flow meter. In this approach, a test sample of the lubricant is placed between two plates with known temperatures, and the rate of heat transfer through the sample is measured, allowing for the determination of its thermal conductivity. In lubrication systems, the thermal conductivity determines how efficiently heat is transferred away from contact surfaces where friction and heat generation occur. The effective heat dissipation is essential to control the operating temperature of machinery. A high thermal conductivity helps to quickly transport heat away from critical components, preventing overheating and potential damage. Also, lubricants with proper thermal conductivity can maintain a stable viscosity across a range of temperatures.

The **specific heat capacity** represents the amount of heat energy required to raise the temperature of a unit mass of a substance by 1 K. In lubrication, it affects how much heat energy a lubricant can absorb before its temperature starts to increase significantly. The specific heat capacity of liquid lubricants can be measured through differential scanning calorimetry (DSC), whereby a small amount of the lubricant is subjected to controlled heating or cooling, and the heat flow required to maintain a constant temperature difference is recorded. Lubricants with higher specific heat capacities can absorb and store more heat energy, helping to stabilize temperature fluctuations in the system. This is particularly important in preventing thermal breakdown of the lubricant, which can lead to reduced effectiveness and even damage. When examining thermal processes in lubricated contacts more closely, assuming constant values can result in inaccuracies when determining temperature distribution, film thickness, and friction. As a result, the thermal conductivity and capacity can also be incorporated into model equations to account for temperature and pressure changes. According to Larsson and Andersson [11], the pressure-dependent thermal conductivity can be described by

$$\lambda(P) = \left(1 + \frac{A_1 \cdot P}{1 + A_2 \cdot P}\right) \cdot \lambda_0, \tag{3.23}$$

where the pressure is inserted in GPa, and the coefficients A_1 and A_2 have to be experimentally determined. Furthermore, the product of specific heat capacity c and density in convective heat transfer can be expressed as

$$\rho \cdot c(T, P) = \left[1 + \beta_0 \cdot \left(1 + B_1 \cdot P + B_2 \cdot P^2\right) \cdot (T - T_0)\right] \cdot \left(1 + \frac{C_1 \cdot P}{1 + C_2 \cdot P}\right) \cdot (\rho \cdot c)_0, \tag{3.24}$$

with the lubricant-specific parameters β_0, B_1, B_2, C_1, and C_2. The term $\rho \cdot c$ is sufficient for calculating convective heat transfer but does not represent the dependence of the heat capacity on pressure and temperature. However, if one divides the aforementioned equation by one of the density equations introduced earlier, this dependence becomes possible to account for. As a result, the specific heat capacity increases with rising temperature and changes only moderately with pressure (both increases and decreases in value) [1].

The **flash point** of a liquid lubricant is the temperature (in °C), at which it produces vapor that can ignite when exposed to an open flame or spark. It is a critical safety parameter for lubricating oils used in machinery, as oils with lower flash points are more prone to catching fire or exploding in high-temperature and high-friction conditions. A higher flash point indicates greater safety and stability under extreme operating conditions.

The **pour point** is the lowest temperature (°C), at which the lubricant remains fluid and can be poured or pumped without forming solid or waxy deposits or becoming too viscous to use. This property is crucial for oils used in cold climates or equipment that operates in low-temperature environments. A lower pour point indicates an improved low-temperature performance, preventing the lubricant from becoming too thick and causing poor lubrication under cold conditions.

Oxidative stability refers to the resistance to oxidation, which occurs when the lubricant reacts with oxygen in the environment over time. Oxidation can lead to the formation of harmful sludge, acids, and other byproducts that can degrade the lubricant's performance and reduce the lifespan of the equipment it lubricates. The oxidative stability is the temperature (°C), at which significant oxidation reactions begin to occur. Lubricants with higher oxidative stability can resist oxidation and maintain their performance and lubrication properties for a more extended period.

The **vapor pressure** is a measure of how readily a lubricant evaporates or turns into vapor at a specific temperature. A higher vapor pressure indicates that the oil evaporates more easily, which can lead to increased oil consumption, emissions, and potential health and environmental concerns. In contrast, lower vapor pressure oils are more stable and tend to retain their properties without significant loss due to evaporation.

Apart from the aforementioned properties, various other factors can be of relevance, depending on the respective application. For instance, **solubility** might be important if lubricants need to be compatible with other fluids in the system. **Demulsibility** is important in applications with water exposure, as the lubricant should separate easily from water, preventing emulsification. **Filtration compatibility** ensures that lubricants do not form deposits that clog filters, reducing system efficiency. **Foam resistance** is crucial as excessive foaming can reduce lubrication effectiveness, so lubricants should resist foaming. **Additive compatibility** is crucial, as lubricants often contain additives; compatibility is vital to maintain performance. **Aging stability** ensures that lubricants maintain their properties over time with minimal degradation or change due to aging or storage. Considering the environmental impact, including **biodegradability** and **toxicity**, is crucial in environmentally sensitive applications.

3.2 LUBRICATION FUNDAMENTALS

The reduction of friction and wear of tribo-technical contacts when employing liquid lubricants can be primarily attributed to the partial or complete separation of the contacting partners through an intermediate medium with low shear resistance. Depending on the macro-geometry, arrangement, kinematics (see Section 1.5), and surface topography (see Section 1.4) of the contacting bodies, as well as the properties of the lubricant (see Section 3.1), various lubrication and friction conditions can be established. This is illustrated in the so-called Stribeck curve[2] (Figure 3.4) by plotting a coefficient of friction μ over the lubricant film height parameter λ

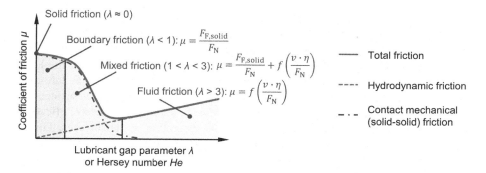

FIGURE 3.4 Stribeck curve: Coefficient of friction versus lubricant gap parameter or Hersey number as well as contributions to overall friction behavior in different regimes of lubricated contacts.

---

[2] Named after Richard Stribeck who reported such characteristic behavior for journal bearings operating at different speeds and loads.

$$\lambda = \frac{h}{R_q}, \tag{3.25}$$

which describes the ratio of the film thickness h to the combined surface roughness of both contacting surfaces.

While friction is relatively high in the case of direct solid contact (see Section 1.7), it significantly decreases through the formation of a natural or artificial adsorption layer or the presence of a solid lubricant (see Chapter 4) under boundary lubrication. The load is borne by the contacting asperities of the contact partners while shearing during relative motion predominantly occurs within the boundary layer. Under mixed lubrication, friction further diminishes as a portion of the load is carried by the lubricant under pressure in the contact area, leading to partial separation of the surfaces. Due to the increasing contribution of fluid friction with the rising lubricant film height parameter and the decreasing contribution of contact mechanics-related friction, a minimum point emerges in the overall friction as the transition is made from mixed to fluid friction. This point is termed the "transition point." Due to the trend toward improving energy efficiency and the use of low-viscosity lubricants, boundary, and mixed lubrication conditions are nowadays becoming increasingly important. However, due to their complex interactions, mathematical description and numerical modeling of these conditions are extremely demanding. When the contact partners are completely separated (full-film lubrication), friction depends solely on the increasing shear resistance of the lubricant as speed rises. With no interaction between surface asperities, wear is almost eliminated. In principle, the resulting lubricant film thickness (Figure 3.5) decreases with normal force and increases with relative velocity and lubricant viscosity. Therefore, instead of the lubricant film parameter λ, the dimensionless Hersey number

$$He = \frac{v \cdot \eta}{F_N} \tag{3.26}$$

is also frequently used for the x-axis of the Stribeck curve.

With the establishment of a supporting hydrodynamic (HD) lubricant film, the friction and wear behavior depend, among other factors, on the gap geometry, relative movements of the surfaces, the viscosity of the lubricating medium, the load, and the adhesion between this medium and the contact partners. When the lubricant is drawn into the contact zone through a lifting movement of parallel surfaces or into a narrowing gap through a tangential movement, and due to adhesive effects, a hydrodynamic pressure builds up, leading to the separation of the surfaces. This is exemplified in Figure 3.6 for a radial journal bearing, for which the maximum lubricant pressure typically reaches up to 100 MPa.

In general, the minimum achievable lubricant film height increases with rotational speed and viscosity as well as decreases with load. As shown in Figure 3.7, higher viscosities lead to lower friction at slower velocities and to an earlier transition from mixed to full-film lubrication. However, it also induces higher COFs when the lubricant's internal fluid friction dominates due to the larger lubricant gap. In turn, higher loads result in higher COFs at lower velocities and a later transition out of the mixed lubrication regime, but to lower friction in the full-film regime due to the smaller lubricant gap that is formed (Figure 3.7).

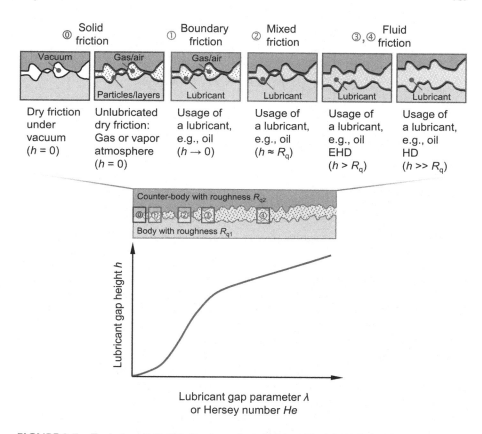

FIGURE 3.5 Evolution of the lubricant gap from height with rising lubricant gap parameter or Hersey number.

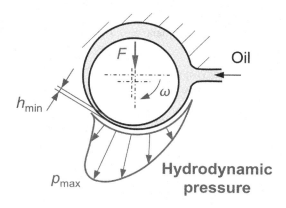

FIGURE 3.6 Pressure build-up and lubricant gap geometry in HD contacts exemplarily shown a radial journal bearing.

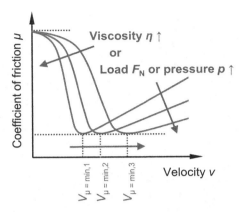

FIGURE 3.7   Influence of increasing velocity, viscosity, and load on friction in HD contacts.

If one of the contact partners is soft or if the pressures increase due to the load or non-conformal contact geometry, the elastic deformation of the surfaces superimposes the HD conditions. This condition is typically referred to as elastohydrodynamics (EHD) or, when considering thermal effects, thermo-elastohydrodynamics (TEHD). The elastic deformation and pressure distribution essentially follow those in dry contacts according to the Hertzian theory (see Section 1.5). However, in the contact entrance region, due to lubricant compression, an earlier pressure rise occurs, and at the contact exit, a sudden pressure drop leads to an elastic constriction, creating a feature known as the Petrusevich spike in the hydrodynamic pressure profile. This is schematically depicted in Figure 3.8 and animated in Video 3.1 in ebook+. The developing lubricant film thickness and the elastic deformation are typically in a similar magnitude. While a nearly parallel lubricant gap with a distance is established over wide ranges, the minimal lubricant film height is found in the constriction region. The degree of the hydrodynamic pressure peak in this region also depends on the contact stresses and might even exceed the pressure in the central contact area if contact loads are high. Under higher loads, the EHD pressure distribution approaches the Hertzian pressure distribution, as the effects of elastic deformation dominate. At lower loads or higher speeds, hydrodynamic processes become more prominent, resulting in a more pronounced pressure

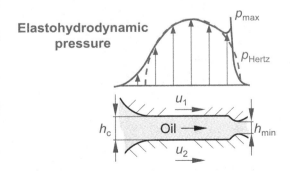

FIGURE 3.8   Pressure build-up and lubricant gap geometry in EHD contacts.

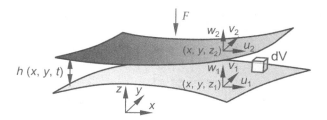

FIGURE 3.9 Spatial directions and velocities to describe fluid flows in lubricated contacts.

peak shifting toward the center of the deformed area. Hydrodynamic lubrication effects can be theoretically simulated for both pure HD contacts and EHD contacts. The fundamentals will be discussed in the following sections.

3.2.1 HYDRODYNAMIC LUBRICATION (HL)

Hydrodynamic lubrication plays a pivotal role in tribological applications, offering crucial advantages over boundary and mixed lubrication as it eliminates the contribution of solid materials and wear.

For a comprehensive mathematical description of the fluid behavior, various continuum mechanics conservation laws for mass, momentum, and energy are required. These can be formulated either in the Lagrangian approach, which focuses on particle-fixed fluid elements, or the Eulerian form, which deals with fixed control volumes. By adapting the fundamental fluid equations with appropriate boundary conditions to specific flow problems, the behavior of lubricants in HD contacts can be also described. In this regard, the lubricating gap between two surfaces moving relative to each other is based on the Cartesian coordinate system depicted in Figure 3.9, with spatial directions x, y, z, and velocity vectors u, v, w. Since the dimensions of the flow field are generally much larger than the scales of interatomic or intermolecular interactions, a uniform continuum can be considered, and the properties of the lubricant (see Section 3.1) are taken into account as averaged quantities.

3.2.1.1 Mass Conservation

When accounting for mass, the starting point is an infinitesimal fluid element with a density ρ and volume V. The mass balance is established at an unchanging volume element dV with edge lengths ∂x, ∂y, and ∂z to derive the continuity equations. In the case of a constant mass, a change in density necessitates a volume change. Instead of mass, density, as the mass per unit volume, can be employed. The temporal alteration of mass within the unchanging volume element is equal to the difference between inflowing and outflowing mass fluxes. The mass flux, denoted as $\dot{m}$, is the product of fluid density, velocity, and the cross-sectional area through which the fluid moves [1]:

$$d\dot{m} = \frac{\partial m}{\partial t} = \dot{m}_{in} - \dot{m}_{out} = \frac{\partial \rho}{\partial t} dV. \tag{3.27}$$

The differences between inflowing and outflowing mass fluxes in the x, y, and z directions can be determined as illustrated in Figure 3.10. This yields the **continuity**

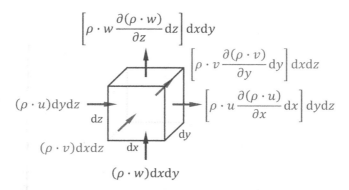

FIGURE 3.10 Mass flows at the volume element (balance volume) according to Ref. [1].

equation (mass conservation) for compressible flows in a differential form and Cartesian coordinates

$$\frac{\partial \rho}{\partial t} + \frac{\partial (\rho \cdot u)}{\partial x} + \frac{\partial (\rho \cdot v)}{\partial y} + \frac{\partial (\rho \cdot z)}{\partial w} = 0, \tag{3.28}$$

according to which the sum of the change of density with time in the volume element and the mass flow in all three spatial directions have to be zero. For incompressible flows, for which the density is constant, we obtain [1]

$$\frac{\partial u}{\partial x} + \frac{\partial v}{\partial y} + \frac{\partial z}{\partial w} = 0. \tag{3.29}$$

3.2.1.2 Momentum Conservation

In accordance with the principle of mass conservation, the conservation of momentum can be also assessed within the volume element. Building upon Newton's second law, which states that force equals mass times velocity, the temporal change in momentum within the volume arises from the disparity between the incoming and outgoing momentum flows $d\dot{I}$, the surface forces acting on the volume element denoted as $dF_{\tau\sigma}$, and the mass forces acting on the mass of the volume element dF_m [1]. The force vectors in the three spatial directions encompass both surface forces (such as pressure forces and friction forces represented by normal stress and shear stress) as well as body forces (including gravity, centrifugal forces, electrical, and magnetic forces). The momentum flux $\dot{I}$ signifies the momentum of the fluid element that traverses a cross-sectional area per unit of time. When density is employed instead of mass (similar to mass conservation), the product of density and velocity corresponds to momentum per unit volume [1]:

$$\frac{\partial (m \cdot v)}{\partial t} = \frac{\partial (\rho \cdot v)}{\partial t} dV = d\dot{I} + dF_{\tau\sigma} + dF_m. \tag{3.30}$$

The momentum flows as well as the normal and shear stresses at the volume element are exemplified for the x-coordinate in Figure 3.11. By employing various equation

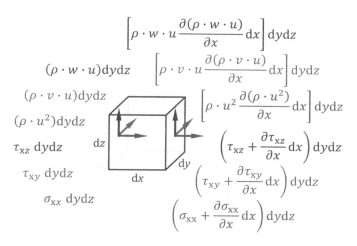

FIGURE 3.11 Momentum flows as well as normal and shear stresses for the x-coordinate at the volume element (balance volume) according to Ref. [1].

transformations (not shown in detail here), we can derive the following fluid-independent and universally applicable momentum equation (Cauchy's equation of motion) for the x-coordinate with the volume force $f_{m,x}$ [1]:

$$\frac{\partial(\rho \cdot u)}{\partial t} + \frac{\partial(\rho \cdot u^2)}{\partial x} + \frac{\partial(\rho \cdot u \cdot v)}{\partial y} + \frac{\partial(\rho \cdot u \cdot w)}{\partial z} = f_{m,x} + \frac{\partial \widehat{\sigma_{xx}}}{\partial x} + \frac{\partial p}{\partial x} + \frac{\partial \tau_{yx}}{\partial y} + \frac{\partial \tau_{zx}}{\partial z}.$$

(3.31)

The equations for the y- and z-coordinates can be obtained similarly (please refer to Ref. [1]). To solve the momentum conservation equations for a specific fluid, it is necessary to introduce fluid-specific relationships between the stress deviator and the velocity field (velocity gradients). The **Navier-Stokes equation** for an unsteady three-dimensional and compressible flow with Newtonian fluid behavior is derived for the x-coordinate as follows [1]:

$$\underbrace{\left(\frac{\partial(\rho \cdot u)}{\partial t} + \frac{\partial(\rho \cdot u^2)}{\partial x} + \frac{\partial(\rho \cdot u \cdot v)}{\partial y} + \frac{\partial(\rho \cdot u \cdot w)}{\partial z} \right)}_{\text{Inertia term}} = \overbrace{f_{m,x}}^{\text{Volume force term}} + \overbrace{\frac{\partial p}{\partial x}}^{\text{Pressure term}} +$$

$$\underbrace{+ \frac{\partial}{\partial x}\left\{ \eta\left[2\frac{\partial u}{\partial x} - \frac{2}{3}\left(\frac{\partial u}{\partial x} + \frac{\partial v}{\partial y} + \frac{\partial w}{\partial z} \right) \right] \right\} + \frac{\partial}{\partial y}\left[\eta\left(\frac{\partial v}{\partial x} + \frac{\partial u}{\partial y} \right) \right] + \frac{\partial}{\partial z}\left[\eta\left(\frac{\partial w}{\partial x} + \frac{\partial u}{\partial z} \right) \right]}_{\text{Friction term}}.$$

(3.32)

For an incompressible flow (with constant density) exhibiting Newtonian fluid behavior, the momentum conservation equation for the x-coordinate can be simplified as follows [1]:

$$\rho \underbrace{\left(\frac{\partial u}{\partial t} + u \frac{\partial u}{\partial x} + v \frac{\partial u}{\partial y} + w \frac{\partial u}{\partial z} \right)}_{\text{Inertia term}} = \overbrace{f_{\text{m,x}}}^{\text{Volume force term}} + \overbrace{\frac{\partial p}{\partial x}}^{\text{Pressure term}} +$$

$$\underbrace{+ \frac{\partial}{\partial x} \left[\eta \left(2 \frac{\partial u}{\partial x} \right) \right] + \frac{\partial}{\partial y} \left[\eta \left(\frac{\partial v}{\partial x} + \frac{\partial u}{\partial y} \right) \right] + \frac{\partial}{\partial z} \left[\eta \left(\frac{\partial w}{\partial x} + \frac{\partial u}{\partial z} \right) \right]}_{\text{Friction term}}.$$

(3.33)

By combining the Navier-Stokes equations with thermodynamic state equations and thermal energy conservation in a system of coupled nonlinear differential equations, the flow in hydrodynamically lubricated contacts can be also computed using specially developed equation solvers or commercial CFD programs. This has been on the rise due to the increase in computational capacities in recent years. Direct numerical simulations (DNS) can be conducted for both laminar and turbulent flows. The latter requires a very high resolution of the flow to capture even small vortices. Static numerical simulations of turbulent flow replace terms in the Navier-Stokes equations with a temporal mean and fluctuation value. These averaged Navier-Stokes equations are referred to as Reynolds-Averaged Navier-Stokes (RANS). The so-called Stokes equations arise when inertia terms are neglected. In frictionless flow considerations, the friction term can be disregarded. When pressure and friction terms dominate, inertia and volume terms might be neglected, yielding the so-called Reynolds equation [1].

3.2.1.3 Similarity Numbers

With dimensionless similarity numbers, it can be easily assessed, which terms dominate and which can be neglected.

The **Reynolds number** indicates the ratio of inertia to frictional force

$$Re = \frac{\text{Inertia force}}{\text{Friction force}} = \frac{v \cdot L}{\upsilon} = \frac{\rho \cdot v \cdot L}{\eta}, \tag{3.34}$$

where L is a characteristic length specific to each application, often the lubricant film thickness h. When Re is below an application-specific limit (e.g., 41.3·(diameter/clearance)$^{0.5}$ for hydrodynamic journal bearings), laminar flow occurs; otherwise, it is turbulent [1]. If $Re \ll 1$, frictional forces dominate, and inertial forces can be neglected [12].

The **Froude number**

$$Fr = \frac{\text{Inertia force}}{\text{Gravity force}} = \frac{v^2}{\rho \cdot v^2}, \tag{3.35}$$

which arises from the ratio of inertia to gravity, allows for an estimation of volume forces [12]. The ratio of gravity to frictional force is provided by the quotient of the Reynolds and Froude numbers. When the values are small, gravity forces can be disregarded.

Furthermore, the **Euler number** represents the ratio of pressure to tangential forces [13]:

$$Eu = \frac{\text{Pressure force}}{\text{Tangential force}} = \frac{\Delta p}{g \cdot L}. \tag{3.36}$$

EXERCISE 3.1

Consider a hydrodynamic journal bearing operating at 40°C. The lubricating oil is of ISO VG 46 class. The shaft with a nominal diameter of $D = 200$ mm rotates with $n = 2000$ min^{-1} and is loaded by a normal force of $F = 50$ kN. Due to the tolerances on the diameter of the shaft and the bearing bushing, expect an average clearance of $c = 200$ μm. For the sake of simplicity, assume the lubricant gap height h to be half the bearing clearance (no eccentricity of the shaft relative to the bushing). Evaluate if the flow can be assumed as laminar.

3.2.1.4 The Reynolds Equation

Reynolds [14] derived the differential equation named after him to calculate the pressure distribution in a given lubrication gap from the incompressible Navier-Stokes equations for Newtonian fluids. This derivation was based on the following **assumptions** [15]:

- Negligible effects of inertia, gravity, and surface tension due to the low mass and acceleration of the lubricant.
- The pressure is assumed to be constant along the lubrication gap height because the contact area greatly exceeds it.
- Validity of Stokes' adhesion condition for the lubricant on the walls of the contacting bodies.
- Laminar flow conditions.
- Lubricant interfaces are arranged at shallow angles or parallel to each other.
- Significant shear stresses and velocity gradients only in the direction of the lubrication gap height.
- Constant density and viscosity across the lubrication gap.
- Newtonian behavior of the lubricant.

The last two mentioned limitations can be partially resolved, allowing for lubricant properties that vary with the gap height to account for thermal effects as well as non-Newtonian behavior [16]. However, through the aforementioned assumptions, the Navier-Stokes equations can be simplified to only include pressure and friction terms. Furthermore, in the latter terms, the velocity gradients in the x- and y-directions, as well as the changes in velocity and pressure in the lubrication film thickness direction, can be neglected. After integrating the equations twice in the lubrication film thickness direction and considering the boundary conditions at the upper ($z = h$: $u = u_1$, $v = v_1$) and lower ($z = 0$: $u = u_2$, $v = v_2$) moving surfaces for the integration constants C_i, the velocity gradients are given as

$$\dot{\gamma}_x = \frac{\partial p}{\partial x}\left(z - \frac{\int_0^h \frac{z}{\eta}\,dz}{\int_0^h \frac{1}{\eta}\,dz}\right) + \frac{u_1 - u_2}{\eta \int_0^h \frac{1}{\eta}\,dz},$$

$$\dot{\gamma}_y = \frac{1}{\eta}\frac{\partial p}{\partial y}\left(z - \frac{\int_0^h \frac{z}{\eta}\,dz}{\int_0^h \frac{1}{\eta}\,dz}\right) + \frac{v_1 - v_2}{\eta \int_0^h \frac{1}{\eta}\,dz}. \tag{3.37}$$

When inserted into the equations for the velocity distribution in the lubricant gap, the following equations can be obtained:

$$u = \frac{1}{\eta}\frac{\partial p}{\partial x}\left(\int_0^z \frac{z}{\eta}\,dz - \frac{\int_0^h \frac{z}{\eta}\,dz}{\int_0^h \frac{1}{\eta}\,dz}\int_0^z \frac{1}{\eta}\,dz\right) + \frac{u_1 - u_2}{\int_0^h \frac{1}{\eta}\,dz}\int_0^z \frac{1}{\eta}\,dz + u_2,$$

$$v = \frac{1}{\eta}\frac{\partial p}{\partial y}\left(\int_0^z \frac{z}{\eta}\,dz - \frac{\int_0^h \frac{z}{\eta}\,dz}{\int_0^h \frac{1}{\eta}\,dz}\int_0^z \frac{1}{\eta}\,dz\right) + \frac{v_1 - v_2}{\int_0^h \frac{1}{\eta}\,dz}\int_0^z \frac{1}{\eta}\,dz + v_2. \tag{3.38}$$

By inserting these expressions into the integrated continuity equation

$$\frac{\partial}{\partial t}(\rho \cdot h) + \frac{\partial}{\partial x}\left(\int_0^h \rho \cdot u\,dz\right) + \frac{\partial}{\partial y}\left(\int_0^h \rho \cdot v\,dz\right) = 0, \tag{3.39}$$

we obtain the **generalized Reynolds equation** according to Peiran and Shizhu [17] in a slightly different notation:

$$0 = -\frac{\partial}{\partial t}\left(\int_0^h \rho\,dz\right) +$$

$$+ \frac{\partial}{\partial x}\left\{\left[\frac{\int_0^z \frac{z}{\eta}\,dz \cdot \int_0^h \rho \cdot \left(\int_0^z \frac{1}{\eta}\,dz'\right)dz}{\int_0^h \frac{1}{\eta}\,dz} - \int_0^h \rho \cdot \left(\int_0^z \frac{z'}{\eta}\,dz'\right)dz\right]\frac{\partial p}{\partial x}\right\} +$$

$$+\frac{\partial}{\partial y}\left\{\left[\frac{\int_{0}^{z}\frac{z}{\eta}\mathrm{d}z\cdot\int_{0}^{h}\rho\cdot\left(\int_{0}^{z}\frac{1}{\eta}\mathrm{d}z'\right)\mathrm{d}z}{\int_{0}^{h}\frac{1}{\eta}\mathrm{d}z}-\int_{0}^{h}\rho\cdot\left(\int_{0}^{z}\frac{z'}{\eta}\mathrm{d}z'\right)\mathrm{d}z\right]\frac{\partial p}{\partial x}\right\}-$$

$$-\frac{\partial}{\partial x}\left[\frac{\int_{0}^{h}\rho\cdot\left(\int_{0}^{z}\frac{1}{\eta}\mathrm{d}z'\right)\mathrm{d}z}{\int_{0}^{h}\frac{1}{\eta}\mathrm{d}z}(u_1-u_2)+\int_{0}^{h}\rho\cdot\mathrm{d}z\cdot u_2\right]-$$

$$\frac{\partial}{\partial y}\left[\frac{\int_{0}^{h}\rho\cdot\left(\int_{0}^{z}\frac{1}{\eta}\mathrm{d}z'\right)\mathrm{d}z}{\int_{0}^{h}\frac{1}{\eta}\mathrm{d}z}(v_1-v_2)+\int_{0}^{h}\rho\cdot\mathrm{d}z\cdot v_2\right]. \tag{3.40}$$

Thereby, the first term describes time-transient squeeze effects due to movement in the normal direction. The second and third terms yield the pressure-induced flow (Poiseuille terms) and the fourth and fifth terms describe the velocity-induced flow (Couette terms), which are all illustrated in Figure 3.12.

If only Newtonian fluid behavior (and isothermal conditions) is considered, the integral terms with variable density and viscosity over the lubrication gap disappear. We obtain the original **Reynolds differential equation**:

$$0=-\overbrace{\frac{\partial}{\partial t}(\rho\cdot h)}^{\text{Squeeze term}}+\overbrace{\frac{\partial}{\partial x}\left(\frac{\rho\cdot h^3}{12\cdot\eta}\frac{\partial p}{\partial x}\right)+\frac{\partial}{\partial y}\left(\frac{\rho\cdot h^3}{12\cdot\eta}\frac{\partial p}{\partial y}\right)}^{\text{Pressures/Poiseuille term}}-\underbrace{\frac{\partial}{\partial x}\left(\rho\cdot h\frac{u_1+u_2}{2}\right)-\frac{\partial}{\partial y}\left(\rho\cdot h\frac{v_1+v_2}{2}\right)}_{\text{Velocity/Couette term}},$$

$$\tag{3.41}$$

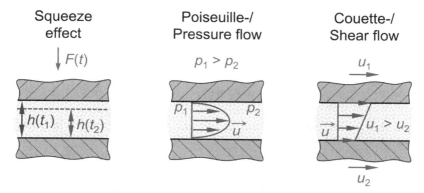

FIGURE 3.12 Hydrodynamic flow proportions of the Reynolds equation according to Ref. [18].

which can be further reduced to

$$0 = \overbrace{\frac{\partial}{\partial x}\left(\frac{h^3}{12\cdot\eta}\frac{\partial p}{\partial x}\right)}^{\text{Pressures/Poiseuille term}} - \overbrace{\frac{\partial}{\partial x}(h\cdot u_{\mathrm{m}})}^{\text{Velocity/Couette term}} \tag{3.42}$$

when neglecting time-transient effects (quasi-stationary assumption) and assuming a constant density as well as pressure gradient and flow in y-direction.

A solution to the differential Reynolds equation requires knowledge of the variable lubrication film thickness h, which is composed of the zero-gap h_0 and the geometries of the contacting bodies. Generally, by integrating the pressure over the contact domain at a fixed geometry, it becomes possible to calculate the **load-carrying capacity** (*LLC*) of the contact, i.e., the normal load that can be supported under given contact conditions. This is a common measure to compare different hydrodynamic contact designs. Alternatively, one can establish a **load balance equation**

$$F_{\mathrm{N}} = \int_{\Omega_c} p(x,y)\,\mathrm{d}x\,\mathrm{d}y, \tag{3.43}$$

where the film thickness $h(x,y)$ has to be iteratively adjusted so that the resulting integral over the pressure distribution equals the externally applied normal load. Thereby, the fluid friction opposing the surface movements can be an interesting measure to compare different designs. For pure flows in the x-direction, this can be determined by integrating the shear stress τ_{zi} at the center of the lubrication gap:

$$F_{\mathrm{F}} = \int_{\Omega_c} \tau_{\mathrm{zi}}\,|_{z=h/2}\,\mathrm{d}x\,\mathrm{d}y, \tag{3.44}$$

with

$$\tau_{\mathrm{zi}} = \frac{\partial p}{\partial x}\left(z - \frac{\int_0^h \frac{z}{\eta}\mathrm{d}z}{\int_0^h \frac{1}{\eta}\mathrm{d}z}\right) + \frac{1}{\int_0^h \frac{1}{\eta}\mathrm{d}z}\left(u_{i,z=h} - u_{i,z=0}\right), \tag{3.45}$$

which simplifies to

$$\tau_{\mathrm{zi}} = \eta\frac{\partial u}{\partial z} \tag{3.46}$$

for non-Newtonian and isothermal conditions, i.e., no variations in gap height directions.

Generally, since the Reynolds equation contains the derivation of the lubricant gap to the coordinate in the flow direction ($\partial h^3/\partial x$ or $\partial h/\partial x$), we can already conclude that a perfectly parallel gap geometry under quasi-stationary conditions (no time-dependent squeeze effects) would not generate a hydrodynamic pressure and, therefore, cannot separate two surfaces. Instead, hydrodynamic pressure generation requires surfaces that are converging in the flow direction and is further enhanced by the degree of convergence as well as fluid viscosity η and the hydrodynamically effective sliding velocity u_m, explaining the increase of film thickness with the Hersey number. To gain further insights into the relationships of actual HL contacts, we discuss some simplified example problems related to the analysis of bearing components assuming two-dimensionality (no pressure gradient or flow in the y-directions) as well as isoviscous and isothermal conditions.

3.2.1.5   Example: Slider Bearing with Linearly Converging Gap Height

In the first example, we take a linear slider bearing with the leading edge height h_1 and the trailing edge height h_0 of the pad, as shown in Figure 3.13. These kinds of bearings are versatile components to provide precise linear (or rotational for axial support when arranged respectively) motion with minimal friction and wear. Their design makes them suitable for a wide range of applications across various industries, particularly where precision, load capacity, and reliability are essential. While the inclined slider is stationary, the other surface moves with u. For this system, the film thickness along the x-axis can be described as follows:

$$h(x) = h_0 + \frac{(h_1 - h_0) \cdot x}{L} = h_0 \cdot \left(1 + \frac{k \cdot x}{L}\right), \tag{3.47}$$

whereby k is the convergence ratio

$$k = \frac{h_1 - h_0}{h_0} \tag{3.48}$$

and L the length of the pad.

The change in the film thickness can be connected to the change in position:

$$dh = \frac{h_0 \cdot k}{L} \, dx \text{ or } dx = \frac{L}{h_0 \cdot k} \, dh. \tag{3.49}$$

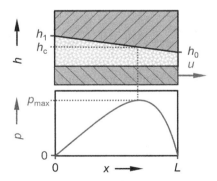

FIGURE 3.13 Linear slider bearing with converging lubricant gap height in flow direction as well as resulting hydrodynamic pressure distribution.

Assuming a constant viscosity, the Reynolds equation (in slightly different notation) can be used to describe the behavior of the fluid between both surfaces:

$$\frac{\partial}{\partial x}\left(h^3\frac{\partial p}{\partial x}\right)=6\cdot\eta\cdot u\frac{\partial h}{\partial x}.\tag{3.50}$$

Assuming that the peak pressure is observed at h_c and integrating the terms to resolve the partial differentials as well as replacing the differentials with Eq. 3.49, we first obtain

$$\frac{h_0\cdot k}{L}\frac{\partial p}{\partial h}=6\cdot\eta\cdot u\frac{h-h_c}{h^3}\tag{3.51}$$

and then after another integration

$$\frac{h_0\cdot k}{6\cdot\eta\cdot u\cdot L}\cdot p=-\frac{1}{h}+\frac{h_c}{2h^2}+C,\tag{3.52}$$

with the integration constant C. Assuming zero pressure at the edges, we can express the constants as follows:

$$C=\frac{1}{h_0(k+2)}\quad\text{and}\quad h_c=\frac{2h_0(k+1)}{(k+2)}.\tag{3.53}$$

The resulting pressure as a function of the height can be therefore expressed as

$$p(h)=\frac{6\cdot\eta\cdot u\cdot L}{h_0\cdot k}\left[-\frac{1}{h}+\frac{h_0}{h^2}\frac{(k+1)}{(k+2)}+\frac{1}{h_0(k+2)}\right],\tag{3.54}$$

and its maximum at h_c is

$$p_{max}=\frac{6\cdot\eta\cdot u\cdot L}{h_0^2\cdot k}\left[-\frac{(k+2)}{4(k+1)}+\frac{1}{k+2}\right].\tag{3.55}$$

This can be used to calculate the LLC per unit slider width B as the integral of the pressure over x

$$\frac{LLC}{B}=\int_0^B p\,dx=\frac{6\cdot\eta\cdot u\cdot L^2}{h_0^2\cdot k^2}\left[-\ln(k+1)+\frac{2k}{k+2}\right],\tag{3.56}$$

as well as the frictional force per unit slider width B

$$\frac{F_f}{B}=\int_0^B\eta\frac{\partial u}{\partial z}\,dx=\frac{\eta\cdot u\cdot L}{h_0}\left[\frac{6}{k+2}-\frac{4\ln(k+1)}{k}\right]\tag{3.57}$$

of the bearing, allowing to calculate the COF:

$$\mu=\frac{F_f}{LLC}=\frac{k\cdot h_0}{L}\left[\frac{3k-2(k+2)\cdot\ln(k+1)}{6k-3(k+2)\cdot\ln(k+1)}\right].\tag{3.58}$$

Interestingly, these expressions suggest that the supportable load is directly proportional to the viscosity, the sliding distance, and the square of the contact length and inversely proportional to the square of the minimum film thickness as well as the convergence ratio. In turn, the COF in these bearings does not depend on the viscosity of the lubricating fluid (however, as mentioned, it does affect LLC). Regarding the bearing design, we can aim at maximizing the LLC, which is achieved for $k = 1.2$. In contrast, the minimum COF would be realized for $k = 1.55$, which is different for the maximized lift conditions. Hence, selecting the convergence ratio always represents a compromise between lower friction to enhance energy efficiency and a higher LLC to avoid solid-solid contact and thus wear.

EXERCISE 3.2

Consider an air thrust bearing consisting of six equal pads with a length $L = 38$ mm and width $B = 114$ mm. The minimum film thickness is $h_0 = 10\ \mu$m and the convergence ratio is $k = 1.3$. The flow moves with a velocity of $u = 0.5$ m/s and the viscosity of air is $\eta = 0.02$ mPa·s.

Neglecting three-dimensional effects, determine
- the lifting capacity LLC of the bearing,
- the LLC if instead of air it is lubricated with oil ($\eta_{oil} = 100$ mPa·s), and
- the resulting COF for oil lubrication.

3.2.1.6   Example: Rayleigh Step Bearing

In the next example, we consider the so-called Rayleigh[3] step bearing, which is a linear slider bearing. Instead of a linearly converging lubricant gap, this configuration features two parallel areas with height h_1 and h_0, respectively, and an abrupt step of h_s at distance x_s from higher to lower film thickness in the flow direction, see Figure 3.14.

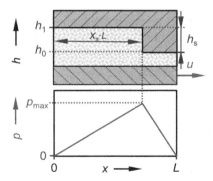

FIGURE 3.14 Linear slider bearing with step in the lubricant gap height as well as resulting hydrodynamic pressure distribution.

---

[3] Named after Lord Rayleigh, a British mathematician and physicist, who reported the bearing type in 1918.

This configuration, when neglecting side leakage, is known to deliver the highest LLC compared to other geometries. The lubricant film thickness can be described as

$$h(x) = \begin{cases} h_1 & \text{for } 0 \le x < x_s L \\ h_0 & \text{for } x_s L \le x \le L \end{cases}. \tag{3.59}$$

Assuming a zero-pressure gradient at the in- and outlet due to a constant film thickness and no discontinuity at the position of the step

$$x_s \cdot L \left(\frac{\partial p}{\partial x} \right)_{\text{at } h_1} = -(L - x_s \cdot L) \left(\frac{\partial p}{\partial x} \right)_{\text{at } h_0} \tag{3.60}$$

as well as a constant mass flow

$$h_1^3 \left(\frac{\partial p}{\partial x} \right)_{\text{at } h_1} + 6 \cdot \eta \cdot u \cdot h_1 = h_2^3 \left(\frac{\partial p}{\partial x} \right)_{\text{at } h_0} + 6 \cdot \eta \cdot u \cdot h_0, \tag{3.61}$$

we obtain linear pressure gradients in both regions:

$$p(x) = \begin{cases} \dfrac{6 \cdot \eta \cdot u \cdot (1 - x_s) \cdot h_s \cdot \dot{x}}{(1 - x_s) \cdot h_1^3 + x_s \cdot h_0^3} & \text{for } 0 \le x < x_s L \\[3mm] -\dfrac{6 \cdot \eta \cdot u \cdot x_s \cdot h_s \cdot x}{(1 - x_s) \cdot h_1^3 + x_s \cdot h_0^3} & \text{for } x_s L \le x \le L \end{cases}, \tag{3.62}$$

resulting in the triangular pressure distribution shown in Figure 3.14. The maximum pressure can be therefore obtained at the step position:

$$p_{\max} = \frac{6 \cdot \eta \cdot u \cdot L \cdot x_s \cdot (1 - x_s) \cdot h_s}{(1 - x_s) \cdot h_1^3 + x_s \cdot h_0^3}. \tag{3.63}$$

This is a function of the bearing geometry and especially the step, approaching zero for an infinite h_0 or a step height h_s of zero (two completely parallel surfaces). The configuration that produces the largest pressure, and thus the greatest LLC per unit width (i.e., the area of the triangular)

$$\frac{LLC}{B} = \frac{p_{\max} \cdot L}{2} = \frac{3 \cdot \eta \cdot u \cdot L^2 \cdot x_s \cdot (1 - x_s) \cdot h_s}{(1 - x_s) \cdot h_1^3 + x_s \cdot h_0^3}, \tag{3.64}$$

i.e.,

$$\frac{\partial p_{\max}}{\partial x_s} = 0 \text{ and } \frac{\partial p_{\max}}{\partial h_s} = 0 \tag{3.65}$$

is achieved for

$$\frac{h_0}{h_s} = 1.155 \text{ and } x_s = 0.7182. \tag{3.66}$$

3.2.1.7 Example: Journal Bearing

In the last example, we consider an infinitely wide hydrodynamic journal bearing as depicted earlier in Figure 3.6. These are mechanical components used to radially support rotating shafts within a stationary bushing. These types of bearings are commonly used in a wide range of applications, including engines, turbines, pumps, and other machinery requiring smooth and efficient shaft rotation. Therein, hydrodynamic journal bearings allow to establish high speed, low noise generation, and good damping of vibrations (stable rotation). Let us consider a bearing with the clearance c between the shaft (radius R_I) and the bushing (radius R_B), i.e., $c = R_B - R_I$. If the shaft is a little off-center (i.e., O_I is displaced from O_B, see Figure 3.15) by the eccentricity e, the representative geometric relations can be described as follows:

$$\overline{O_IC} + \overline{CA} = \overline{O_IB} + \overline{BA} \, , \tag{3.67}$$

$$e \cdot \cos\alpha + R_B \cdot \cos\beta = R_I + h \, , \text{ therefore} \tag{3.68}$$

$$h = e \cdot \cos\alpha + R_B \cdot \cos\beta - R_I. \tag{3.69}$$

Assuming the angle β to be small, i.e., the bearing clearance and film thickness are small compared to the radii, the film thickness can be "unwrapped" from the shaft and viewed as a periodic stationary profile [19] (neglecting the curvature of the lubricant film; see Figure 3.16, left), thus $\cos\beta \approx 1$ and

$$h = e \cdot \cos\alpha + R_B - R_I = d \cdot \cos\alpha + c = c \cdot (\varepsilon \cdot \cos\alpha + 1), \tag{3.70}$$

where ε is the eccentricity ratio, which can assume values between 0 and 1. For instance, when $\varepsilon = 1$, then $e = c$, and the maximum film height at $\alpha = 0$ is $h = 2c$.

The Reynolds equation in an integrated form can be written as

$$\frac{\partial p}{\partial x} = \frac{6 \cdot \eta \cdot R_I \cdot \omega \cdot (h - h_m)}{h^3}, \tag{3.71}$$

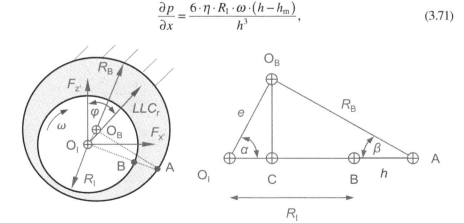

FIGURE 3.15 Radial journal bearing and geometrical relations between geometries and film thickness.

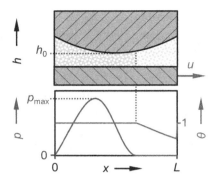

FIGURE 3.16 Infinitely wide parabolic-shaped lubricant gap height as well as resulting hydrodynamic pressure and fractional film content distribution.

where ω denotes the rotational speed and h_m stands for the film thickness where $\partial p/\partial x = 0$, Furthermore, considering $\partial x = R_I \cdot \partial \alpha$ and the "unwrapped" film thickness, we can write

$$\frac{\partial p}{\partial \alpha} = \frac{6 \cdot \eta \cdot R_I^2 \cdot \omega}{c^2}\left[\frac{1}{(\varepsilon \cdot \cos\alpha + 1)^2} - \frac{h_m}{c \cdot (\varepsilon \cdot \cos\alpha + 1)^3}\right]. \tag{3.72}$$

The integral for the pressure can be resolved using the Sommerfeld substitution

$$1 + \varepsilon \cdot \cos\alpha = \frac{1 - \varepsilon^2}{1 - \varepsilon \cdot \cos\gamma}, \tag{3.73}$$

where γ is the so-called Sommerfeld variable, which, in turn, depends on the eccentricity ratio and the angle [19]:

$$\sin\alpha = \frac{\sqrt{1 - \varepsilon^2}\sin\gamma}{1 - \varepsilon \cdot \cos\gamma}, \quad \cos\alpha = \frac{\cos\gamma - \varepsilon}{1 - \varepsilon \cdot \cos\gamma},$$

$$\sin\gamma = \frac{\sqrt{1 - \varepsilon^2}\sin\alpha}{1 + \varepsilon \cdot \cos\alpha}, \quad \cos\alpha = \frac{\cos\gamma - \varepsilon}{1 - \varepsilon \cdot \cos\gamma},$$

$$\partial\alpha = \frac{\sqrt{1 - \varepsilon^2}}{1 - \varepsilon \cdot \cos\gamma}\partial\gamma. \tag{3.74}$$

Assuming zero pressure at the point of maximum film thickness, we can formulate the pressure as

$$p = \frac{6 \cdot \eta \cdot R_I^2 \cdot \omega \cdot \varepsilon \cdot \sin\alpha \cdot (2 - \varepsilon \cdot \cos\alpha)}{c^2 \cdot (2 + \varepsilon^2) \cdot (1 + \varepsilon \cdot \cos\gamma)^2}, \tag{3.75}$$

which yields a point symmetric pressure distribution (the so-called Sommerfeld solution, see Figure 3.15 on the right) with positive pressures for $0 \le \alpha \le \pi$ and equally negative pressures for $\pi \le \alpha \le 2\pi$. Deriving this, we first obtain

$$h_{\mathrm{m}} = \frac{2c \cdot \left(1 - \varepsilon^2\right)}{2 + \varepsilon^2},$$

(3.76)

providing the angle where $\partial p / \partial x = 0$

$$\alpha_{\mathrm{m}} = \cos^{-1} \frac{-3\varepsilon}{2 + \varepsilon^2},$$

(3.77)

i.e., the pressure becomes maximum [19]. Making use of

$$\sin \alpha_{\mathrm{m}} = \sqrt{1 - \cos^2 \alpha_{\mathrm{m}}} = \frac{\sqrt{4 - 5\varepsilon^2 + \varepsilon^4}}{2 + \varepsilon^2},$$

(3.78)

the maximum pressure can be determined as

$$p_{\max} = \frac{\eta \cdot R_{\mathrm{I}}^2 \cdot \omega}{c^2} \cdot \frac{3\varepsilon \cdot \sqrt{4 - 5\varepsilon^2 + \varepsilon^4} \left(4 - \varepsilon^2\right)}{2 \left(2 + \varepsilon^2\right) \left(1 - \varepsilon^2\right)}.$$

(3.79)

From this, we can assess the *LLC* by integration and various transformations (see Ref. [19] for more details). Considering the negative pressures behind the minimum film thickness ($\pi \le \alpha \le 2\pi$) would mean a normal load at a right angle to the line of centers (attitude angle φ of 90°), i.e., a shaft center moving away from the bearing center in a right angle to the load. By simply "cutting off" the negative pressures, i.e., setting them to zero, we can obtain a physically more meaningful solution (the so-called half-Sommerfeld solution [20]) with the load components in the two Cartesian directions:

$$F_{x'} = \int_0^{\pi} p \cdot R_{\mathrm{I}} \cdot \sin\left(\pi - \alpha\right) d\alpha = \cdots = \frac{6 \cdot \eta \cdot R_{\mathrm{I}} \cdot \omega \cdot \pi \cdot \varepsilon}{c^2 \cdot \sqrt{\left(2 + \varepsilon^2\right)\left(1 - \varepsilon^2\right)}},$$

$$F_{z'} = \int_0^{\pi} p \cdot R_{\mathrm{I}} \cdot \cos\left(\pi - \alpha\right) d\alpha = \cdots = \frac{12 \cdot \eta \cdot R_{\mathrm{I}} \cdot \omega \cdot \varepsilon^2}{c^2 \cdot \left(2 + \varepsilon^2\right)\left(1 - \varepsilon^2\right)},$$

(3.80)

which ultimately yields the radial *LLC*

$$LLC_{\mathrm{r}} = \sqrt{F_x^2 + F_z^2} = \frac{6 \cdot \eta \cdot R_{\mathrm{I}} \cdot \omega \cdot \varepsilon \cdot \sqrt{\pi^2 - \varepsilon^2 \left(\pi^2 - 4\right)}}{c^2 \cdot \left(2 + \varepsilon^2\right)\left(1 - \varepsilon^2\right)}$$

(3.81)

and the attitude angle

$$\varphi = \tan^{-1} \left(\frac{F_x}{F_z}\right) = \tan^{-1} \left(\frac{\pi}{2\varepsilon} \sqrt{1 - \varepsilon^2}\right).$$

(3.82)

Generally, we can observe that for an eccentricity ratio of zero, the pressure and the supported load become zero since we no longer have a converging lubricant gap and the attitude angle is 90°. With increasing load, the shaft first moves at a right angle to the applied load and, at higher loads, approaches the bearing along the load vector. Finally, for the eccentricity ratio of 1, we would have an attitude angle of 0° and infinite pressure and *LLC*. This implies that load increases on hydrodynamic journal bearings can be accommodated by higher eccentricities.

3.2.1.8   Comment on the Considered Examples

Please note that in all three exemplary studies of the hydrodynamic use cases discussed above, several assumptions and simplifications have been made to create a tractable model and facilitate analysis. Firstly, the established consideration was for fixed lubricant gap profiles, which did not vary depending on the external stress conditions (no load balance). Additionally, we assumed a two-dimensional analysis (one-dimensional Reynolds equation), neglecting three-dimensional effects that could influence the behavior of the system. Nevertheless, we can learn some general relations of general hydrodynamic contact characteristics. Furthermore, we did not consider side leakage, assuming a perfect sealing between the moving components and the surrounding environment. Also, we did not analyze pressure differences at the inlet and outlet of the bearing, i.e., an additionally acting hydrostatic pressure, which could have practical implications for the overall system. We did also not account for variations in the viscosity and density of the lubricating fluid, which can vary under different operating conditions and affect the system's performance (see Section 3.1). Lastly, we did not consider cavitation effects (see next paragraph). Due to the complexity of the problem, finding closed-form exact analytical solutions for this kind of problem is challenging or impossible, thus requiring numerical methods or specialized software for practical applications. For more information regarding solutions of representative hydrodynamic cases, the interested reader is referred to Ref. [19].

3.2.1.9   Cavitation

As previously described, the first solutions of the Reynolds equation did not consider cavitation effects, which resulted in a point-symmetrical pressure profile, with the maximum positive pressure corresponding to the minimum negative pressure. This would assume that the lubricant can transmit relatively high tensile stresses, which is not likely the case under the high pressures, tangential shear, and thermal effects in tribological contacts. Instead, it leads to the tearing of the lubricant film and the formation of gas or vapor bubbles, known as cavitation effects. This results in a two-phase flow of lubricant and gas or vapor in the cavitation region, generally where the diverging lubricant gap is located. The prevailing pressure corresponds to the cavitation pressure, which is the solubility or vapor pressure of the lubricant. Various cavitation models have been developed for consideration of cavitation effects. The simplest correction is the above-mentioned half-Sommerfeld solution, simply setting negative pressure to zero [21]. However, this leads to a discontinuity in the transition between the pressure and cavitation regions, which, in turn, contradicts the continuity equation or mass conservation. Some improvement has been achieved by the so-called Reynolds boundary conditions [22], whereby during the solution of the Reynolds equation, it is ensured that the pressure gradient becomes zero at the transition from the pressure to the cavitation region, and the pressure corresponds to the cavitation pressure. This continuous transition satisfies mass conservation in the pressure region but continues to be violated in the cavitation region. Despite errors in the calculation results due to the violation of the continuity equation, these **non-mass-conserving cavitation models** have gained wide popularity in literature due to their relatively simple implementation. These errors can be overcome by **mass-conserving cavitation models**. The so-called JFO theory [23–25] assumes a fluid transport between the emerging gas and vapor bubbles, which can also reform. The numerical solution of the Reynolds equation thus involves dynamic boundary conditions, which are complex to implement. Therefore, Elrod [26] developed a more efficient approach that considers the boundary condition of the JFO cavitation theory without solving it directly by introducing a switching variable for density ratio or gap fill level into the Reynolds equation, making it valid in both the pressure and cavitation regions and efficiently solvable using numerical models. Thereby, the so-called fractional film content

$$\theta = \frac{h_{liq}}{h} = \frac{\rho_{mix}}{\rho} = \frac{\eta_{mix}}{\eta} \tag{3.83}$$

represents to which extent the gap is filled with lubricant. The distribution of hydrodynamic pressure as well as fractional film content for an infinitely wide, parabolic-shaped lubricant gap (as is the case in the previously discussed radial journal bearing) is depicted in Figure 3.16. Thereby, $\theta = 1$ in the pressurized ($p > 0$), converging region, i.e., the gap is completely filled with lubricant. Subsequently, the film cavitates in the diverging area, which leads to the pressure dropping down to zero and $0 < \theta \leq 1$.

3.2.1.10 Other Factors to Consider

Micro-geometry/roughness can significantly influence hydrodynamic pressure in lubricated contacts, either reducing or increasing macro-hydrodynamic pressure development based on surface topography characteristics. These **micro-level effects** can be incorporated directly (deterministic) or indirectly (stochastic). In **deterministic coupling** of micro- and macro-hydrodynamics, surface topography is mathematically integrated into the lubricant gap equation, necessitating fine resolution of computational approaches and resulting in extended calculation times. Alternatively, an indirect approach has been developed, involving stochastic methods and an averaged form of the Reynolds equation. By employing flow factors such as those proposed by Patir and Cheng [27,28], the mean volume flow between two rough surfaces in a control volume is balanced directionally for a given mean gap height, considering defined boundary conditions. These flow factors account for differences between rough and smooth gaps in terms of pressure flow Φ_p (pressure-induced flow through locally varying film thickness) and shear flow Φ_s (flow due to relative motion of the surface topography), thereby impacting the Poiseuille and Couette terms in the Reynolds equation:

$$0 = \frac{\partial}{\partial x} \left(\Phi_p \frac{h^3}{12 \cdot \eta} \frac{\partial p}{\partial x} \right) - \frac{\partial}{\partial x} (h \cdot u_m) - u_m \cdot R_q \frac{\partial \Phi_s}{\partial x}. \tag{3.84}$$

For an infinite lubricant gap parameter λ, i.e., very high film thickness or very low roughness, $\Phi_p \rightarrow 1$ and $\Phi_s \rightarrow 0$. Otherwise, a pressure flow factor greater than one corresponds to reduced pressure build-up, while a pressure flow factor less than one corresponds to increased pressure build-up. The shear flow factor can only take values greater than zero and enhances the pressure build-up. The local contact area distribution when two rough surfaces come into contact and the resulting fluid flow are abstractly illustrated in Figure 3.17. It becomes evident that flow is impeded by the transverse orientation of asperities, which implies that fluid molecules need to travel a longer distance. This is also reflected in the flow factors determined by Patir and Cheng, which are plotted in Figure 3.17 for different Peklenik factors (see Section 1.4.2). With increasing transverse orientation, the pressure build-up caused by surface texture and additional shear flow also increases.

In **mixed lubrication** scenarios, both hydrodynamic effects and solid-solid asperities resulting from surface roughness come into play. When determining the actual contact area and pressures, various methods can be employed, either directly (deterministically) or indirectly (stochastically). In the indirect approach, the total *LLC* is the sum of contributions from the hydrodynamic pressure generation and the direct solid asperity contact. The latter can be considered as a function of the average fluid film height through integrated solid asperity contact pressure profiles. These profiles can be computed for specific surface pairings using simplified models that assume idealized roughness geometries and statistical averaging, e.g., the Greenwood–Williams model (see Section 1.5.3).

Furthermore, the generation of **temperature gradients** can be a result of frictional heating arising from the internal shear resistance within the lubricant or the contact of solid asperities, which in term influence the solid and fluid properties (see Section 3.1). To this end, heat conduction and convection within the lubricant gap and contacting bodies, along with heat sources attributed to lubricant compression/expansion and shearing, especially in the context of mixed lubrication, where additional heat sources originate from contacting asperities (flash temperatures), can be considered.

Finally, despite the presence of parallel or conformal contact geometries, the application of greater external forces, inertial forces, hydrodynamic pressures, and elevated temperatures can result in macro-scale deformations of the bearing components, shafts, or adjacent components. These deformations directly impact the shape

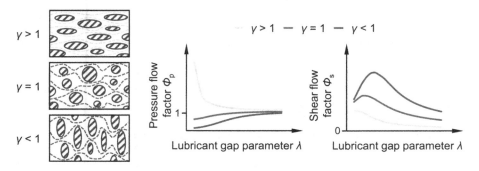

FIGURE 3.17 Fluid flow for different contact area orientations (left) and pressure and flow factor evolution in dependency of Peklenik factor and lubricant gap height parameter (right).

of the lubricant gap and the associated pressure distribution and might have to be considered in respective numerical considerations. Frequently, this also is referred to as **elastohydrodynamics**, even though it refers to hydrodynamic contacts with more macroscopic deformation.

3.2.2   Elastohydrodynamic Lubrication (EHL)

The term elastohydrodynamics or elastohydrodynamic lubrication (EHL) has already been introduced in the 20th century, referring to concentrated contacts in various applications, including roller bearings, gears, and cam/follower systems. Thereby, the hydrodynamic pressure generation as described in the previous section is superimposed with the elastic deformation, see Figure 3.8. In contrast to conformal, hydrodynamic contacts, concentrated contacts involve non-conforming surfaces and can generate pressure on the order of magnitude of several hundred MPa to some GPa, which is why such contacts often represent critical points within tribo-technical systems. A lubricated contact can be categorized as elastohydrodynamic if the elastic deformations of the contacting bodies are equal to or greater than the emerging lubricant film thickness [29].

When deformations are of a similar magnitude, it is referred to as **hard EHL** (contact bodies have high elasticity modules), and when deformations are significantly greater, it is termed **soft EHL** (contact bodies have lower elasticity modules). In cases of negligible deformation, lubrication operates in a hydrodynamic manner (see Section 3.2.1). Under high loads or for soft materials, the EHL pressure distribution approximates the Hertzian pressure distribution, dominated by elastic deformation effects. Conversely, at lower loads, higher speeds, and/or harder materials, hydrodynamic processes gain prominence, leading to a more pronounced pressure peak shifted toward the outlet, potentially surpassing the central pressure. This dynamic interplay between load, speed, and contact shape significantly influences the lubricant film's shape and pressure distribution. In more line-like contacts, such as those in roller bearings or spur gears, the lubricant film in these contacts also takes on an elongated shape, being largely constant transversely to the flow (despite potential edge effects at the extremes of the geometries). In circular or elliptical point-shaped contacts, the lubricant film thickness typically features a horseshoe-shaped constriction, see Figure 3.18.

The theoretical consideration of an EHL contact generally requires the **coupled solution** of the

- **Navier-Stokes equations** or the **Reynolds equation** (as usually valid simplification) accounting for the **lubricant's hydrodynamics,**[4]
- **solids' elastic deformation,**[5]
- **lubricant gap equation,**[3]

---

[4] Similarly to HL contacts, micro-hydrodynamic effects induced by the surface topography can be considered deterministically or stochastically by employing an averaged Reynolds equation and the flow-factor concept.

[5] Similarly to HL contacts, mixed lubrication effects, i.e., the deformation and LLC contribution of asperity contact, can be considered deterministically or stochastically.

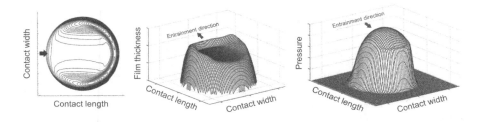

FIGURE 3.18 Contour (left) and 3D plot of the film thickness (middle) and pressure (right) distribution in a circular point contact. (Reprinted from Ref. [30] with permission.)

- **load balance equation,**
- **lubricant's rheological behavior** (see Section 3.1), and
- **energy equation** to account for thermal effects (if applicable).

Therefore, besides the external load, the pressure distribution and lubricant gap geometry of EHL contacts are generally influenced by the contact geometry (the solids' radii R_i, which can be combined to the effective radius R'), the velocities u_i, the elastic properties of the materials (elastic modulus E_i and Poisson's ratio v_i, which can be combined for both bodies to the reduced Young's modulus $E' = 1/E^*$), the fluid density ρ and viscosity η (including the viscosity-pressure coefficient α) as well as (if applicable) the thermal properties (thermal conductivity and spec. heat capacity), see Figure 3.19.

In the simplest form (neglecting non-Newtonian and thermal effects), EHL contacts demand a coupled solution of the hydrodynamics and the elastic deformation. When flow dynamics and structural mechanics interact, it is commonly referred to as **Fluid–Structure Interaction** (FSI). Given that both aspects influence each other, an effective coupling method is essential. One approach involves solving

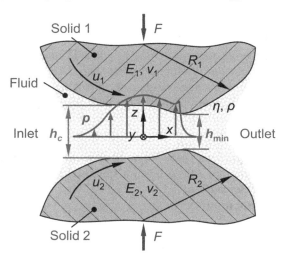

FIGURE 3.19 Schematic illustration of an infinite EHL line-contact with the relevant variables. (Adapted from Ref. [31] with permission by CC BY 4.0.)

hydrodynamics first and then addressing deformation caused by pressure, known as **weak coupling**. If this step occurs only once, it is called unidirectional coupling. In this basic approach, the interaction between fluid and structure is inadequately considered. In weak coupling, the two steps are typically repeated sequentially until a defined convergence criterion is met. In contrast, **strong coupling** combines the solution of the two discretized differential equations into a single joint algebraic system of equations. While it is possible to use the full Navier-Stokes equations to address the EHL contact problem, this requires discretizing the lubricant gap in the gap height direction, making normalization and mesh transformation more challenging. Instabilities may also arise due to an excessive response of elastic deformation to the previous iteration step's surface pressure. Consequently, hydrodynamics is often represented in the form of the Reynolds differential equation.

Historically, both weak and strong couplings have been applied to address EHL problems. Weak coupling, in the form of sequential solutions, was the first and most widely used approach. The initial works employed an inverse method, solving the Reynolds equation for the lubrication gap at a given pressure, often using a half-space approach for computing elastic deformation. Another method is the direct approach, which solves the Reynolds equation for pressure. The introduction of the multi-grid method accelerated calculations, and the use of Multi-Level Multi-Integration (MLMI) methods and relaxation techniques enabled the calculation of higher pressures, with notable contributions from researchers like Venner and Lubrecht [32,33]. In contrast, using strong coupling with a half-space model for elastic deformation results in a fully populated matrix within the nonlinear system of equations to be solved. This places a significant memory and computational burden on the system. In case of a point contact, these limitations were often dictated by the hardware of the era. This is a key reason why weakly coupled solutions prevailed historically. To address these limitations and achieve a sparse matrix, reducing computational and memory demands, Habchi [18,34] proposed replacing elastic deformation with a Finite Element (FE) formulation. Additionally, the Reynolds differential equation is discretized using an FE approach. Strong coupling exhibits excellent convergence behavior, requiring relatively few iterations for a solution. Furthermore, commercial software algorithms can be readily applied in this context. For more details on numerical EHL modeling, the interested reader is referred to Ref. [19].

Simulating EHL contacts is inherently intricate and computationally intensive. These simulations can be prohibitively demanding, making them inaccessible to a wider audience and challenging to integrate into more comprehensive multibody system dynamics simulations. To extend the applicability of EHL calculations and their findings, researchers introduced **non-dimensional similarity groups**. Dowson and Higginson [35,36] established a set of material

$$G = \alpha_{\mathrm{p}} \cdot E', \tag{3.85}$$

velocity

$$U = \frac{\eta_0 \cdot u_{\mathrm{m}}}{E' \cdot R'}, \tag{3.86}$$

load

$$W_{2D} = \frac{F}{l \cdot E' \cdot R'}, \; W_{3D} = \frac{F}{E' \cdot R'^2}, \tag{3.87}$$

and lubricant film thickness parameters

$$H_c = \frac{h_c}{R'}, \; H_{min} = \frac{h_{min}}{R'} \tag{3.88}$$

and introduced empirically derived regression equations (henceforth referred to as **"proximity equations"**) that correlate these parameters based on numerical predictions, e.g.

$$H_{min,VE} = f(U,G,W) \approx 2.53 \cdot G^{0.55} \cdot U^{0.7} \cdot W_{2D}^{-0.125}. \tag{3.89}$$

It becomes evident that a higher load parameter leads to a higher film thickness (enters the equation with a negative exponent), however, exerts a relatively minor impact. In contrast, the material and velocity parameters, particularly the product of velocity and viscosity, significantly influence the film thickness, whereby higher values lead to higher film thickness (positive exponents).

Moes [37] demonstrated that the aforementioned parameters can be unidirectionally converted into three parameters, effectively introducing what is referred to as the modified load

$$M_{2D} = W_{2D} \cdot (2U)^{-\frac{1}{2}}, \; M_{3D} = W_{3D} \cdot (2U)^{-\frac{3}{4}}, \tag{3.90}$$

the viscosity

$$L = G \cdot (2U)^{\frac{1}{4}}, \tag{3.91}$$

and the film thickness parameter

$$H_{min} = h_{min} \cdot U^{-\frac{1}{2}}, \tag{3.92}$$

thus, allowing to estimate integral film thickness parameters:

$$H_{min} = f(M,L) \approx 1.56 \cdot M_{2D}^{-0.125} \cdot L^{0.55}. \tag{3.93}$$

The relation between the minimum film thickness and the load as well as the viscosity parameters is illustrated in Figure 3.19. It can be seen that the film thickness decreases with the load parameter and increases with the viscosity parameter. Thereby, the aforementioned regression equations are essentially approximations and possess finite validity. When the specified range does not apply (Martin-Gümbel or Moes conditions as in Figure 3.20), there might be a tendency to underestimate the resulting lubricant film heights.

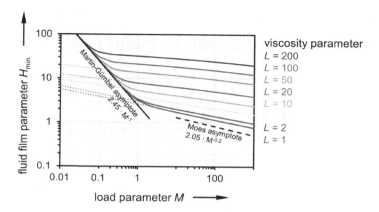

FIGURE 3.20 Minimum lubricant gap H_c versus load parameter M at different viscosity parameters L. (Adapted from Ref. [31] with permission by CC BY 4.0.)

To adjust to different dominating effects, Johnson [38] suggested the elasticity

$$P_E = \sqrt{\frac{F^2}{\eta_0 \cdot u_m \cdot l^2 \cdot R'}} \tag{3.94}$$

and the pressure-viscosity

$$P_\alpha = \sqrt{\frac{\alpha_p^2 \cdot F^3}{\eta_0 \cdot u_m \cdot l^3 \cdot R'^2}}.$$

Thus, we can classify EHL contacts into different regimes, where the piezo-viscous fluid behavior (i.e., pressure dependence) has to be considered or whether isoviscous fluid behavior is sufficient and where the elastic behavior dominates or if the solid is rather rigid, see Figure 3.21. Therefore, depending on the regime, we can define

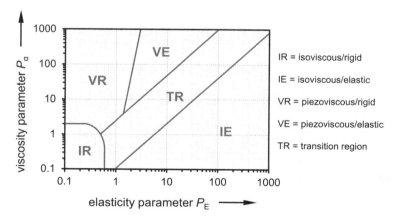

FIGURE 3.21 Lubrication regimes in an infinite line-contact. (Adapted from Ref. [31] with permission by CC BY 4.0.)

different proximity equations [29]. For piezo-viscous and elastic contacts (VE), i.e., large M and large L, we can stick to Eq. 3.93 introduced above. In the case of isoviscous fluid and easily deformable contacts, i.e., large M and $L \to 0$ (soft EHL), the film thickness can be estimated by

$$H_{\min,\,\mathrm{IE}} \approx 2.05 \cdot M_{2\mathrm{D}}^{-1} \,. \tag{3.95}$$

In turn, the minimum film thickness for piezo-viscous and rigid contacts (small M and large L) can be expressed as

$$H_{\min,\,\mathrm{VR}} \approx 1.05 \cdot L^{\frac{2}{3}} \,, \tag{3.96}$$

and for isoviscous and rigid contacts (small M, $L \to 0$) approximated by

$$H_{\min,\,\mathrm{IR}} \approx 2.45 \cdot M_{2\mathrm{D}}^{-1}. \tag{3.97}$$

Over the last decades, these proximity equations have been continuously refined by new simulation results to broaden the range of validity, e.g., [39]

$$H_{\min} \approx \left\{ \left[\left(0.99 \cdot M_{2\mathrm{D}}^{-0.125} \cdot L^{0.75} \cdot t \right)^{r} + \left(2.05 \cdot M_{2\mathrm{D}}^{-0.2} \right)^{r} \right]^{\frac{s}{r}} + \left(2.45 \cdot M_{2\mathrm{D}}^{-1} \right)^{s} \right\}^{\frac{1}{s}} , \text{ with} \tag{3.98}$$

$$s = 3 - e^{-2 \cdot M_{2\mathrm{D}}^{-1}}, r = e^{1 - \frac{4}{L+5}}, t = 1 - e^{-3.5 \cdot M_{2\mathrm{D}}^{-0.125} \cdot L^{-0.25}} ,$$

and expanded to also account for the central film thickness, e.g., [40]

$$H_{\mathrm{c}} \approx \left[\left(H_{\mathrm{c,IR}}^{\frac{7}{3}} + H_{\mathrm{c,IE}}^{-\frac{7}{3}} \right)^{\frac{3}{7} \cdot s} + \left(H_{\mathrm{c,VR}}^{\frac{7}{2}} + H_{\mathrm{c,VE}}^{\frac{7}{2}} \right)^{-\frac{2}{7} \cdot s} \right]^{s^{-1}} , \text{ with} \tag{3.99}$$

$$s = \frac{1}{5} \left(7 + 8 \cdot e^{-2 \cdot \frac{H_{\mathrm{c,IE}}}{H_{\mathrm{c,IR}}}} \right),$$

$$H_{\mathrm{c,IR}} \approx 3 \cdot M^{-1}, H_{\mathrm{c,IE}} \approx 2.62105 \cdot M^{-\frac{1}{5}}, H_{\mathrm{c,VR}} \approx 1.28666 \cdot L^{\frac{2}{3}}, H_{\mathrm{c,VE}} \approx 1.31106 \cdot M_{2\mathrm{D}}^{-\frac{1}{8}} \cdot L^{\frac{3}{4}}.$$

Based on Figure 3.22, the central film thickness generally decreases with the viscosity parameter L as well as initially with an increase load parameter M. However, please note that it is only marginally influenced by the load in a broad range.

In addition, approaches for circular or elliptical point contacts were developed whereby the equations from Hamrock and Dowson are [41]

$$H_{\min} \approx 3.68 \cdot G^{0.49} \cdot U^{0.68} \cdot W_{3\mathrm{D}}^{-0.073} \left(1 - e^{-0.68 \cdot \frac{a}{b}} \right). \tag{3.100}$$

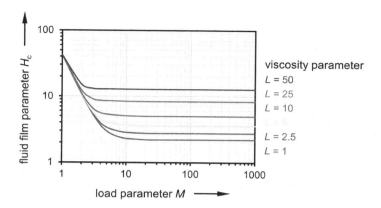

FIGURE 3.22 Central lubricant gap H_c versus load parameter M at different viscosity parameters L. (Adapted from Ref. [31] with permission by CC BY 4.0.)

Chittenden et al. [42] further elaborated this as

$$H_{\min} \approx 3.68 \cdot G^{0.49} \cdot U^{0.68} \cdot W_{3D}^{-0.073} \left(1 - e^{-0.67 \cdot \left(\frac{R_s}{R_e} \right)^{\frac{2}{3}}} \right),$$

$$H_c \approx 4.31 \cdot G^{0.49} \cdot U^{0.68} \cdot W_{3D}^{-0.073} \left(1 - e^{-1.23 \cdot \left(\frac{R_s}{R_e} \right)^{\frac{2}{3}}} \right), \text{ with}$$

$$\frac{R_s}{R_e} = \frac{\frac{R_y}{R_x} \cdot \cos^2 \xi + \sin^2 \xi}{\cos^2 \xi + \frac{R_y}{R_x} \cdot \sin^2 \xi},$$

(3.101)

even accounting for different inflow directions, as well as Nijenbanning et al. [43]

$$H_c \approx \left\{ \left[H_{c,IR}^{\frac{3}{2}} + \left(H_{c,IE}^{-4} + H_{00}^{-4} \right)^{-\frac{3}{8}} \right]^{\frac{2}{3} s} + \left(H_{c,VR}^{-8} + H_{c,VE}^{-8} \right)^{-\frac{s}{8}} \right\}^{s^{-1}}$$

$$s = 1.5 \cdot \left(1 + e^{-1.2 \cdot \frac{H_{IE}}{H_{IR}}} \right),$$

(3.102)

$$H_{00} = 1.8 \cdot \left(\frac{R_x}{R_y} \right)^{-1},$$

$$H_{c,IR} \approx 145 \cdot \left[1 + 0.796 \cdot \left(\frac{R_x}{R_y} \right)^{\frac{14}{15}} \right]^{-\frac{15}{7}} \cdot \left(\frac{R_x}{R_y} \right)^{-1} \cdot M_{3D}^{-2},$$

$$H_{c,IE} \approx 3.18 \cdot \left[1+0.006 \cdot \ln\left(\frac{R_x}{R_y}\right)+0.63 \cdot \left(\frac{R_x}{R_y}\right)^{\frac{4}{7}}\right]^{-\frac{14}{25}} \cdot \left(\frac{R_x}{R_y}\right)^{-\frac{1}{15}} \cdot M_{3D}^{\frac{2}{15}},$$

$$H_{c,VR} \approx 1.29 \cdot \left[1+0.691 \cdot \left(\frac{R_x}{R_y}\right)\right]^{-\frac{2}{3}} \cdot L^{\frac{2}{3}},$$

$$H_{c,VE} \approx 1.48 \cdot \left[1+0.006 \cdot \ln\left(\frac{R_x}{R_y}\right)+0.63 \cdot \left(\frac{R_x}{R_y}\right)^{\frac{4}{7}}\right]^{-\frac{7}{20}} \cdot \left(\frac{R_x}{R_y}\right)^{-\frac{1}{24}} \cdot M_{3D}^{-\frac{1}{12}} \cdot L^{\frac{3}{4}}$$

are widely adopted in literature due to their good agreement with experimental observations. It is important to note that all aforementioned approaches have their specific applicability ranges. For more information regarding non-dimensional similarity groups as well as proximity equations to estimate film thickness, the interested reader is referred to Ref. [31].

EXERCISE 3.3

Assume a purely radial loaded deep groove ball bearing, whereby the balls have a diameter of $D_B = 3.5$ mm and the raceway has a diameter of $D_R = 22$ mm and a curvature radius of $R_R = 1.8$ mm, corresponding to a reduced radius of $R' = 1.51$ mm. The inner ring rotates at $n_R = 6000$ min^{-1}, which yields a ring speed at contact with the inner ring of $u_R = 7$ m/s. Assume no slip between balls and raceway, allowing the balls to rotate at $n_B = -21900$ min^{-1}. The inner ring is made of steel ($E_R = 2.08 \cdot 10^5$ N/mm², $v_R = 0.3$), and the balls of ceramic ($E_B = 3.18 \cdot 10^5$ N/mm², $v_B = 0.26$). The highest loaded contact between the ball and raceways experiences a force of $F = 430$ N, resulting in a Hertzian contact pressure of ca. 3.4 GPa as well as the semi-axes of the contact ellipse $a \approx 820$ μm and $b \approx 75$ μm (neglecting inertia and friction forces). The contact is lubricated by an oil with a dynamic viscosity of $\eta_0 = 50$ mPa·s and a pressure-viscosity coefficient of $\alpha_p = 1.9 \cdot 10^{-8}$ Pa^{-1}. The averaged roughness of the raceway and the ball are $R_{qR} = 0.10$ μm and $R_{qB} = 0.05$ μm, respectively.

- Estimate the resulting minimum film thickness in the highest loaded contact using the approach from Hamrock and Dowson (Eq. 3.100).
- Determine the lubricant gap parameter.
- In which lubrication regime is the contact operating?

It is important to note that EHL contacts are profoundly affected by various factors, which might be only considered by certain correction factors for the aforementioned analytically solvable approaches [31] or complex numerical simulations [19]. For instance, at elevated speeds and variations in surface velocities (commonly referred to as slip), there is the potential for substantial local temperature increases within the lubricating fluid. This phenomenon can lead to fluid temperatures surpassing the bulk temperatures of the contacting components. These temperature rises are primarily attributable to the compression of the lubricant as it enters the contact zone and the shearing that occurs within the contact's core. As a result, the lubricant's viscosity and density tend to reduce. Similarly, the shear stresses inside the fluid due to the high-velocity difference of the contacting bodies might lead to a decrease in viscosity (see Section 3.1). Consequently, thermal and shear-thinning effects may also decrease the thickness of the lubricating film. Furthermore, an EHL contact can encounter issues related to insufficient lubrication, often triggered by factors such as high velocities, the use of highly viscous lubricants, or limited lubricant availability. Thus, the initiation of pressure build-up cannot initiate at any distance away from but closer to the central contact point, resulting in a notably thinner lubricating film compared to situations where the contact is fully flooded with lubricant. It is also well known that surface topography plays a pivotal role in determining the formation of EHD lubricating films. When waviness or roughness is oriented perpendicular to the direction of motion, it can, in specific cases, have a beneficial impact on the overall height of the lubricant film. However, structures aligned in the direction of motion tend to have adverse effects. Moreover, the presence of rough surfaces often leads to localized, significantly elevated pressure levels. As lubricant film thickness diminishes, the simultaneous occurrence of a solid asperity contact and the development of hydrodynamic pressure (mixed lubrication) can take place, effectively sharing the normal load.

3.3 LIQUID LUBRICANTS

3.3.1 Oil-Based Lubricants

One of the most common classes of liquid lubricants considered in industry connects to oils, which are viscous substances made up of long polymer molecules that interact with each other, lubricated surfaces, and the surrounding environment. Oil-based lubricants are used in many applications, from automotive and industrial machinery to medical and consumer products. They provide many advantages, including reduced friction and protection against wear, an excellent thermal and oxidation stability, and a general protection against corrosion. Though they are more expensive to produce, transport, and store than their water-based competitors, the operation regime of oil lubricants is still substantially larger. At the same time, the usage of oil-based lubricants can pose certain problems due to their ability to absorb dirt, dust, and other particles distributed on surfaces or in the atmosphere. Moreover, they show a pronounced dependency of pressure, temperature, and shearing (see Section 3.1). In a general sense, the used oil-based lubricants can be subdivided into three classes: mineral, synthetic, and bio-oils.

Mineral oils are petroleum-derived lubricants that form a major fracture of lubricants used in applications (ranging from lubrication for gears, bearings, chains, slides, and connections, among others) due to their chemical and physical properties. Mineral oils are mixtures of different hydrocarbon molecules obtained from refining crude oil. Consequently, the corresponding size and share variations of the molecules in mineral oils can be quite large. Most of the oil consists of paraffins, for which C atoms are straight or branched chains, and naphthenes, in which some of the C atoms form rings (Figure 3.23). If the amount of carbon present in paraffin chains is much higher than in naphthene rings, the oil is called paraffinic, and vice versa for naphthenic oil. Mineral oils are inexpensive, demonstrate good lubrication characteristics and compatibility with various materials, and can be used from −17°C and up to a maximum temperature of 130°C, while super-refined oils can reach operating temperatures of up to 200°C (e.g., straight paraffin oils). As mineral oils are non-renewable and their toxicity poses a threat to the environment, their use becomes more and more constrained due to legislative regulations. For instance, in some industries such as agriculture or offshore drilling, the use of petroleum-based lubricants poses the risk of environmental and ecosystem contamination as much as it would be in case of the improper disposal of the lubricants.

Synthetic oils, in turn, are intentionally created through further refining of mineral oils or chemical synthesis. Some examples are illustrated in Figure 3.24. Since the process of making synthetic oils is quite expensive, they are usually developed for certain applications requiring superior physical and chemical oil properties, such as resistance to thermal changes, efficiency under high contact pressure conditions, stability at low temperatures, or even compatibility with vacuum or humid conditions. In this case, the advanced physical properties, such as viscosity, density, and pour point, are induced by structural modifications of the lubricant base stock. In this context, synthetic oils can be used up to ~400°C and even above in some cases. The needed chemical properties of the finished or formulated lubricant, such as thermal and oxidation stability and interactions with the surfaces, are provided primarily by the additives used with the base stock.

As the thermal stability of polymeric molecules depends on the structure of the hydrocarbon chains, an improved high-temperature compatibility and stability can be achieved by the replacement of weakly bonded fractions of the molecule with branched hydrocarbons and by using various oxidation inhibitors as additives. For instance, viscosity-temperature and volatility properties of **polybutene** (low molecular weight) and **alphaolefin oligomers** of decene-1 are optimized by polymerizing olefin monomers.

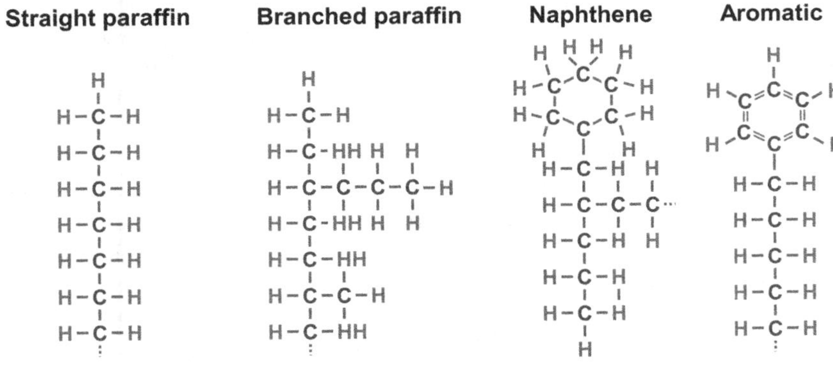

FIGURE 3.23 Schematic illustration of different types of mineral oils with the respective chain arrangement and building blocks.

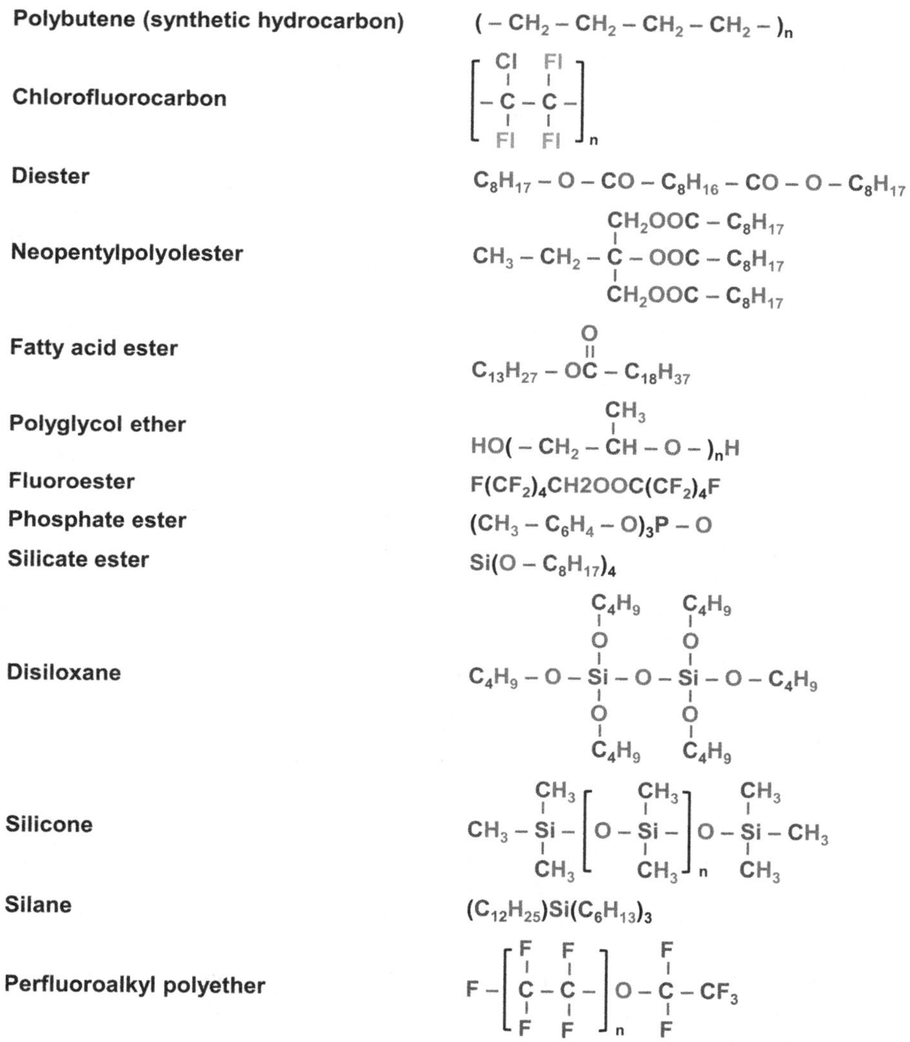

Polybutene (synthetic hydrocarbon) $(-CH_2-CH_2-CH_2-CH_2-)_n$

Chlorofluorocarbon
$$\left[\begin{array}{c} Cl \quad Fl \\ | \quad | \\ -C-C- \\ | \quad | \\ Fl \quad Fl \end{array}\right]_n$$

Diester $C_8H_{17}-O-CO-C_8H_{16}-CO-O-C_8H_{17}$

Neopentylpolyolester
$$\begin{array}{l} CH_2OOC-C_8H_{17} \\ | \\ CH_3-CH_2-C-OOC-C_8H_{17} \\ | \\ CH_2OOC-C_8H_{17} \end{array}$$

Fatty acid ester
$$\begin{array}{c} O \\ \| \\ C_{13}H_{27}-OC-C_{18}H_{37} \end{array}$$

Polyglycol ether
$$\begin{array}{c} CH_3 \\ | \\ HO(-CH_2-CH-O-)_nH \end{array}$$

Fluoroester $F(CF_2)_4CH2OOC(CF_2)_4F$

Phosphate ester $(CH_3-C_6H_4-O)_3P-O$

Silicate ester $Si(O-C_8H_{17})_4$

Disiloxane
$$\begin{array}{ccc} C_4H_9 & & C_4H_9 \\ | & & | \\ O & & O \\ | & & | \\ C_4H_9-O-Si-O-Si-O-C_4H_9 \\ | & & | \\ O & & O \\ | & & | \\ C_4H_9 & & C_4H_9 \end{array}$$

Silicone
$$\begin{array}{ccccc} CH_3 & & CH_3 & & CH_3 \\ | & & | & & | \\ CH_3-Si- & \left[O-Si-\right. & & \left.\right]_n & O-Si-CH_3 \\ | & & | & & | \\ CH_3 & & CH_3 & & CH_3 \end{array}$$

Silane $(C_{12}H_{25})Si(C_6H_{13})_3$

Perfluoroalkyl polyether
$$F-\left[\begin{array}{c} F \quad F \\ | \quad | \\ C-C- \\ | \quad | \\ F \quad F \end{array}\right]_n O-\begin{array}{c} F \\ | \\ C-CF_3 \\ | \\ F \end{array}$$

FIGURE 3.24 Schematic illustration of different examples of synthetic oils with the respective chain arrangement and building blocks.

In the polymerized molecules, the low bond energy of the C-C linkage remains a drawback. Therefore, further improvements are possible by substituting the H atoms with a mixture of Cl and F in the hydrocarbon chain to protect the C-C bond as demonstrated in **chlorinated and fluorinated compounds**. The resulting fluorinated hydrocarbons and chlorofluorocarbons are generally good lubricants, chemically inert, and have high oxidation and thermal stability. Unfortunately, their synthesis is difficult and costly, and the introduced modifications lead to high volatility as well as reduced viscosity.

Another approach to extend the application of synthetic lubricants is to use **compounds containing ester linkages**: The product of reacting alcohol ($R'-CH_2OH$)

with an (in)organic acid (R-COOH), in which organic groups R' come from alcohol used and R from the acid used. This synthesis became one of the most widely used in industry. Due to the higher bond energy compared to C-C linkages in hydrocarbons of mineral oils, the ester linkages are more stable against heat. Therefore, the ester lubricants show excellent viscosity-temperature and volatility properties as well as good mixability with additives aiming at improving their oxidation and lubrication behavior. The main disadvantage of ester-based lubricants is the acid-catalyzed degradation reactions that can proceed by oxidative or thermal mechanisms. Ester-based compounds are designed by various combinations of acids and alcohols, creating an extensive library of lubricants with desired characteristics.

For instance, **fatty acid esters** (with low molecular weight), used as internal lubricants in floppy disks and magnetic tapes, offer excellent lubrication at low temperatures with metals and metal oxides, which are reactive to fatty acids.

Diesters (higher molecular weight than fatty acid esters), common synthetic lubricants for aircraft engines, show an improved volatility than fatty acids. Polyglycol esters and fluoroesters, derived from organic carboxylic acids and fluoroalcohols, are used as hydraulic fluids. Phosphate esters such as tricresyl phosphate (TCP, $(CH_3C_6H_4O)_3PO$) provide excellent lubricity and are widely used as anti-wear additives for petroleum.

Silicate esters' reaction product of silicic acid $[SiO_x(OH)_4]_n$ and an alkyl group alcohol $[CH_3$-OH Methyl alcohol] is often used as low-temperature lubricants. The presence of the Si-O-C bonds distinguishes the silicate esters from silanes and silicones (Si-C direct bond). They have high thermal stability, low viscosity, and fair lubricity, but poor hydrolytic (water removing) stability.

Silicones (Si-O and Si-C linkages) or **polysiloxanes**, such as dimethyl silicone polymers, offer an excellent chemical inertness, thermal stability, low volatility, and low surface tension. However, these silicon-based fluids are expensive and are, therefore, applied for applications involving extreme operating temperatures (−50°C to 370°C), such as in grease formulations used in space and military applications. Due to their chemical inertness, they tend to have a poor boundary lubrication behavior, especially with steel, thus preventing their use in high-load-bearing applications. However, it is important to note that they show a good (elasto-)hydrodynamic lubrication behavior.

Silanes (Si-C linkages), another type of synthetic lubricant, differ from silicones as they lack the familiar Si-O linkage that provides the central backbone of the silicone polymer fluid. Silanes are formed from $SiCl_4$ and organo-metallic groups. They are typically used as high-temperature lubricants in the absence of air or oxygen, which makes them particularly interesting for space applications.

Perfluoropolyethers (PFPEs) can be prepared with different molecular weights affecting their range of viscosity range. They show oxidative stability up to 320°C and thermal stability up to 370°C. In addition, they exhibit a low surface tension, and can easily attach different polar groups. Therefore, PFPEs are widely used as oils in greases for high-temperature, high vacuum, and chemical-resistant applications, as hydraulic fluids, gas turbine engine oils, and lubricants in satellite instruments or magnetic rigid disks in hard drives.

A summary of different characteristics of commonly used synthetic lubricants in comparison to mineral oils is presented in Table 3.1.

TABLE 3.1

Typical Properties of Commonly Used Classes of Synthetic Oil Lubricants [44–48]

| | Therm. Stability in °C | Kin. Viscosity in cSt at | | | | | Spec. Gravity at 20°C | Therm. Cond. in cal/h·°C | Spec. Heat Cap. at 38°C in cal/(g·°C) | Flash Point in °C | Pour Point in °C | Oxidative Stability in °C | Vapor Pressure at 20°C in Torr |
		−20°C	0°C	40°C	100°C	200°C							
Mineral oils	135	170	75	19	5.5	n/a	0.86	115	0.39	105	−57	n/a	10^{-6}–10^{-2}
Diesters	210	193	75	13	3.3	1.1	0.90	132	0.46	230	−60	n/a	10^{-6}
Neopentyl polyol esters	230	16	16	15	4.5	n/a	0.96	n/a	n/a	250	−62	n/a	10^{-7}
Phosphate esters	240	85	38	11	4		1.09	109	0.42	180	−57	n/a	10^{-7}
Silicate esters	250	115	47	12	4	1.3	0.89	n/a	n/a	185	−65	n/a	10^{-7}
Disiloxanes	230	200	100	33	11	3.8	0.93	n/a	n/a	200	−70	n/a	30
Silicones													5×10^{-8}
Phenyl methyl	280	850	250	74	25	22	1.03	124	0.34	260	−70	240	
Fluoro	260	20000	n/a	190	30	24	1.20	n/a	n/a	290	−50	220	
Polyphenylesters													10^{-8}
4P-3E	430	n/a	2500	70	6.3	1.4	1.18	133	0.43	240	−7	290	
5P-4E	430	n/a	n/a	363	13.1	2.1	n/a	n/a	n/a	290	+4	290	
Perfluoropolyethers													
Fomblin YR	370	n/a	8000	515	35	n/a	1.92	82	0.24	none	−30	320	10^{-9}
Fomblin Z-25	370	1000	440	150	41	n/a	1.87	n/a	0.20	none	−67	320	3×10^{-12}

Source: Adapted from Ref. [49].

Please note that these are just reference values to indicate the order of magnitude. Actual oils, even though they are of the same type, might vary to some extent.

Bio-lubricants have attracted lots of attention as alternatives to petroleum-based oils to address sustainability concerns as they are biodegradable and have low toxicity toward the ecosystem and humans [50]. Historically, bio-lubricants were presented by animal fats, such as shark and whale oils. However, since the use of these animal fat-based lubricants is limited to ~120°C, with time the focus has shifted toward plant-based lubrication, which resembles a more sustainable and effective source of renewable oils. While most mineral oils and synthetic oils have a biodegradability below 40% and above 70%, respectively, vegetable oils feature a biodegradability of 90%–98%. Among the currently widely used plant-based bio-lubricants are castor, palm, and rapeseed oil, while fish oil and lanolin oil (derived from wool-bearing animals, such as sheep) reflect the commonly applied animal-based lubricants. Bio-based lubricants tend to provide good lubrication characteristics but are still much less oxidation and thermally stable than mineral or synthetic oils since they break down and generate sticky deposits. Moreover, due to the limited availability of plant-based oils and their cost being higher than for mineral and synthetic oils, bio-oils are mostly used in food and pharmaceutical industries.

Most bio-lubricants consist of fatty acids promoting the formation of a thin protective film on the involved rubbing surfaces, which helps to reduce wear and enhances oxidation resistance during operation [51]. Castor oils, for instance, are notably investigated by different industries due to its excellent lubrication performance attributed to its high viscosity and the domination of ricinoleic acid, which plays a major role in the oil's lubricity [52]. Jojoba oil, one of the bio-lubricants used in pharmaceutical applications, consists of a double bond on each side of the chain thus enhancing its stability during exposure to mechanical stresses [53,54]. Rapeseed oil has been evaluated as an additive to synthetic lubricants, and pennycress oil, also known as fatty acid methyl ester, has been tested as a potential biodiesel and jet fuel [55].

Different reports compared the performance of bio-oils to commonly used synthetic oils (Table 3.2). Recent efforts demonstrated beneficial characteristics of certain lipid molecules. Specifically, it has been shown that triacylglycerols (TAGs), wax esters, and erucic acids, all of which contain long-chain fatty acids, improved the resulting oxidation stability and lubrication efficiency. This knowledge proposed new approaches toward the controlled design of bio-lubricants using genetic modifications routes or synthetic routes for the modifications.

3.3.2 Additive Systems

Since base oils, regardless if they are mineral, synthetic, or bio-oils, provide only a certain range of characteristics, additional pathways for enhancing their performance are needed. This is usually possible with **additive systems** such as **anti-wear (AW)** and **extreme pressure (EP)** additives as well as **friction modifiers**. In this regard, the used additives are chemical compounds that are added to the base oils in a range of 0.1–0.5 wt.-% (in some cases up to 30 wt.-%) to improve their performance and enhance their lubrication properties by tailoring their viscosity and oxidation resistance as well as reducing foaming of the oils. Commonly used additives include detergents, dispersants, and viscosity index improvers as well as rust and corrosion inhibitors, friction modifiers, and pour point depressants. To obtain the most

TABLE 3.2
Qualitative Comparison of the Characteristics of Various Bio-Lubricants to Reference Lubricants Based upon [56]

Bio-Lubricant	Reference Lubricant	Qualitative Comparison
Castor oil	Mineral oil	Larger viscosity index, lower deposit-forming tendencies, lower volatility, higher concentrations of antioxidants
	SAE 20W40	Friction and wear reduction, lower volatility, biodegradability
Coconut oil	SAE20W50	Friction and wear reduction
Jatropha oil	SAE20W40	Friction and wear reduction
Palm oil	SAE 20W40	Friction and wear reduction, lower volatility, biodegradability
	SAE20W50	Friction reduction, higher stability of the lubricating film, better protection against oxidation and corrosion, higher reactivity of unsaturated hydrocarbon chains
	SAE 40	Higher viscosity, friction reduction
Pongamia oil	SAE 20W40	Friction reduction, lower brake-specific energy consumption, higher brake thermal efficiency, elimination of emissions
Soybean oil	Mineral oil	Friction reduction, non-toxicity, cheaper
	Synthetic oil	Larger viscosity
Rapeseed oil	SAE20W40	Improved oxidative stability, better cold flow behavior, friction reduction

desirable characteristics needed for certain complex systems, the number of additives in the package can sometimes exceed a dozen. The appropriate additive system used in oils can vary depending on the type of oil being used, the manufacturer's specifications, and the desired lubricant's performance. Apart from directly improving the friction and wear performance, they can also help to enhance the useful lifetime and reduce maintenance costs (oil changes) and the overall carbon footprint.

Viscosity index improvers (VIIs) are one of the most common additives in lubricants [57] and help to improve the oil's viscosity index, i.e., maintaining the viscosity at high and low temperatures that would be challenging with the pure base oil only. The primary usage of VIIs connects with multigrade engine oils, gear oils, automatic transmission fluids, power steering fluids, greases, and hydraulic fluids. These applications are mostly relevant in automobiles because of the substantial temperature fluctuations that vehicles experience. Since the oil itself generally shows a decreasing viscosity with increasing temperature (see Section 3.1), the VIIs should show minimal impact at low temperatures, while increasing the viscosity at higher temperatures. Consequently, VII modifiers are usually made from polymeric molecules exhibiting a unique behavior. The molecule chains contract at low temperatures, leaving the fluid viscosity unaffected, while at high temperatures the chains relax, increasing the fluid viscosity (Figure 3.25). The longer the polymer chain is (higher molecular weight), the better the thickening properties the polymer exhibits. Unfortunately, longer polymeric chains suffer from degradation induced by mechanical shearing. Therefore, the needed improvement with longer stability can be achieved by using larger quantities of lower molecular weight polymers instead. The most

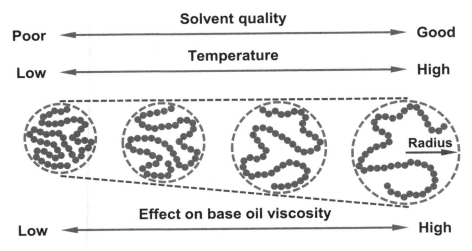

FIGURE 3.25 Schematic representation of the coil expansion mechanism leading to an increased solvent quality with temperature and thus to polymer expansion and viscosity improvement.

common VIIs currently used in industry are polyalkyl methacrylate (PAMA), olefin copolymer (OCP), polyisobutylene (PIB), and hydrogenated styrene-diene (HSD). Apart from the coil mechanism to affect viscosity, some polymers, such as HSD, are believed to raise viscosity through the association/aggregation of macro-molecules, resulting in micelle formation. Furthermore, polymers can influence the viscosity through secondary mechanisms, such as knot formation and their interaction with adjacent solvent molecules.

Friction modifiers also resemble common additives in lubricants with the overall goal of reducing friction between the involved moving parts, increasing the life of the lubricant, and improving the overall efficiency of the system compared to lubrication with pure base oil. The most common examples of friction modifiers are long-chain alcohols, amines, and fatty acids. Friction modifiers enable easier shearing of the surfaces, which is especially important under boundary lubrication. They (condensation polymerized oxidation products) usually have a polar group (-OH) at one end of the molecule that reacts with the contacting metallic and oxidic surfaces through adsorption. This leads to the formation of thin layers (partially monolayers) with low shear strength, so-called **tribo-films**. The latter are effective only at relatively low loads and moderate temperatures (80°C–150°C). Possible friction modifiers that are efficient in engine oils and can be dissolved in oil include organomolybdenum compounds such as molybdenum dithiophosphate/dialkyldithiocarbamate (MoDTP and MoDTC) and ashless organic ester- or acid-based additives like stearic acid or octadecanol.

The underlying mechanisms for the attachment of these modifiers to the involved rubbing surfaces connect with **physisorption** and **chemisorption**. In the case of physisorption, the molecules attach to the metal surfaces by electrostatic forces. An example of this type of interaction is schematically presented in Figure 3.26 (left) for

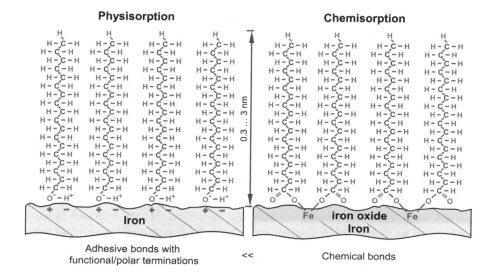

FIGURE 3.26 Physisorption (left) and chemisorption (right) mechanisms of surface attachment of friction modifiers.

octadecanol ($C_{18}H_{37}OH$) molecules lubricating steel surfaces. Since the octadecanol molecules have -OH groups at one of the ends, these -OH groups cause a charge redistribution on the metal surface that facilitates van der Waals bonding between the additives and sliding materials. Due to the weak bonding, several issues should be taken into consideration:

- The presence of surface oxides, in the form of native oxides or oxides formed during sliding, can inhibit the necessary charge redistribution on the metal surfaces, thus suppressing the physisorption of the films.
- The physisorbed films cannot sustain high contact loads. As their maximum shear stresses are on the order of 10–20 MPa, they easily degrade under boundary lubrication reaching contact pressures above 100 MPa.
- These protective films generally are not durable due to the secondary bond nature between the metal surfaces and molecules, resulting in the removal of the molecules from the sliding interfaces and loss of protection. An increase in additive concentration only partially solves the problem of coverage, since this goes hand in hand with other drawbacks, such as an increased tendency of agglomeration.

An example of chemisorbed protective films to protect the involved surfaces and to reduce friction is presented in Figure 3.26 (right) based on the use of stearic acid ($C_{18}H_{36}O_2$) on steel surfaces. In this case, the molecules of the friction modifiers can be anchored by reacting with the surface oxygen groups to form monolayers of iron stearate. In this regard, it is important to emphasize that this anchoring mechanism is universal for AW additives, such as sulfur and phosphorous-based compounds, which will be reviewed in the next paragraph. For friction modifiers, however, the major concept remains to facilitate easy shearing between the sliding interfaces, rather than prevent the surface from wearing off. Therefore, a minimum chain length of the fatty acids is required. As the chain length increases, the interactions between surfaces decrease, or longer molecules provide better self-support. Another key factor for the polar group adsorption is the similar polarity of the workpiece surface, which is typically achieved in metallic materials through the formation of an oxide layer. However, when dealing with ceramic materials, issues can arise in terms of adsorption. For instance, fatty acids with polar end groups are easily adsorbed onto aluminum oxide through ionic bonding, leading to reduced friction beyond a certain chain length. In contrast, there is almost no adsorption on silicon carbide with covalent bonding, thus resulting in only a marginal influence of the resulting COF.

Even though the same additive can serve both as a friction modifier and AW component, their mechanisms are conceptually different. **AW additives** are needed to prevent wear, even if friction is not affected. Therefore, the AW additives function by forming a relatively thick coating, in the form of organic, metal-organic, or metal salt films, on the sliding surfaces that cannot be easily removed through shearing. These additives form a protective layer on the metallic surfaces, reducing wear and extending the lifespan of the component. Common examples are zincdialkyldithiophosphate (ZDDP), molybdenum dithiocarbamate (MoDTC), tricresyl phosphate (TCP), and ethyl stearate mostly used for esters and other synthetic lubricants [58–60].

Furthermore, the creation of reaction layers with low shear resistance through tribochemical reactions between lubricating oil constituents and the metallic material surface offers an enhanced thermal and mechanical resilience compared to layers formed via physical or chemisorptive means. To achieve this, lubricating oils incorporate **EP additives** like sulfur-phosphorous [61,62], chlorinated, or phosphorus compounds. The efficacy of these additives hinges on the rate at which reaction layers form, a process influenced by factors such as activation energy through extreme pressures and/or surface temperature ($> 150°C$). A critical aspect is the concentration of these additives at the interface between the metal and the lubricating oil. Elevating their adsorbability – in other words, their surface concentration – augments their impact, whereas displacement of these additives from the metal surface by other polar compounds has the opposite effect.

In addition to the aforementioned additives with functions commonly important for the application needs, there are other types of additives:

- **oxidation inhibitors** prevent the early formation of aging residues thus extending the life of the lubricants,
- **corrosion protection agents** prevent corrosion of metallic surfaces,
- **anti-rust agents** prevent rust on metal parts during downtime periods,
- **metal deactivators** avoid catalytic influences of metals on the oxidation process,
- **dispersants** help to break down sludge-forming, insoluble contaminants and keep them in suspension to avoid deposition on the surfaces,
- **pour point improvers** lower the pour point, and
- **foam inhibitors** reduce the shearing-induced formation of foam due to better adhesion of the lubricant and thus prevent lubrication failure.

Instead of using single additives, additive packages are typically employed to combine a carefully balanced mixture of various additives, each designed to address different aspects related to lubrication. This multi-faceted approach ensures that lubricants can adapt to a wide array of operating conditions and requirements, making them more versatile and reliable.

3.3.3 OTHER LIQUID LUBRICANTS

Even though oils make the most efficient and widely used approach for lubricating moving systems, **water-based lubricants** (WBLs) become increasingly popular among those looking for a sustainable alternative to traditional lubricants, that are environmentally friendly and much easier to clean off. Made mostly of water, these lubricants are safe to use on almost any surface and will not damage some materials in the same way as an oil-based lubricant could (for instance, silicone, which is not compatible with synthetic lubricants). WBLs are of particular interest for hydropower plants, the food industry, and pharmaceuticals, where contaminations originating from oil-based lubrication cause major safety concerns. WBLs are also traditional low-cost lubricants used in metalworking processes (such as cutting and polishing), where they do not only act as friction and wear modifiers but also as a coolant [63].

The performance of WBLs can be optimized by the addition of different additive packages. Alternatively, WBLs can be created using biomass-derived ethylene glycol and glycerol that already contain significant amounts of water [64]. The main disadvantages of WBLs relate to their low viscosity and corrosion acceleration and that they dry up relatively quickly and need to be reapplied during use. Therefore, to extend their lifetime and to improve their lubrication characteristics, similar approaches as for traditional oil-based lubricants are used. For instance, incorporating additives (such as tetrabutylphosphonium benzotriazole or alkylglucopyranosides) or through galvanic couplings using electrical charge to limit corrosion processes.

The viscosity and low-temperature range are meanwhile improved by the addition of **ionic liquids** (ILs, salts made of cations and anions with melting points below 100°C). In particular, ILs provide additional benefits of lubrication improvement (at conditions of high pressure and temperature) originating from their high thermal stability, non-flammability, high conductivity, and low vapor pressure assisting them in forming strong adsorption films. Unfortunately, the use of ILs can also accelerate the corrosion of metals.

Nanoparticles can be also employed as additives to improve the tribological characteristics of WBLs while suppressing corrosion that is ascribed to the surface charge distribution suppressing the charge exchange in the metallic surfaces. It has been shown that the addition of silicon oxide nanoparticles, typically carrying a negative surface potential, reduces corrosion of copper surfaces exposed to NaCl solution [65]. The adsorbed nanoparticles are drawn by electrostatic forces to a positively charged Cu surface, impeding the access of chlorine ions from the solution to the surface. This resulted in a considerable deceleration of corrosion processes. Meanwhile, the adsorption of the nanoparticles on the sliding surfaces helps to separate the sliding contact and reduce the induced damage. Similar effects have been demonstrated for other nanomaterials, such as titanium oxide, titanium carbide, zinc oxide, nanodiamonds, etc. [66–68].

Another alternative lubricant is **glycerol** (also called glycerin), a sugar alcohol derived from animal fat or vegetable oil. Glycerol is a clear, odorless, and non-toxic viscous liquid characterized by a low volatility, low freezing point, and ability to hold moisture and dissolve in water. It is widely used in marine environments, the production and processing of food, pharmaceuticals, and cosmetics as a moisturizer, thickener, and sweetener. It has also found application in medical devices, such as catheters syringes, and dental restorations. The excellent lubrication performance of glycerol originates from its physical and chemical characteristics. Glycerol has a high viscosity index thus allowing it to maintain its lubricating performance across a wide range of operating conditions. The thermal degradation of glycerol occurs by dehydration and the formation of acrolein at temperatures above 250°C [69]. Apart from being non-toxic and biodegradable, glycerol demonstrated an excellent load-carrying capacity. However, glycerol used as a lubricant does also present some challenges and bottlenecks. One of the main drawbacks of glycerol is its hygroscopic nature, which implies that it readily absorbs water from the atmosphere. While it is important in some range of applications, such as in the cosmetic industry, it can also lead to increased oxidation and corrosion in applications involving metallic surfaces. The rather low temperature of the onset of its degradation also presents a drawback in some applications.

ILs are salts that exist in a liquid state at or near room temperature. ILs have the potential to provide an improved lubrication performance than conventional liquid lubricants due to their low volatility, high thermal stability, and unique interfacial properties while being non-toxic and non-flammable. The structure of ILs consisting of bulky cations and small anions (Figure 3.27) provides them with low vapor pressure and excellent thermal stability which are critical characteristics for high-temperature lubrication applications. Moreover, ILs are chemically stable and can be tailored to have specific properties, such as viscosity, lubricity, and solubility. The tribological properties of ILs result from their unique surface activity and ability to form boundary films on metallic surfaces. The lubricity of ILs can be further improved by the addition of surface-active agents or friction modifiers. Alternatively, ILs can be used as additives themselves

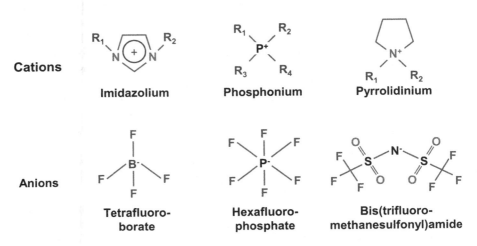

FIGURE 3.27 Schematic representation of typical molecular structures of ionic liquids. (Redrawn from Ref. [70] with permission by CC BY 4.0.)

to improve the water resistance, shear stability, and wear resistance of conventional oil-based lubricants. Unfortunately, the high cost of production has limited the commercial use of ILs. It also should be kept in mind that the compatibility of ILs with materials needs to be carefully evaluated before their use in specific applications.

Furthermore, there has recently been a great interest in minimizing the need for oil-based lubrication in combustion engine components by providing the needed lubrication characteristics directly from the used **fuels** [71–73]. Since most fuels are highly chemically reactive and can contain traces of water and other impurities as a result of the synthesis process, the fuel system surfaces are often affected by corrosion and oxidation. Over time, fuel injectors, valves, and other engine components can accumulate deposits, affecting their performance and efficiency. Addressing the tribological challenges, in this case, can be done from two perspectives: the deposition of appropriate coating systems for surface protection (such as carbides, nitrides, and oxides) or by incorporating fuel additives to improve their performance and efficiency by reducing the wear of the fuel lubricated contacts. The additives used in fuels are designed to possess detergent, inhibit corrosion and oxidation, and stabilize the properties. Usually, the additive packages are expected to tailor unique requirements and challenges of each fuel type and, therefore, differ for gasoline versus diesel, and alternative fuels.

3.3.4 Superlubricity with Liquid Lubricants

In the quest to reduce and optimize frictional losses in lubricated systems, specific focus has been placed on the complete elimination of friction, known as superlubricity [74,75]. Thereby, a COF below 0.01 is considered as the superlubricity threshold. Despite the ubiquitous use of oil-based lubricants in automotive and engineering/manufacturing fields and incremental improvements over time, achieving and maintaining macro-scale friction coefficients below 0.01 across lubrication regimes under real-life operating conditions has been rather difficult. There are several tribological applications, for which these deficient lubricating technologies have been a limiting factor in the design and realization of next-generation moving mechanical systems. Pertinent examples include vehicles, gears and bearing systems of wind turbines, and a myriad of other types of moving mechanical assemblies that exist in manufacturing, aerospace, power generation, and other cross-cutting industrial sectors. Fundamental physical aspects of common oil-based lubricants, such as viscosity and its strong dependence on temperature as well as the stability of the additive packages, affect their lubrication properties in the long run. The different lubrication regimes exert different levels of shear forces and contact pressures within the tribological contact zone, demanding a robust dynamic response from the lubricant itself to remain in contact and ensure its efficiency across different applied conditions.

Polar and ionic liquids have emerged as favorable choices for achieving superlubricity among various liquid media options. Notably, exemplary cases, such as glycerol and glycol, have displayed a promising performance in superlubricity and offer room for further improvement. These liquids exhibit reduced friction, particularly when in contact with surfaces protected by coatings like diamond-like carbon (DLC), sapphire, and alumina, or when interacting with bearing steels, hardened steels, carburized steels, and basic silicate glasses. Moreover, these materials have garnered substantial attention in the engineering community due to their adaptability for further enhancements through the incorporation of low-dimensional materials and the inclusion of liquid additives.

Because of its readily availability, good miscibility, low cost, and extremely low viscosity, water presents an excellent option for reducing the viscosity of glycerol, which can be effective in passivating sliding interfaces, and thus serve as a friction modifier. The liquid lubricant-enabled superlubricity usually relies on the formation of a tribo-film, a hydration layer, or a molecular brush. However, even for EHL conditions, superlubricity has been achieved when the intercalated liquid formed a very thin layer thus enabling easy shearing between the sliding surfaces [76].

3.3.5 LIQUID LUBRICANT AGING

Liquid lubricants, such as oils, are the lifeblood of machinery, vital for ensuring their efficiency and longevity. Yet, the aging of liquid lubricants is an unavoidable reality. These lubricants, meticulously selected and maintained, cannot evade the effects of time. Lubricant aging stems from multiple factors. Mechanical and thermal stress[6] or UV radiation can lead to the degradation of lubricant molecules and the formation of detrimental byproducts. Furthermore, the interaction of lubricants with oxygen, heat, and contaminants initiates oxidation. Contaminants, like water (also from ambient humidity), dirt, and (catalytic) metal particles, accelerate aging by promoting chemical reactions [77]. This can lead to aging products, such as sludge or carbon in the oil. Furthermore, lubricant additives deplete over time (i.e., are "consumed"), diminishing their protective qualities and effectiveness. As lubricants age, their properties undergo alterations that impact machinery. Viscosity changes can make lubricants too thin, reducing their ability to provide an adequate film thickness, or too thick, increasing friction and heat generation. Oxidation generates acidic compounds that corrode metal surfaces, potentially leading to component damage. Depleted additives weaken the lubricant's protective capacity, and contaminants accumulate, further accelerating deterioration.

Addressing lubricant aging requires vigilance and proactive strategies. Routine oil analysis, examining viscosity, sludge and acid content, saponification, chemical composition, and additive as well as wear particle concentration provides insights into the lubricant's condition and guides replacement decisions. In this regard, Filtration systems can be effective to remove contaminants, thus extending lubricant life. Maintaining machinery within recommended temperature ranges reduces thermal stress on lubricants. In some cases, additives can be replenished, reinvigorating the lubricant's protective properties.

3.4 GREASES

Greases are semi-solid lubricants made up of

- **base oil**,
- **thickeners**, and
- **additives**.

---

[6] Actually, every rise of 10°C above 70°C might halve the oil service life.

They are mostly applied in the automotive and aerospace industries as well as construction, manufacturing, and agriculture to machinery and equipment experiencing large contact pressures, for which the use of liquid lubricants is limited due to the potential transition to boundary lubrication. Alternatively, greases are a solution to localize the lubricant in applications that cannot sustain immersion into a bath of liquid lubricant. The base oil that makes up about 65–95 wt.-% of the grease can be derived from natural, such as mineral and vegetable oils, or synthetic sources, such as synthetic esters, or silicone (see Section 3.3.1). The general piezo-viscous and non-Newtonian characteristics of lubricating oils (see Section 3.1) are generally maintained in greases, which make greases an ideal choice for high-load applications. The additives used in greases usually take up to 5 wt.-% and are a combination of different chemicals serving different purposes in improving the performance of the base grease, such as their resistance to water, oxidation, corrosion, extreme temperatures, or pressure. They also help to adjust their viscosity and enhance their wear and friction-reducing characteristics. Thickeners make up 5–30 wt.-% and are used to maintain the additives in the contact interface by forming a fibrous structure and cross-linking the molecules. Thickeners are composed of fatty acid soaps, such as calcium, lithium, aluminum, and sodium-based, and non-soaps, such as clay, PTFE, polyurea, and silica. For instance, lithium soaps are composed of salts of fatty acids and lithium hydroxide, forming a colloidal soap suspension when combined with oil. The fibers of the thickener can range lengths of about 1–100 μm and length-to-diameter ratio from 10 to 100, affecting the resulting consistency of the grease at a given thickness and application conditions. Currently, most used greases are based on lithium and calcium soaps, while some additional efforts have been dedicated to develop natural thickeners for the design of 100% biocompatible and biodegradable greases. In this regard, nanomaterials such as cellulose and nano-clays have been considered.

To some extent, we can imagine (the thickeners of) grease like a sponge that can soak up and release the base oil. Just as a sponge soaks up liquid when you press it against a surface, grease can absorb and retain the base oil. The thickening agent in grease holds the base oil within its structure. When pressure or force is applied, the sponge releases the absorbed liquid. Similarly, in the context of machinery and lubrication, when components move and exert pressure, the grease releases minimal amounts of the base oil to lubricate the moving parts. Over time, the sponge can release the liquid it had absorbed, and it might be necessary to replenish it. Similarly, grease can gradually release its base oil, and maintenance or re-greasing is required to ensure that the lubrication remains effective.

The exact mechanism of grease lubrication when inside the tribo-contact itself still is a matter of ongoing scientific discussion as the entire analysis of grease-lubricated interfaces is challenging. As greases are classified as non-Newtonian substances with a yield shear stress, they show different rheological responses to both the shear rate and shear duration. Greases tend to show a viscoelastic behavior, which depends on the applied shear rates. They demonstrate a solid, or very highly viscous behavior for low shear rates and a viscous behavior with potential shear thinning at higher shear rates, approaching the rheology of the base oil. Furthermore, greases

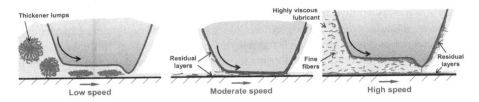

FIGURE 3.28 Schematic illustration of the different lubrication states experienced by greases with increasing velocity (from left to right). (Adapted from Ref. [83] with permission by CC BY 4.0.)

are thixotropic[7] materials, which implies that the measured stress also depends on the time of shearing [78]. Due to the constant exposure to high stresses, the grease thickener structure continuously degrades and is transformed from a viscoplastic material that behaves as a rigid body at low stresses to a viscous material that flows as a fluid at elevated high stresses. The shear stress τ in the grease can be described as

$$\tau = \tau_y + \eta_a \cdot \dot{\gamma}, \tag{3.103}$$

with the yield stress of the grease τ_y (above which the grease starts to flow) as well as the apparent viscosity

$$\eta_a = a + b \cdot \dot{\gamma}^{n-1}. \tag{3.104}$$

The yield stress of the grease strongly depends on temperature, e.g., it drops from 500 Pa at 20°C to 100 Pa at 60°C for lithium grease [79]. In the aforementioned equations, it is assumed that the grease flow is composed of two regimes (Newtonian and non-Newtonian) and the total shearing stress required to produce the same shear rate for both regimes is the sum of the shear stresses required for each [80]. At stresses below the yield shear stress ($0\text{--}2 \times 10^3$ s^{-1}), the grease is retained in the contact, thus sealing the interfaces from moisture and environmental effects. At high shear rates (above 2×10^3 s^{-1}), greases show almost Newtonian flow characteristics, and their viscosities are only slightly greater than those of their base oils. Similarly, the grease follows similar lubrication film formation trends as the lubricating base oil. However, the film thickness initially tends to decrease with velocity, since the acting shear causes the viscosity to decrease thus leading to the film decay and starvation of the contact, which is followed by an increase when the base oil properties start to dominate (Figure 3.28). Please note that the structure of the thickener fibers may degrade with mechanical aging, which impacts the lubrication effectiveness of the grease, since the permeability of the damaged grease fibers changes and suppresses the efficiency of the oil lubrication film formations. This structural degradation results in a loss of its apparent viscosity (softening), which can be described by

$$\eta_G = \eta_{BO}(1 - B \cdot \Phi), \tag{3.105}$$

---

[7] Isothermal decrease in the structural (apparent) viscosity during shearing (at both constant and variable shear rates) followed by an increase in the viscosity and the re-solidification once shearing has ended.

where η_{BO} is the viscosity of the base oil, B represents a constant, and Φ stands for the volume fraction of the thickener. In practice, e.g., for rolling element bearings,[8] this implies that the grease film may be short-lived and the thickener plays a minor role while the service life is mainly driven by the base oil viscosity [81,82].

Apart from mechanical stresses, the structural integrity can be also negatively impacted by exposure to environmental pollutants such as water, process liquids, or other contaminants, as well as to temperature variations. For instance, thermally induced failures of greases result from the melting of the thickeners or oil thinning, which negatively affects the resulting surface tension and capillary action, thus leading to the effect that the oil cannot be retained in the thickener matrix. At lower temperatures, the major failure is caused by the oil component, while both the onset of oxidation of the oil and the softening of the soap deteriorate the grease performance at elevated temperatures. The lower temperature limit of greases is determined by their bleeding properties or the pour point of the base oil. The upper temperature limit is defined by the grease drop point (droplet falls from a standardized cup). Hence, greases are typically engineered to provide a high dropping point. This characteristic enables them to retain the base oil within the thickener matrix for extended periods, ensuring prolonged operational effectiveness. In any case, for the selection of a suitable lubricating grease, the operating conditions (operating temperature, load, speed, installation, environmental conditions, maintenance frequency, etc.) should be known as precisely as possible.

3.5 LUBRICANT STORAGE, HANDLING, AND MAINTENANCE

Proper storage and handling are fundamental in preserving lubricants. Inadequate storage can expose lubricants to contaminants, moisture, and extreme temperatures, potentially leading to degradation and inducing a performance deterioration. To this end, dedicated storage facilities with controlled temperature and humidity should be established, and sealed containers should be used to prevent contamination and moisture ingress. Moreover, cross-contamination when handling lubricants should be avoided by the usage of clean and dedicated transfer equipment. Moreover, regular sampling and analysis of lubricants (e.g., by spectroscopy, particle counting, and viscosity measurement) are crucial to evaluate both the lubricant's condition and the machinery it serves. These methods help to detect potential issues early and make informed maintenance decisions. Maintaining clean lubricants is vital to prevent wear and contamination, which is why filtration and cleaning methods play a significant role in the overall lubricant maintenance strategy. It is possible to either employ in-line filters to remove particulate contaminants from circulating lubricants or offline filtration units to clean lubricants before reintroducing them into the system, especially in critical applications. Also, centrifugation can be used for large systems to effectively separate solid contaminants and water from the lubricant. Finally, lubricants are designed for longevity, but there comes a time when they must be replaced. Either routine replenishment or replacement schedules should be developed. For this, monitoring lubricant condition and equipment performance is needed to determine when to change the lubricant.

---

[8] Of which roughly 90% in the market are lubricated by greases, while oil or solid lubrication only make up the remaining 10%.

3.6   CHECK YOURSELF

- What are the primary functions of liquid lubricants in tribological systems, and how do they contribute to reducing friction and wear?
- What are the key properties of liquid lubricants that affect their performance in tribological contacts, and how do these properties change with temperature, pressure, and shear rate?
- How can the rheological behavior of liquid lubricants be categorized, and what are the implications of different rheological models for lubricant performance?
- What are some important thermal properties of liquid lubricants, and why are they significant in controlling temperature and preventing thermal breakdown in machinery?
- What are the safety and stability-related properties of liquid lubricants, such as flash point, pour point, oxidative stability, and vapor pressure, and why are they crucial for various industrial applications?
- What are the primary factors that contribute to the reduction of friction and wear in tribo-technical contacts when using liquid lubricants?
- How is the lubrication and friction behavior in tribological contacts categorized based on the Stribeck curve, and what are the key lubrication conditions associated with different regions on the curve?
- What is the significance of the Hersey number, and how does it relate to the lubricant film thickness and friction behavior in tribological contacts?
- What role do hydrodynamic lubrication and elastohydrodynamics play in minimizing friction and wear in contacts, and how do factors such as viscosity, load, and relative velocity influence these lubrication regimes?
- What are the fundamental conservation laws in continuum mechanics that are essential for mathematically describing fluid behavior in hydrodynamic lubrication?
- What is the significance of the Reynolds number in determining whether the flow in hydrodynamic lubrication is laminar or turbulent?
- What is the Reynolds equation, and how does it relate to the behavior of the fluid in the bearing?
- What assumptions and simplifications are made when deriving the Reynolds equation?
- How is the load-carrying capacity of a hydrodynamic contact calculated, and what is its relevance in comparing different contact designs?
- What is cavitation, and how does it affect lubricated contacts?
- What is the role of micro-geometry or roughness in influencing hydrodynamic pressure, and how can it be incorporated into lubrication models?
- Explain the concept of mixed lubrication and how it involves both hydrodynamic effects and solid-solid asperities in the presence of surface roughness.
- What is elastohydrodynamic lubrication, and how does it differ from hydrodynamic lubrication?
- What are the key parameters and equations used to estimate minimum film thickness in EHL contacts, and why are they important?
- In the context of EHL, how do the material, velocity, and load parameters affect the lubricant film thickness?

- What are other factors that can influence EHL contacts?
- What are the primary advantages of using oil-based lubricants, and how do they compare to water-based lubricants?
- What are the three main classes of oil-based lubricants, and how do they differ in terms of their composition and properties?
- How do synthetic oils differ from mineral oils, and what are some of the key characteristics of commonly used synthetic lubricants?
- Why are bio-lubricants gaining attention as alternatives to petroleum-based oils, and what are some examples of plant- and animal-based bio-lubricants?
- What role do additives play in enhancing the performance of lubricants, and what are some common types of additives used in lubricant formulations?
- What are the main functions of viscosity index improvers in lubricants, and how do they help improve the performance of lubricating oils?
- How do friction modifiers and anti-wear additives differ in their mechanisms and purposes within lubricating oils?
- What are some common examples of additives used in lubricants to prevent wear, reduce friction, inhibit oxidation, and protect against corrosion, and how do they enhance the overall performance of lubricants?
- What are some key advantages of water-based lubricants compared to traditional oil-based lubricants, and in which industries are they particularly beneficial?
- How can the performance of water-based lubricants be optimized, and what are the main challenges associated with using them?
- What is the role of ionic liquids in lubrication, and what advantages and challenges do they present compared to traditional liquid lubricants?
- What is superlubricity, and why is it a significant focus in reducing friction in lubricated systems?
- How do polar and ionic liquids, such as glycerol and glycol, contribute to achieving superlubricity, and what role do they play in passivating sliding interfaces?
- What strategies can be employed to address the aging of liquid lubricants, and why is routine oil analysis essential in this context?
- What are the main components of greases, and how do they differ from liquid lubricants in terms of application and characteristics?
- Describe the role of thickeners in greases and how they help maintain the lubricant's properties. What are some common types of thickeners used in greases?
- How does the rheological behavior of greases vary with shear rate and duration, and what is the significance of their yield stress in lubrication?
- What factors can impact the structural integrity of greases, leading to their failure, and how does temperature influence grease performance?
- Why is proper storage, handling, and maintenance of lubricants essential, and what methods can be employed to ensure the longevity and effectiveness of lubricants in machinery?

3.7   SOLUTIONS FOR THE EXERCISES

EXERCISE 3.1

$$Re = \frac{v \cdot L}{v} = \frac{\pi \cdot D \cdot n \cdot \dfrac{h}{2}}{v} = \frac{\pi \cdot 200 \text{ mm} \cdot \dfrac{1500}{60 \text{ s}} \cdot \dfrac{0.2 \text{ mm}}{2}}{46 \text{ mm}^2/\text{s}}$$

$$= 34.15 \ll 41.3 \cdot \sqrt{\frac{D}{c}} = 1306$$

→ The flow in the lubricant gap can be assumed to be laminar.

EXERCISE 3.2

- LLC of each pad: $LLC_{\text{pad}} = \dfrac{6 \cdot \eta \cdot u \cdot B \cdot L^2}{h_0^2 \cdot k^2}\left[-\ln(k+1) + \dfrac{2k}{k+2}\right] = 2.5 \text{ N}$

  → LLC of six pads: $LLC = 6 \cdot LLC_{\text{pad}} = 15 \text{ N}$

  Note: Misalignment would not create the same conditions for all pads in the application working under realistic conditions.

- $LLC_{\text{oil}} = \dfrac{\eta_{\text{oil}}}{\eta} LLC = 5000 \cdot LLC = 75 \text{ kN}$

  Note: This force is equivalent to the weight of 100 average adults.

- $\mu = \dfrac{k \cdot h_0}{L}\left[\dfrac{3k - 2(k+2)\cdot\ln(k+1)}{6k - 3(k+2)\cdot\ln(k+1)}\right] = 0.0012$

  Note: This is below the superlubricity threshold (< 0.01), which is desirable for bearing systems, thus indicating very efficient lubrication.

EXERCISE 3.3

- $G = \alpha_{\text{p}} \cdot E' \approx 5138$, with $E' = \dfrac{2}{\left(\dfrac{1-v_R^2}{E_R} + \dfrac{1-v_B^2}{E_B}\right)} = 273673 \dfrac{\text{N}}{\text{mm}^2}$

$$U = \frac{\eta_0 \cdot u_{\text{m}}}{E' \cdot R'} = \frac{\eta_0 \cdot \dfrac{\left(\dfrac{D_B}{2} \cdot 2\pi \cdot n_B + u_R\right)}{2}}{E' \cdot R'} \approx 1.8 \cdot 10^{-10},$$

$$W_{3D} = \frac{F}{E' \cdot R'^2} \approx 6.9 \cdot 10^{-4},$$

$$h_{\min} = H_{\min} \cdot R' \approx 3.68 \cdot G^{0.49} \cdot U^{0.68} \cdot W_{3D}^{-0.073} \left(1 - e^{-0.68 \cdot \frac{a}{b}} \right) R' \approx 15\,\mu m$$

- $\lambda = \dfrac{h_{\min}}{\sqrt{R_R^2 + R_B^2}} = 1.3$

- $1 < \lambda\, 3 \rightarrow$ The contact operates in the mixed lubrication regime with a significant amount of solid asperity contact.

REFERENCES

1. D. Bartel, *Simulation von Tribosystemen: Grundlagen und Anwendungen.* Vieweg+ Teubner research, Vieweg Verlag, Friedrich & Sohn Verlagsgesellschaft mbH, Braunschweig, 2010.
2. D. Dowson, G.R. Higginson, *Elasto-Hydrodynamic Lubrication: International Series on Materials Science and Technology,* Elsevier, Cambridge, MA, 2014.
3. B. Bode, Modell zur Beschreibung des Fließverhaltens von Flüssigkeiten unter hohem Druck, *Tribologie und Schmierungstechnik* 36 (1989) 182–189.
4. H. Vogel, Das temperaturabhangigkeitsgesetz der viskositat von flussigkeiten, *Physikalische Zeitschrift.* 22 (1921) 645–646.
5. C. Barus, Note on the dependence of viscosity on pressure and temperature, *Proceedings of the American Academy of Arts and Sciences, JSTOR* 27 (1891) 13–18.
6. H. Eyring, Viscosity, plasticity, and diffusion as examples of absolute reaction rates, *The Journal of Chemical Physics* 4 (1936) 283–291.
7. S. Bair, W.O. Winer, A rheological model for elastohydrodynamic contacts based on primary laboratory data. *Transactions of the ASME Journal of Lubrication Technology* 101 (1979): 25.
8. M.M. Cross, Rheology of non-Newtonian fluids: A new flow equation for pseudoplastic systems, *Journal of Colloid Science* 20 (1965) 417–437.
9. P.J. Carreau, Rheological equations from molecular network theories, *Transactions of The Society of Rheology* 16 (1972) 99–127.
10. K. Yasuda, R. Armstrong, R. Cohen, Shear flow properties of concentrated solutions of linear and star branched polystyrenes, *Rheologica Acta* 20 (1981) 163–178.
11. R. Larsson, O. Andersson, Lubricant thermal conductivity and heat capacity under high pressure, *Proceedings of the Institution of Mechanical Engineers, Part J: Journal of Engineering Tribology* 214 (2000) 337–342.
12. H. Herwig, *Strömungsmechanik*, Springer, Verlag, 2006.
13. H. Kruggel-Emden, S. Wirtz, V. Scherer, A study on tangential force laws applicable to the discrete element method (DEM) for materials with viscoelastic or plastic behavior, *Chemical Engineering Science* 63 (2008) 1523–1541.
14. O. Reynolds, IV. On the theory of lubrication and its application to Mr. Beauchamp tower's experiments, including an experimental determination of the viscosity of olive oil, *Philosophical Transactions of the Royal Society of London* 177 (1886) 157–234.
15. R. Gohar, *Elastohydrodynamics Ellis Horwood Ltd.(England)*, John Wiley &Sons, Chichester, 1988.
16. B. Najji, B. Bou-Said, D. Berthe, New formulation for lubrication with non-Newtonian fluids *Journal of Tribology* 111(1) (1989) 29–34.

17. Y. Peiran, W. Shizhu, A generalized Reynolds equation for non-Newtonian thermal elastohydrodynamic lubrication, *Journal of Tribology* 112(4) (1990) 631–636.

18. W. Habchi, *Finite Element Modeling of Elastohydrodynamic Lubrication Problems*, John Wiley & Sons, Chichester, 2018.

19. B.J. Hamrock, S.R. Schmid, B.O. Jacobson, *Fundamentals of Fluid Film Lubrication*, CRC Press, Boca Raton, FL, 2004.

20. A. Sommerfeld, Zur hydrodynamischen theorie der schmiermittelreibung, *Zeitschrift für Mathematik und Physik* 50 (1904) 155.

21. L. Gümbel, Verleich der Ergebnisse der rechnerischen Behandlung des Lagerschmierungsproblem mit neueren Versuchsergebnissen, *Monatsblätter Berliner Bezirk (VDI)* (1921) 125–128.

22. H.W. Swift, The stability of lubricating films in journal bearings. (Includes appendix), in: *Minutes of the Proceedings of the Institution of Civil Engineers*, Thomas Telford-ICE Virtual Library, 1932, pp. 267–288.

23. B. Jakobsson, L. Floberg, The finite journal bearing, considering vaporization (Das Gleitlager von endlicher Breite mit Verdampfung), Report No. 3 from the Institute of Machine Elements, Chalmers Univ. of Technology, Gothenburg, Sweden (1957).

24. K.-O. Olsson, Cavitation in dynamically loaded bearing, *Transaction of Chalmers University of Technology, Sweden* 308 (1957).

25. L. Floberg, *On Journal Bearing Lubrication Considering the Tensile Strength of the Liquid Lubricant*, Lund Technical University, Lund, 1973.

26. H.G. Elrod, A cavitation algorithm, *ASME Journal of Lubrication Technology* 103(3) (1981) 350–354.

27. N. Patir, H. Cheng, An average flow model for determining effects of three-dimensional roughness on partial hydrodynamic lubrication, *Journal of Lubrication Technology* 100(1) (1978) 12–17.

28. N. Patir, H. Cheng, Application of average flow model to lubrication between rough sliding surfaces, *Journal of Lubrication Technology* 101(2) (1979) 220–229.

29. M. Schouten, H. Van Leeuwen, Die Elastohydrodynamik: Geschichte und Neuentwicklungen, Gleitwälzkontakte: Grundlagen und Stand der Technik bei Wälzlagern, Zahnrädern und Nockenstösseln sowie stufenlos übersetzenden Getrieben (CVT), in: *Tagung Fulda*, 4–5 Oktober 1995, VDI-Verlag 1995, pp. 1–47.

30. A. Jolkin, R. Larsson, Film thickness, pressure distribution and traction in sliding EHL conjunctions, *Tribology Series*, (1999) 505–516.

31. M. Marian, M. Bartz, S. Wartzack, A. Rosenkranz, Non-dimensional groups, film thickness equations and correction factors for elastohydrodynamic lubrication: A review, *Lubricants* 8 (2020) 95.

32. C.H. Venner, A. Lubrecht, Multigrid techniques: A fast and efficient method for the numerical simulation of elastohydrodynamically lubricated point contact problems, *Proceedings of the Institution of Mechanical Engineers, Part J: Journal of Engineering Tribology* 214 (2000) 43–62.

33. C. Venner, A. Lubrecht, *Multilevel Methods in Lubrication*, Elsevier, Amsterdam, Netherlands, 2000.

34. W. Habchi, I. Demirci, D. Eyheramendy, G. Morales-Espejel, P. Vergne, A finite element approach of thin film lubrication in circular EHD contacts, *Tribology International* 40 (2007) 1466–1473.

35. D. Dowson, G. Higginson, The effect of material properties on the lubrication of elastic rollers, *Journal of Mechanical Engineering Science* 2 (1960) 188–194.

36. D. Dowson, G. Higginson, A. Whitaker, Elasto-hydrodynamic lubrication: A survey of isothermal solutions, *Journal of Mechanical Engineering Science* 4 (1962) 121–126.

37. H. Moes, Discussion on paper D1 by D, Dowson. *Proceedings of the Institution of Mechanical Engineers* 180 (1966) 244–245.

38. K. Johnson, Regimes of elastohydrodynamic lubrication, *Journal of Mechanical Engineering Science* 12 (1970) 9–16.

39. H. Moes, Optimum similarity analysis with applications to elastohydrodynamic lubrication, *Wear* 159 (1992) 57–66.

40. H. Moes, *Lubrication and Beyond-University of Twente Lecture Notes code 115531*, University of Twente, Enschede, The Netherlands, 2000.

41. B.J. Hamrock, D. Dowson, Ball bearing lubrication: The elastohydrodynamics of elliptical contacts, *Journal of Lubrication Technology* 104(2) (1981) 279–281.

42. R. Chittenden, D. Dowson, J. Dunn, C. Taylor, A theoretical analysis of the isothermal elastohydrodynamic lubrication of concentrated contacts. I. Direction of lubricant entrainment coincident with the major axis of the Hertzian contact ellipse, *Proceedings of the Royal Society of London. A. Mathematical and Physical Sciences* 397 (1985) 245–269.

43. G. Nijenbanning, C.H. Venner, H. Moes, Film thickness in elastohydrodynamically lubricated elliptic contacts, *Wear* 176 (1994) 217–229.

44. M. Torbacke, E. Kassfeldt, *Lubricants: Introduction to Properties and Performance*, John Wiley & Sons (Chichester, England), 2014.

45. J.W. Miller, Synthetic lubricants and their industrial applications, *Journal of Synthetic Lubrication* 1 (1984) 136–152.

46. C. Kajdas, Industrial lubricants, in: *Chemistry and Technology of Lubricants*, Springer, Verlag, 1992, pp. 228–263.

47. V. Stepina, V. Vesely, *Lubricants and Special Fluids*, Elsevier, Cambridge, MA, 1992.

48. T. Bartels, W. Bock, Gear lubrication oils, in: T. Mang, W. Dresel, (Eds), *Lubricants and Lubrication*. 3rd ed., Wiley-VCH Verlag GmbH & Co. KGaA, Weinheim, 2017, 293–344.

49. B. Bhushan, *Principles and Applications of Tribology*, John Wiley & Sons, Chichester, 1999.

50. B.K. Sharma, K.M. Doll, S.Z. Erhan, Ester hydroxy derivatives of methyl oleate: Tribological, oxidation and low temperature properties, *Bioresource Technology* 99 (2008) 7333–7340.

51. R. Al Sulaimi, A. Macknojia, M. Eskandari, A. Shirani, B. Gautam, W. Park, P. Whitehead, A.P. Alonso, J.C. Sedbrook, K.D. Chapman, D. Berman, Evaluating the effects of very long chain and hydroxy fatty acid content on tribological performance and thermal oxidation behavior of plant-based lubricants, *Tribology International* 185 (2023) 108576.

52. Q. Zeng, The lubrication performance and viscosity behavior of castor oil under high temperature, *Green Materials* 10 (2021) 51–58.

53. H. Saleh, M.Y. Selim, Improving the performance and emission characteristics of a diesel engine fueled by jojoba methyl ester-diesel-ethanol ternary blends, *Fuel* 207 (2017) 690–701.

54. K. Suthar, Y. Singh, A.R. Surana, V.H. Rajubhai, A. Sharma, Experimental evaluation of the friction and wear of jojoba oil with aluminium oxide (Al2O3) nanoparticles as an additive, *Materials Today: Proceedings* 25 (2020) 699–703.

55. E. Durak, A study on friction behavior of rapeseed oil as an environmentally friendly additive in lubricating oil, *Industrial Lubrication and Tribology* 56 (2004) 23–37.

56. H. Mobarak, E.N. Mohamad, H.H. Masjuki, M.A. Kalam, K. Al Mahmud, M. Habibullah, A. Ashraful, The prospects of biolubricants as alternatives in automotive applications, *Renewable and Sustainable Energy Reviews* 33 (2014) 34–43.

57. A. Martini, U.S. Ramasamy, M. Len, Review of viscosity modifier lubricant additives, *Tribology Letters* 66 (2018) 1–14.

58. A. Shirani, S. Berkebile, D. Berman, Promoted high-temperature lubrication and surface activity of polyolester lubricant with added phosphonium ionic liquid, *Tribology International* 180 (2023) 108287.

59. J. Zhang, H. Spikes, On the mechanism of ZDDP antiwear film formation, *Tribology Letters* 63 (2016) 1–15.
60. H. Spikes, Low-and zero-sulphated ash, phosphorus and sulphur anti-wear additives for engine oils, *Lubrication Science* 20 (2008) 103–136.
61. A. Papay, Antiwear and extreme-pressure additives in lubricants, *Lubrication Science* 10 (1998) 209–224.
62. N. Canter, Special report: Trends in extreme pressure additives, *Tribology and Lubrication Technology* 63 (2007) 10.
63. M.H. Rahman, H. Warneke, H. Webbert, J. Rodriguez, E. Austin, K. Tokunaga, D.K. Rajak, P.L. Menezes, Water-based lubricants: Development, properties, and performances, *Lubricants* 9 (2021) 73.
64. H. Ji, X. Zhang, T. Tan, Preparation of a water-based lubricant from lignocellulosic biomass and its tribological properties, *Industrial & Engineering Chemistry Research* 56 (2017) 7858–7864.
65. M. Barisik, S. Atalay, A. Beskok, S. Qian, Size dependent surface charge properties of silica nanoparticles, *The Journal of Physical Chemistry C* 118 (2014) 1836–1842.
66. Y. Cui, M. Ding, T. Sui, W. Zheng, G. Qiao, S. Yan, X. Liu, Role of nanoparticle materials as water-based lubricant additives for ceramics, *Tribology International* 142 (2020) 105978.
67. H.T. Nguyen, K.-H. Chung, Assessment of tribological properties of Ti3C2 as a water-based lubricant additive, *Materials* 13 (2020) 5545.
68. T. Liu, C. Zhou, C. Gao, Y. Zhang, G. Yang, P. Zhang, S. Zhang, Preparation of Cu@ SiO2 composite nanoparticle and its tribological properties as water-based lubricant additive, *Lubrication Science* 32 (2020) 69–79.
69. X. Wang, F. Zhao, L. Huang, Low temperature dehydration of glycerol to acrolein in vapor phase with hydrogen as dilution: From catalyst screening via TPSR to real-time reaction in a fixed-bed, *Catalysts* 10 (2019) 43.
70. A.E. Somers, P.C. Howlett, D.R. MacFarlane, M. Forsyth, A review of ionic liquid lubricants, *Lubricants* 1 (2013) 3–21.
71. A. Shirani, Y. Li, O.L. Eryilmaz, D. Berman, Tribocatalytically-activated formation of protective friction and wear reducing carbon coatings from alkane environment, *Scientific Reports* 11 (2021) 20643.
72. A. Shirani, Y. Li, J. Smith, J. Curry, P. Lu, M. Wilson, M. Chandross, N. Argibay, D. Berman, Mechanochemically driven formation of protective carbon films from ethanol environment, *Materials Today Chemistry* 26 (2022) 101112.
73. D. Berman, A. Erdemir, Achieving ultralow friction and wear by tribocatalysis: Enabled by in-operando formation of nanocarbon films, *ACS Nano* 15 (2021) 18865–18879.
74. A. Ayyagari, D. Berman, K.I. Alam, A. Erdemir, Progress in superlubricity across different media and material systems-a review, *Frontiers in Mechanical Engineering* 8 (2022).
75. D. Berman, A. Erdemir, A.V. Sumant, Approaches for achieving superlubricity in two-dimensional materials, *ACS Nano* 12 (2018) 2122–2137.
76. X. Chen, T. Kato, M. Kawaguchi, M. Nosaka, J. Choi, Structural and environmental dependence of superlow friction in ion vapour-deposited aC: H: Si films for solid lubrication application, *Journal of Physics D: Applied Physics* 46 (2013) 255304.
77. K. Jacques, T. Joy, A. Shirani, D. Berman, Effect of water incorporation on the lubrication characteristics of synthetic oils, *Tribology Letters* 67 (2019) 105.
78. P.M. Lugt, A review on grease lubrication in rolling bearings, *Tribology & Lubrication Technology* 66 (2010) 44–48, 50–56.
79. T.E. Karis, R.N. Kono, M.S. Jhon, Harmonic analysis in grease rheology, *Journal of Applied Polymer Science* 90 (2003) 334–343.
80. A.E. Yousif, Rheological properties of lubricating greases, *Wear* 82 (1982) 13–25.

81. V. Wikström, E. Höglund, Starting and steady-state friction torque of grease-lubricated rolling element bearings at low temperatures-part II: Correlation with less-complex test methods, *Tribology Transactions* 39 (1996) 684–690.
82. G. Dalmaz, R. Nantua, An evaluation of grease behavior in rolling bearing contacts, *Lubrication Science* 43 (1987) 905–915.
83. X. Li, F. Guo, G. Poll, Y. Fei, P. Yang, Grease film evolution in rolling elastohydrodynamic lubrication contacts, *Friction* 9 (2021) 179–190.

4 Surface Engineering

Tribological properties of common engineering materials such as metals (in particular steels) or polymers can be significantly improved by tailoring their surface properties while retaining or keeping their bulk unchanged. This approach is widely used in industry since the tribological performance is a surface phenomenon. Therefore, treating the surface up to a few micrometers (or even as thin as nanometers in some cases) can substantially improve the application worthiness at a fraction of the cost relative to the entire bulk of the material being modified with an expensive or exotic material. The newly created or modified boundary layer thus provides functional aspects, such as wear resistance, friction control, corrosion resistance, (fatigue) strength increase, temperature resistance, biocompatibility, among others, as well as some decorative aspects, such as color, gloss, flatness, among others. As already outlined, the properties that affect the tribological performance of mechanical components are surface roughness, hardness, chemical composition or activity, and applied stresses. To tailor the needed characteristics and to control these properties as well as the resulting tribological performance, the processes used for surface modifications can be achieved

- without changing the chemical composition in the boundary layer of the material,
- with a change of the chemical composition in the boundary layer of the material, but with the original material still being present, and
- by applying another material.

Furthermore, typically employed approaches differ if they form a layer or not as well as in how they affect materials in terms of depth (Figure 4.1). The depth of the modification is an important aspect as it determines how far beneath the surface the properties, both physical and chemical, change. For example, in Chapter 1, we learned that the contact size and pressure depend on the mechanical characteristics of materials in contact. At the same time, the friction-induced temperature increase is dependent on the thermal conductivity of the surfaces and bulk material. And further, the thickness of the modifications affects shearing of the material, thus friction, while hardness, contributes to its wear resistance.

Surface engineering methods can be further divided into three categories:

- **Surface Finishing and Texturing**: Modifying the surface topography by polishing, patterning, dimpling, etching, or other means.
- **Surface Hardening**: Surface heat treatments, such as carburizing, nitriding, and boriding followed by hardening and tempering; and surface melt-freezing processes such as electron beam or laser melting.
- **Surface Coatings**: Applying metallurgical, ceramic, or polymeric-based coatings as well as their composites.

DOI: 10.1201/9781003397519-4

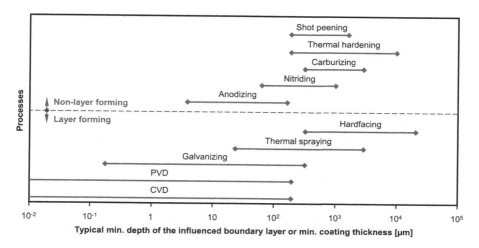

FIGURE 4.1 Thickness of the influenced boundary layer or coating thickness from various non-layer or layer-forming surface engineering processes.

These approaches can be alternatives for an application (see Video 4.1 in ebook+), but can also be partly overlapping and various methods are frequently combined. In the following, we discuss each approach in detail and summarize the common solutions and practices involved.

4.1 SURFACE FINISHING AND TEXTURING

Surface finishing refers to a set of processes and techniques applied to the surface of a material to improve its properties, appearance, or both. The goal is to achieve specific functional or aesthetic qualities, making the material suitable for various applications. Surface finishing methods vary widely and can be categorized into mechanical, chemical, and electrochemical processes. The latter, such as **chemical etching** or **electropolishing**, harness chemical reactions and electrical potentials to precisely alter the involved material surfaces, resulting in improvements ranging from smoother surfaces, corrosion resistance to aesthetic aspects. These methods can also be employed for cleaning or degreasing samples or parts.

Mechanical surface finishing methods (Figure 4.2) focus on altering the topography of a material through physical means. Thereby, mechanical techniques are typically ablative, path determined manufactured processes with geometrically undefined cutting edges that basically are tribological systems that rely on wearing-off in a controlled way by harder counter-bodies. In this sense, **grinding** employs chemically inert abrasive particles with a high hardness to remove material. The latter can be controlled by the selection of grain size and distribution on the sandpaper, however strongly depends on the hardness and brittleness of the processed material. **Polishing** also utilizes abrasives and polishing compounds to create smooth, reflective surfaces in metals, ceramics, and polymers. Frequently, various grinding and polishing stages are combined to achieve the desired surface qualities.

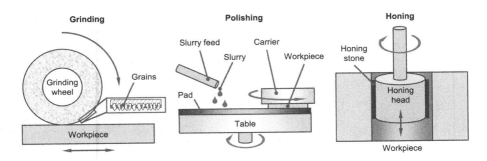

FIGURE 4.2   Schematic illustrations of various mechanical surface finishing processes.

While grinding rather leads to anisotropic surfaces with a preferential orientation (in the main direction of abrasive grain movement), polishing can achieve isotropic surfaces (Section 1.4.2). **Lapping** is also a precision method using abrasive slurries and plates for achieving high-precision flat surfaces, crucial for precision components, optical lenses, and semiconductor wafers. **Honing** and **superfinishing** employ abrasive grains to create a fine surface texture and improve geometric form, commonly used in automotive and aerospace industries for cylinders. These processes also allow to tailor the orientation of roughness and fabricate, e.g., cross-hatched patterns (stochastic surface textures).

Similarly, surface texturing is a technique used to modify the surface topography of a material to enhance its performance, which particularly holds true for adsorption enhancement as well as friction and wear reduction. The primary goal of surface texturing is to create micro-scale discrete and/or periodic features that can alter the interaction between two contacting surfaces. These features can be isotropic, involving the distribution of uniformly spaced micro-dimples or micro-cavities across the surface, or anisotropic, consisting of aligned structures with preferential orientation. Typically, surface textures are deeper compared to roughness features induced by finishing processes (e.g., grinding, polishing, or honing). Moreover, the lateral texture dimensions are at least one order of magnitude larger than the roughness features. Common methods include laser surface texturing (LST), micro-milling, chemical etching, and electrochemical machining. The proper choice of the texturing technique depends on factors such as precision requirements, material properties, cost-effectiveness, and scalability as illustrated Figure 4.3.

The advantages of surface texturing to reduce friction and wear are multifaceted and depend on the underlying lubrication regime. In the case of dry friction, surface textures help to reduce the area of contact and adhesion thus reducing friction. Moreover, surface textures act as a trap for generated wear debris thus removing it from the tribological contact zone, which is beneficial for friction and wear. Under boundary lubrication, surface textures can help to generate pressure-induced tribo-films due to locally increased contact pressures. The lubricant held within the textured features also forms a protective film that prevents metal-to-metal contact, reducing wear and extending the lifespan of the material. As a result, this allows for the reduction of direct contact between surfaces, improving the lubrication

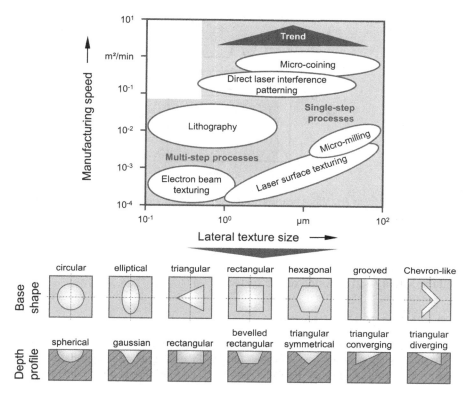

FIGURE 4.3 Classification of different surface texturing techniques regarding the achievable lateral texture size and fabrication speed (top) as well as typical base shapes and bottom profiles of surface textures (bottom). (Redrawn and extended from Ref. [1].)

efficiency, and minimizing friction. Concerning mixed lubrication, surface textures reduce friction and wear by reducing the real area of contact, trapping generated wear debris, and acting as a lubricant reservoir to be easily resupplied when needed. Under lubricated conditions, surface textures may induce changes in the respective flow pattern thus creating more laminar flows instead of turbulent ones. Moreover, textures can build-up an additional hydrodynamic pressure thus generating an additional load-bearing capacity, even for nominally parallel surfaces (Figure 4.4).

In the case of anisotropic textures, such as grooves or patterns with preferential orientation, they can also influence the directionality of the contact forces and guide the lubricant flow. This directional control of lubrication aids in minimizing friction and wear by optimizing the distribution and retention of the lubricant film between the involved surfaces.

The applicability of surface texturing is wide, since it can be easily adapted to various materials including metals and alloys, polymers, ceramics, and composites, thus providing benefits for multiple industries. For instance, in automotive applications, surface texturing is used in engine components, such as piston rings and cylinder liners, to reduce friction and improve fuel efficiency. In manufacturing processes,

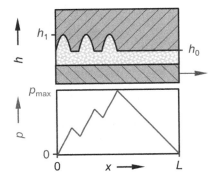

FIGURE 4.4 Linear HL slider bearing with dimpled contact inlet and the resulting hydrodynamic pressure distribution.

texturing is utilized on cutting tools to reduce the tool wear and enhance the machining performance. In biomedical systems, surface texturing allows for control of the cellular activity and adhesion on the surfaces.

4.2 SURFACE HARDENING

As already demonstrated, an enhanced hardness generally imparts a greater wear resistance,[1] such as abrasive, adhesive, and surface fatigue, and degradation under sliding or rolling conditions. However, an increased hardness could be only required for a few to several hundred microns from the surface, where the maximum Hertzian contact pressures are experienced. At the same time, hardening the whole component is not desirable to prevent changes in the mechanical performance of the system since an elevated hardness reduces the material's toughness, ductility, and ability to arrest crack growth. Even more, bulk hardening is more expensive from a processing point of view since working with the hardened parts requires extra steps. Therefore, imparting or increasing the superficial hardness only to a certain extent is extensively adopted in industry. Generally, available methods can be clustered in

- **Mechanical bombardment**, e.g., shot peening,
- **Thermal hardening**, e.g., flame hardening, induction hardening,
- **Thermochemical diffusion processes**, e.g., carburizing, chromium plating, boriding, nitriding, carbonitriding,
- **Plasma processes**, e.g., carburization, polymerization, nitration, and
- **Beam processes**, e.g., laser, electron, or ion radiation.

Shot peening/blasting (Figure 4.5) involves bombarding the surface of a material with small, high-velocity spherical particles, such as steel shots or ceramic beads. This process induces compressive residual stresses in the surface layer, which enhances the material's resistance to fatigue and stress corrosion cracking. Shot peening also induces work hardening, inducing an increased surface hardness and improved wear resistance.

---

[1] This might, however, imply an increased wear of the counter-body.

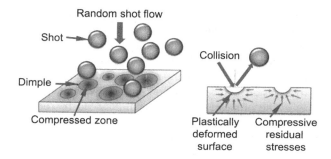

FIGURE 4.5 Schematic illustration of shot peening. (Reprinted and adapted from Ref. [2] with permission.)

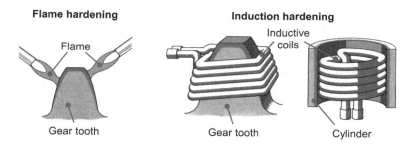

FIGURE 4.6 Schematic illustration of flame and induction hardening. (Adapted from Ref. [3].)

Flame hardening (Figure 4.6) involves heating the surface of a material using an oxyfuel flame and then rapidly quenching it. The localized heating and subsequent quenching create a hardened layer. Flame hardening is performed on parts made of mild steel, alloy steel, medium carbon steel, and cast iron. This method is commonly used for large components, such as shafts, rolls, and machine tool guides. Among the disadvantages of the process are that flame hardening cannot be applied as precisely as other hardening techniques and the process may lead to oxidation and decarburization of the material. **Induction hardening** utilizes the principle of electromagnetic induction to heat the surface of a material rapidly, see Figure 4.6. This localized heating is followed by quenching, a rapid cooling process that creates a hardened layer. This approach is often used for parts with complex shapes, such as gears and camshafts, as it allows selective hardening of specific areas while preserving the properties of the underlying material. In this regard, the parts to be treated are placed inside a copper coil and heated above their phase transformation temperature by applying alternating current to the coil that induces an alternating magnetic field within the part leading to the heating of the surface.

Case hardening is achieved by the formation of a surface layer with a different chemical composition, which is similar to the overall approach of depositing coatings. These techniques result in retaining a softer core with enhanced toughness and ductility properties (bulk) while having, at the same time, a hard case that is wear-resistant with a superior load-bearing ability, realizing the best of two materials in one. A harder case on a softer core also results in residual compressive stresses that can be beneficial in tribological contacts. Thereby, atoms diffuse into the base material, forming a hardened layer, while maintaining a relatively softer and more ductile core thus improving the fatigue strength and wear resistance. Since the driving forces for diffusion processes are thermodynamic potential differences and concentration gradients, the processes are typically carried out at elevated temperatures in a controlled atmosphere. The processes are further influenced by the type of involved substances, atomic diameters, lattice structures, and diffusion paths. A prerequisite for the occurrence of diffusion is that the involved substances/materials can form solid solutions. The type of solid solution simultaneously determines the possibilities of movement of the foreign atoms in the lattice. A distinction is made between intercalation and substitutional solid solutions, see Figure 4.7.

Intercalation involves the insertion of foreign atoms or molecules between existing atomic layers without significantly disrupting the original crystal structure. The solubility is generally limited to the ‰ range and irregular. The atom's diameters of the intercalated substances must be smaller than those of the host lattices, e.g., He, H, B, C, B, N, and O. The intercalated material alters the material's properties, such as hardness and wear resistance, by interacting with the lattice. One example is **carburizing**, for which low-carbon steels (due to the required concentration difference) or iron-based alloys used for gears, camshafts, etc., are subjected to a carbon-rich gas atmosphere (methane or propane) or immersed in a molten bath of carburizing material (cyanide or carbonate salt) at temperatures between 900°C and 950°C and carbon atoms are intercalated between lattice planes. After sufficient carbon has diffused into the surface layer, the component is rapidly cooled or quenched, potentially followed by tempering. An exemplary hardness versus depth profile is depicted in Figure 4.8.

Substitutional solid solutions occur when foreign atoms replace some of the host material's original atoms in the crystal lattice, which results in changes to the material's properties, such as hardness and alloy composition. The atomic radii must be of a similar order of magnitude (about ±14%) and the lattice structure should be the same

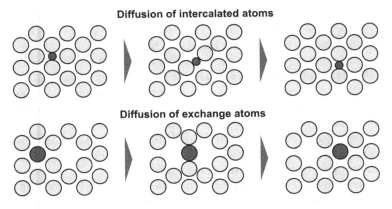

FIGURE 4.7 Illustration of diffusion processes of foreign atoms to form solid solutions, which depend on the respective size of the diffusing atoms.

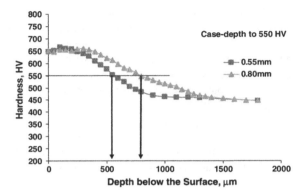

FIGURE 4.8 Typical hardness-depth profiles for case-carburized steel with case-depth of 0.55 and 0.80 mm, respectively. (Reprinted from Ref. [4] with permission.)

or similar (Hume-Rothery rules). This can happen as ring exchange, in which atoms exchange positions within a closed loop or ring of atoms, or missing/void diffusion, in which atoms move through vacancies in the crystal lattice. Examples connect with the replacement of iron atoms in the crystal lattice by elements like chromium or nickel. For chromium plating, the components (typically made of low-carbon steel) are packed in a mixture of inert mineral filler and a chromium donor (e.g., chromium halide) or chromium powder. At a treatment temperature of around 1000°C, the chromium compound evaporates, reaches the surface of the workpiece as a gas dissociates and diffuses into the surface to form a chromium-rich surface layer (up to 35%), typically up to a depth of 50–200 μm, providing an excellent corrosion resistance. In **boronizing**, metal components are exposed to a boron-rich environment (boron-rich powder or boron-containing gases like boron trifluoride or boron trichloride) and heated to temperatures between 800°C and 1100°C. The diffused boron atoms combine with the metal atoms at the surface to form hard borides, such as iron borides, which increase the surface hardness and wear resistance of the material. However, borided surface layers cannot withstand high pressures under rolling loads and the corrosion resistance is low in oxidizing media. In the case of **nitriding**, the component is placed in a nitriding atmosphere, typically composed of ammonia or nitrogen gas. At temperatures of around 500°C–600°C, nitrogen atoms from the atmosphere diffuse into the surface of the material. These atoms combine with the metal atoms at the surface to form hard and wear-resistant nitrides, typically iron nitrides. Before nitriding, the components should be finished, quenched, and tempered already, since the thin iron nitride layers do not permit reworking. These nitriding treatments find applications in various industries, including automotive (gears, crankshaft), aerospace, and tool manufacturing (dies). The diffusion of nitrogen atoms can be combined with carbon atoms (**nitrocarburizing**[2]) to create a balanced surface layer with carbides and nitrites in a single process.

---

[2] Not to be confused with carbonitriding, where primarily carbon diffuses into the material, which is accelerated by the nitrogen.

Plasma processes, utilizing ionized gases to introduce specific elements, can also be employed to form diffusion layers, e.g., **plasma nitriding**. These processes tend to be more eco-friendly, allow to treat a wide range of materials, and typically operate at lower temperatures compared to conventional heat treatments, thus minimizing thermal distortion and preserving material properties.

Beam-based processes have emerged as highly effective methods for surface hardening. **Laser surface hardening** involves directing a focused laser beam onto the material, rapidly heating the surface to critical temperatures, and then quenching to create a hardened layer. In turn, **electron beam processes** utilize high-energy electron beams to selectively heat the material's surface, and **ion beam techniques** involve bombarding the material with high-energy ions, leading to controlled surface modifications. These methods offer several advantages, including precision, minimal thermal distortion, and rapid processing. Laser surface hardening, in particular, allows for localized treatment, making it suitable for intricate geometries. Electron beam processes provide deep penetration, making them effective for higher hardening depths. Ion beam techniques offer versatility in tailoring material properties, enabling fine control over hardness, wear resistance, and other characteristics. These beam-based processes find applications in diverse industries, from automotive and aerospace to tool manufacturing.

4.3   COATINGS

Surface hardening approaches utilizes the principle of modifying the material's superficial layers to improve its mechanical/tribological. This, however, still largely depends on the materials' composition and the modified layer is rather gradual, while it is difficult to control its thickness. A more general approach to control friction and wear relates to the application of coatings as permanently adhering uniform films, both thick and thin, or as powders of lubricating particles, allowing to fulfill the required properties in different locations of tribological contacts (Figure 4.9).

Thus, coatings help to reduce friction thus minimizing the involved energy losses due to the notably changed mechanical properties and shear resistance of the coating compared to the bulk substrate. Due to the enhanced hardness of the coating, this approach tends to enhance the performance and lifespan of components and reduce the need for frequent maintenance. The proper selection of an appropriate coating depends on the specific application requirements, operating conditions, and the properties of the underlying substrate. In the contact of two surfaces, either one or both of which are coated, there are four key parameters that govern the tribological behavior:

- Hardness relationship between the coating and substrate
- Thickness of the coating
- Surface roughness
- Size and hardness of any debris present in the contact zone (due to wear or external sources)

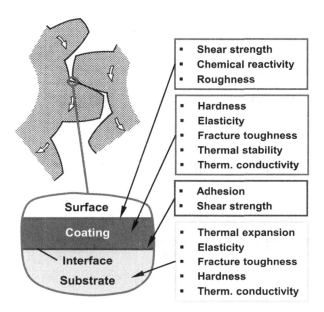

FIGURE 4.9 Tribologically important properties in different zones of the coated surface. (Reprinted and adapted from Ref. [5] with permission.)

The interplay among these aspects gives rise to various contact conditions, each characterized by specific tribological contact mechanisms. Some typical tribological conditions influencing the friction and wear mechanisms of a hard spherical slider on a coated flat surface are schematically illustrated in Figure 4.10.

Some further important points that need to be taken into consideration when selecting appropriate coatings and their deposition strategy include the following:

- Deposited coatings should have necessary protective properties (resist wear, high temperature, corrosion, etc.).
- The coating-surface treatment should not impair the properties of the bulk material.
- Special attention must be paid to the appropriate design of the coating-substrate interface (adhesion, compatibility of thermal and mechanical properties, among others).
- The deposition process should be capable of coating the desired parts, in terms of both size and shape.
- Some changes in substrate material specifications may be necessary to accommodate the coating.
- The surface treatment should be cost-effective, but this judgment should include factors the improved end-product characteristics in addition to the coating costs.
- It is important to consider several coatings techniques to solve the problem before a final selection of the treatment is made.

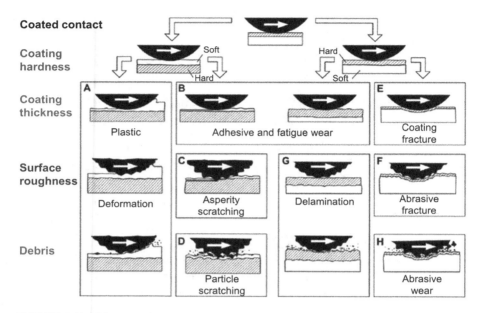

FIGURE 4.10 Macromechanical contact conditions for different mechanisms that influence friction, when a hard spherical slider moves on a coated flat surface. (Reprinted and adapted from Ref. [5] with permission.)

In this regard, the available potential material options for coatings are broad with more and more materials being developed on an everyday basis and highly depend on the selected deposition processes.

4.3.1 Deposition Methods

The available deposition techniques for each material vary largely and can be divided into three sub-groups, which include the deposition from the **gaseous state**, **solution state**, and **molten or semi-molten state** (Figure 4.11). Usually, the selection of the appropriate coating method is made in agreement with the application requirements

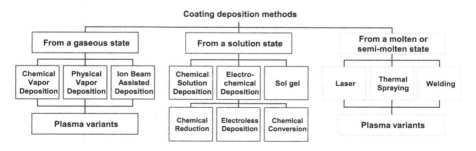

FIGURE 4.11 Different coating deposition methods summarized based on their conceptual approach.

and compatibility with the substrate materials. The methods can be combined to gain better control of the structure and composition.

In **Chemical Vapor Deposition (CVD)**, a precursor gas or vapor containing the desired elements is introduced into a reaction chamber together with the substrate. The interaction between the precursor gas and the substrate is pivotal, triggering chemical reactions at the substrate surface, which manifest as pyrolysis, decomposition, or reactions with co-reactants. The outcome of these reactions is the generation of solid-phase constituents, forming a thin film on the substrate. CVD allows for the deposition of diverse materials, including metals, semiconductors, and ceramics. Even complex geometries and undercuts can be coated uniformly. However, CVD might require the presence of a catalyst on the substrate surface (e.g., Cu, Ni, Fe), and the high temperatures involved (700°C–1000°C) might lead to component distortion. This disadvantage can partly be overcome by plasma assistance (PE-CVD), which provides the activation energy at lower temperatures (down to 200°C). A schematic of the process is shown in Figure 4.12 (left). **Physical Vapor Deposition (PVD)** takes place in a vacuumized reaction chamber, ensuring a controlled environment free from contaminants. In **thermal evaporation**, a solid material, often in the form of a wire or pellet, is heated to a vaporization point. This vaporized material then condenses on the substrate, forming a thin film. This method is particularly effective for materials with lower melting points. In **electron beam evaporation**, an electron beam is used to heat the material, promoting vaporization, see Figure 4.12 (middle). This technique offers precise control over the evaporation process and is suitable for a wide range of materials, including metals and ceramics. **Sputtering**, another PVD method, involves bombarding a target material with high-energy ions, typically argon. This bombardment dislodges atoms from the target, and these atoms then deposit on the substrate. Unbalanced Magnetron Sputtering (UBM), a variant depicted in Figure 4.12 (right), enhances the process by using magnetic fields to confine the plasma and improve coating uniformity. To further enhance PVD processes, reactive sputtering introduces reactive gases (e.g., nitrogen or oxygen) during deposition to create compound films. This enables the deposition of materials like nitrides and oxides. The choice between evaporation and sputtering depends

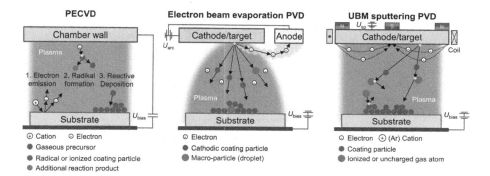

FIGURE 4.12 Schematics of the processes in PE-CVD (left) as well as in electron beam evaporation (middle) and UBM sputtering PVD (right).

on factors like the material, desired film properties, and application requirements. **Ion Beam–Assisted Deposition** (IBAD) is a thin-coating deposition technique that involves the simultaneous use of an ion beam and traditional deposition methods. In this process, a deposition source, such as an electron beam evaporator or sputtering system, is utilized to deposit material onto the substrate. Concurrently, an ion beam source introduces energetic ions, typically inert gases like argon, which bombards the growing film on the substrate, influencing its properties by enhancing adatom mobility, promoting densification, and improving adhesion. In deposition methods from the gaseous state, the deposition rate, coating structure, and properties are controlled by adjusting parameters such as temperature, time, pressure, and gas or target composition. Additionally, they can be used to produce thin coatings in the micro- or even nanometer range as well as multilayered films, enabling the creation of complex structures with specific functionalities. Furthermore, these techniques have been proven to produce tribologically effective coatings on a commercial scale.

   Chemical Solution Deposition (CSD) is a method used to produce coatings by dissolving precursor materials in a liquid solution, forming a thin film on the substrate through chemical reactions. The solution, containing metalorganic compounds, is deposited onto the substrate, and the subsequent heat treatment drives the decomposition and formation of the desired coating. **Electrochemical deposition**, also known as electroplating, involves the electrodeposition of a metal coating onto a conductive substrate through the use of an electrolyte solution and an electric current. As illustrated in Figure 4.13, the substrate acts as a cathode, and the metal ions from the electrolyte are reduced and deposited onto its surface, creating a conformal coating with controlled thickness. This requires surfaces to be electrically conductive. **Sol-gel** involves the transformation of a colloidal suspension of nanoparticles into a gel and then into a solid thin film. The sol is applied to the substrate, and controlled hydrolysis and condensation reactions occur, resulting in a gel. Subsequent drying and heat treatment yields the final coating, which enables the control of composition, thickness, and porosity.

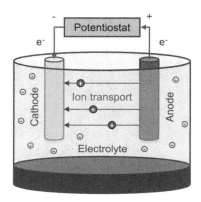

FIGURE 4.13   Schematic of the electrochemical deposition process.

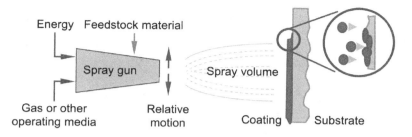

FIGURE 4.14 Schematic illustration of the thermal spraying.

Laser Cladding is a coating technique that involves a high-power laser to melt and fuse a coating material onto a substrate. The laser beam is precisely controlled, allowing for localized and well-defined coatings. **Welding** for coating involves the application of molten or partially molten material onto a substrate through fusion. This is achieved by melting the coating material and the substrate simultaneously, resulting in a metallurgical bond. Both, laser cladding and welding, can also be employed to repair or remanufacture metal components after having experienced wear. **Thermal Spraying** involves the projection of molten or semi-molten particles onto a substrate, see Figure 4.14. Various methods, such as flame spraying, plasma spraying, and High-Velocity Oxygen Fuel (HVOF) spraying, can be used. Thermal spraying allows for the deposition of coatings with diverse materials, including metals, ceramics, and polymers, making it suitable for a wide range of applications. These challenges can be addressed by the search for alternative processing approaches or combining different techniques. In this context, recent studies have demonstrated that the porosity can be beneficially used for the incorporation of 2D solid lubricant powders, thus creating composite coatings with advanced multifunctional properties [6,7].

4.3.2 TRIBOLOGICALLY-EFFECTIVE COATINGS

The most commonly used coating candidates can be divided into different material classes including

- oxides,
- carbides,
- nitrides,
- diamond-like carbon (DLC), and
- friction-adaptive or solid lubricants (which are reviewed in the following section).

Typically reported COFs of these coatings as well as their maximum operation temperatures and mechanical characteristics are summarized in Figure 4.15.

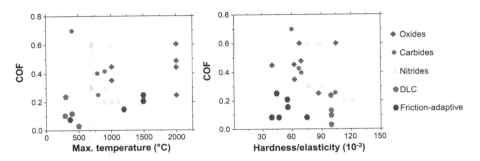

FIGURE 4.15 Ashby plots summarizing literature data on common friction vs. maximum operating temperature as well as mechanical characteristics of different coating classes.

Oxide coatings commonly use protective materials such as aluminum oxide (sapphire), zinc oxide, titania, zirconia, among others. While the friction characteristics of oxides are generally not the greatest, their stability under elevated temperatures makes them ideal candidates for high-temperature applications. Consequently, these coatings are implemented as thermal barrier coatings to protect systems and functional components like turbine blades.

Carbide coatings are composed of carbon and a metallic element, typically tungsten (WC), titanium (TiC), or chromium (Cr_3C_2), and applied by thermal spraying as well as PVD or CVD. These material systems possess an exceptional hardness, which provides an excellent resistance to abrasion, erosion, and adhesion, thereby minimizing wear and extending the lifespan of the coated component. Carbide coatings also exhibit low COFs, reducing the amount of dissipated energy and enabling a smoother operation. Another benefit of carbide coatings connects to their high thermal stability, thus enabling their operation at elevated temperatures without significant degradation, making them suitable for applications, in which frictional heat can reach extreme levels. These coatings act as a protective barrier, preventing surface damage, and preserving the integrity of the underlying material. As a result, their common applications can be found in cutting tools, engine components, like piston rings, valve trains, and cylinder liners, and bearings. Recent studies also demonstrated their excellent resistance of carbide coatings to scuffing and galling of the surfaces thus further expanding their versatility and range of application.

Nitride coatings, such as titanium nitride (TiN) and chromium nitride (CrN), are usually deposited by PVD and offer a high microhardness, often comparable to or even greater than that of carbide coatings. This hardness results in an excellent resistance to abrasion, adhesion, and deformation, making them suitable for applications involving high mechanical stresses. The friction behavior of nitride coatings usually highly depends on the surrounding environment with lower COFs (as low as 0.01) observed for vacuum and dry nitrogen conditions, while higher COFs (as high as 0.2–0.4) are measured under ambient conditions due to the humidity adsorption on the nitride surface [8]. Increasing the temperature allows for a significant decrease in the resulting COF in the air since adsorbed water can be effectively removed. Components coated with nitrides, such as cutting tools, punches, and dies, can withstand heavy loads, and maintain their sharp edges. The coatings' wear resistance and low friction performance help to enhance the performance and reliability of these components under high loads and extreme operating conditions.

Before the discovery of **DLC coatings**, diamonds were used in Swiss watches as a friction-reducing solution for the bearing systems, so-called jewel bearings. The major problem with diamond is that it does not work in vacuum due to its high surface reactivity as a consequence of interactions between non-terminated carbon dangling bonds. To resolve this issue, the coating community proposed hydrogen passivation of the surfaces, which demonstrated a high effectiveness thus enabling even superlubricious states [9]. In DLC coatings (Figure 4.16), the two crystal lattice variants of elemental carbon, diamond, and graphite, connect the individual carbon atoms within the layer through sp^3- and sp^2-hybridizations. The carbon atoms are arranged in a random (amorphous) network without crystalline long-range order, with varying bond distances and angles. DLC coatings can generally be deposited by means of (PE)-CVD or PVD processes. Depending on the predominant hybridization state of carbon atoms, hydrogen-free amorphous carbon films can be distinguished into relatively soft, predominantly sp^2-hybridized amorphous carbon (a-C) coatings or comparatively hard, primarily sp^3-hybridized tetrahedral amorphous carbon (ta-C) coating. Due to the unavoidable addition of hydrogen in form of process gases or through other feedstocks, amorphous carbon films are only classified as hydrogenated above a hydrogen content of about 3 at.-% (a-C:H, ta-C:H). With increasing hydrogen content, the degree of cross-linking between the carbon atoms and thus the resulting hardness of the amorphous carbon layer tend to decrease. At a hydrogen content of > 40 at.-%, these layers are usually referred to as plasma polymers. Since carbon can form bonds with both electropositive and electronegative elements, the coatings can be doped by other non-metallic or metallic elements, such as hydrogen or tungsten. DLC coatings have been reported to exhibit an exceptional hardness and low friction characteristics, supporting their resistance to wear, corrosion, and chemical attack. DLC coatings are widely used as protective coatings for automotive components, cutting tools, and medical devices. Through graphitization of DLC coating, during tribo-testing as well as the transfer to the counter-body, a contact between two surfaces with hydrogen-saturated C-bonds can be formed, leading to extremely low friction.

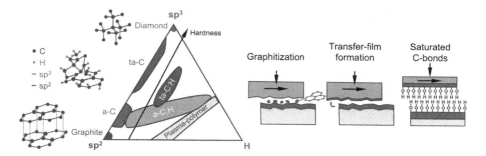

FIGURE 4.16 Classification of DLC coatings depending on the ratio of sp^2- and sp^3-hybridized C-bonds and the hydrogen content (left) as well as low-friction and low-wear mechanism of DLC coatings during sliding (right).

FIGURE 4.17   Different architectures for tribologically-effective coatings.

Polymer-based coatings, among which the most prominent material is polytetrafluoroethylene (PTFE; Teflon), involve the application of a liquid polymer onto the surface, followed by a curing or baking process that transforms it into a solid, adherent coating. Different methods, such as spraying, dipping, or electrostatic deposition, can be utilized to achieve the desired coating thickness and uniformity. PTFE coatings show a list of advantages such as low friction, vacuum compatibility, chemical inertness, and relatively high thermal stability (above 300°C) [10]. The origin of the excellent friction characteristics of PTFE lies in its structure, made of long chains of polymeric $(C_2F_4)_n$ molecules, while the carbon-fluorine bond is highly stable. During sliding against another material, PTFE chains can experience scission, leading to the formation of reactive groups facilitating the creation of a cohesive transfer film on the counter-body. At the same time, the intermolecular forces in PTFE are low, since PTFE is usually not cross-linked and offers a low surface energy, thus resulting in an easy shearing ability between the PTFE layers. However, the structure of PTFE and its low surface energy pose some challenges for coating deposition. Specifically, poor adhesion to the substrates can lead to delamination and high wear of the coating with patchy transfer films. Other commonly employed polymer coatings are polyetheretherketone (PEEK), polyimide, polyurethane, polyphenylene sulfide (PPS), polyvinylidene fluoride (PVDF), polyetherimide (PEI), and polyamide (PA; nylon).

Monolayer coatings consist of a single, uniform layer of any of the aforementioned coating materials, which can be one phase, alloyed (doped), or an isotropic multiphase. However, it should be noted that the **coating architecture** can be adjusted (Figure 4.17) to optimize the desired mechanical and tribological properties. As such, multilayer coatings involve the deposition of multiple layers of different materials. This approach allows for tailored properties, such as an improved adhesion to the substrate, corrosion resistance, and mechanical strength, making them versatile in various applications. Gradient coatings exhibit a gradual variation in composition or thickness across the coating surface. This design provides specific functionalities at different points and helps also to improve adhesion to the substrate, thus preventing delamination. Composite (disperse) coatings involve the dispersion of particles within a matrix, creating a heterogeneous structure. This approach is employed to achieve specific characteristics such as improved wear resistance, or thermal stability.

4.3.3 Layered Materials as Solid Lubricants

The most prominent examples of lamellar materials finding their application potential for tribological needs are 2D layered structures. While the approaches for their deposition and incorporation can mimic the other coating techniques, we wish to separate them to put a particular emphasis on their importance. Solid lubricant

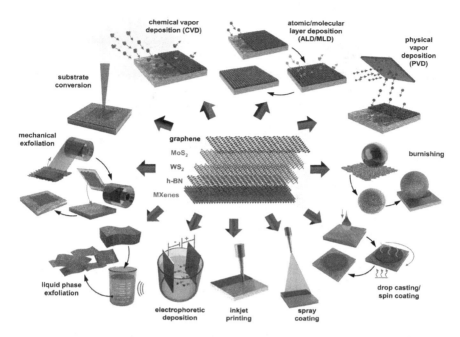

chemical vapor deposition (CVD)

atomic/molecular layer deposition (ALD/MLD)

physical vapor deposition (PVD)

substrate conversion

mechanical exfoliation

graphene

MoS₂

WS₂

h-BN

MXenes

burnishing

liquid phase exfoliation

electrophoretic deposition

inkjet printing

spray coating

drop casting/ spin coating

FIGURE 4.18 Different deposition approaches for 2D materials. (Reproduced with permission from Ref. [11].)

coatings, such as molybdenum disulfide (MoS_2) and graphite, offer excellent lubricating properties even under extreme conditions. These coatings form low-friction layers on the surface, reducing the direct contact between mating surfaces and minimizing wear. Solid lubricant coatings are often applied to components operating in high-temperature environments or where liquid lubricants may not be suitable (vacuum conditions, clean environments). Nowadays, the approaches for the design of 2D materials vary a lot offering flexibility in the adaptation of their use in various mechanical systems (Figure 4.18).

EXERCISE 4.1

What are the key characteristics affecting adhesive friction and how they can be improved by coatings?

Graphite became one of the most conceptually standard solid lubricants, which is widely used in various applications, such as bike chains, key locks, etc. We use graphite every day when writing with pencils since its ability to easily shear (similar to the shearing of a deck of cards) is responsible for the drawn lines. In this case, the interaction between the graphitic layers is through Van-der-Waals forces, which are significantly lower than the primary metallic, ionic, or covalent in-plane atomic forces thus enabling easy sliding and shearing, see Figure 4.19a.

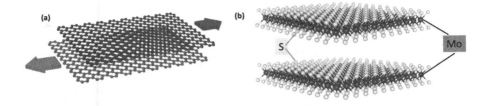

FIGURE 4.19 Shearing of (a) graphitic sheets and (b) MoS_2 sheets as model approaches for solid lubrication.

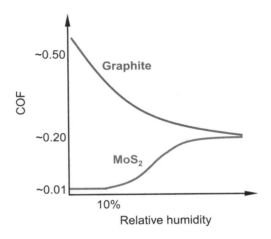

FIGURE 4.20 Humidity dependence of friction for graphite and MoS_2.

Graphite shows an excellent tribological performance with low COFs under ambient conditions. However, the easy shearing capability of graphite is largely affected by the presence of humidity as illustrated in Figure 4.20. In this regard, the presence of water is critically important to terminate the reactive edges of graphene flakes, thus ensuring and maintaining their easy shearing. Once water is eliminated from the material, the flakes accumulate into large chunks producing dusty wear debris. It is important to note that the transition from a dry to a humid environment requires a notable number of molecules to terminate the flakes, thus keeping the shearing ability of the material.

MoS_2 represents another widely used solid lubricant. Even though being considered a 2D material, its flakes are composed of three atomic layers (Figure 4.19b). In this structure, the ionic bonding leads to a charge rearrangement, while Mo atoms become positively charged ions (due to their lower electronegativity) and S atoms become negatively charged ions (due to their higher electronegativity). When MoS_2 is used as a solid lubricant, the actual sliding occurs between the layers of the sulfur atoms. The negative charges of the sulfur ions lead to an electrostatic repulsion between the layers thus decreasing friction significantly.

Water molecules tend to absorb in MoS_2 layers through diffusion and attachment to the edges and defects present, which prevents very low friction [11] (Figure 4.20). Although the resulting COFs are still sufficiently low to be interesting for applications, the high cost of MoS_2 production (in comparison to graphite) and its susceptibility to degrade in contact with water make it less favorable than graphite. The sensitivity to the environment is further affected by temperature [12]. For instance, in a nitrogen and water mixture atmosphere, a temperature increase results in a reduced COF as a result of water removal from the coating. In contrast, for oxygen-enriched environments, a temperature increase accelerates oxygen termination of the MoS_2 flakes, which induced their degradation thus resulting in a loss of lubrication.

Apart from graphite and MoS_2, tungsten disulfide (WS_2), hexagonal boron nitride (h-bN), or MXenes can be named as emerging 2D solid lubricants [11]. As outlined, each material system possesses its advantages and disadvantages. Some of them work better under vacuum and dry conditions, while others function best under high relative humidities. In the same context, some systems tend to induce an excellent friction performance, but once the first signs of wear occur, they tend to lose their beneficial tribological systems. Some materials may not be capable of sufficiently lowering friction to be interesting for applications in terms of energy dissipation, but they can help to induce an excellent wear resistance thus extending their useful lifetime. Therefore, the combination of different solid lubricants is a highly promising alternative to simultaneously induce a beneficial friction and wear performance.

4.3.4 COMPOSITE COATINGS

While coatings can be directly applied as the major components of the lubricating materials, they can be also incorporated into metallic, polymeric, and ceramic matrices. When properly designed, the huge benefit of this approach connects with the homogenous distribution of the lubricant across the bulk matrix, which ensures a continuous release of lubricious material during operation, thus reducing friction and wear. Self-lubricating coatings are commonly used in applications where continuous lubrication is challenging or undesirable, such as in high-temperature environments or vacuum conditions. The design principle of composite coatings is based on the use of materials that already demonstrate improved chemical or physical characteristics, thus enhancing their performance with the incorporation of tribologically beneficial materials. The selection of a composite coating depends on the specific requirements of the application, including operating conditions, load, speed, and environmental factors. Thereby, composite coatings combine different types of reinforcing materials to achieve superior tribological properties. For instance, a coating might incorporate a combination of solid lubricants, ceramic particles, and metallic reinforcements to optimize the resulting friction reduction and wear resistance simultaneously. The most common design of composite coatings employs the incorporation of particles, most often of nanoscale dimensions, such as metal oxides, nitrides, or carbides, or fibers, such as carbon, ceramic, or polymers, into a polymeric or metallic matrix material. The nanoparticles act as reinforcing agents, enhancing the coating's hardness, and wear resistance, as well as reducing friction. Meanwhile, the metal or polymer matrix provides the necessary ductility and toughness as well as improves the adhesion to the substrate.

4.3.5 Advanced Lubrication Solutions

Recently, significant advances in the design of lubrication solutions focus on new adaptive, self-healing, and self-lubrication concepts. Among the most deserving of attention are

* **Chameleon coatings**
* **Self-healing coatings**
* **Tribo-catalytic coatings**

Chameleon coatings are designed to adapt on demand to changes in environment and temperature. A typical example of this concept is a mixture of WS_2 or MoS_2 and graphite powders [13,14], see Figure 4.21. In dry environments, which are advantageous for WS_2 and MoS_2 but difficult for graphite, hexagonal MoS_2 (or WS_2) patches prevail at the contacting surfaces, while their basal planes align parallel to the substrate and the sliding motion. The shear usually occurs in the most oxidation-resistant, inert, and slippery interface, between densely packed (002) basal planes. When humidity increases, the MoS_2 (or WS_2) layers on the counter faces start to degrade and form non-lubricious oxides. As a result of the wear, graphite layers are exposed to sliding, replacing MoS_2 (or WS_2) and making the surface inert and slippery since the presence of water molecules is beneficial for graphite but not for MoS_2 [13]. Upon transitioning back to the dry cycle, the desorption of water molecules promotes the wearing off of graphite sheets, revealing fresh MoS_2 (or WS_2) grains and causing MoS_2 (or WS_2) layers to recoat the surface. In the successful demonstration of the

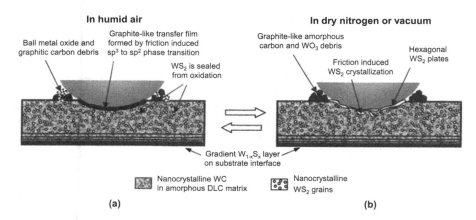

FIGURE 4.21 (a and b) Example of WS_2 + graphite chameleon coating adaptation mechanism. (Reprinted from Ref. [14] with permission.)

chameleon coating concept, MoS_2 (or WS_2) and graphite were also combined with Sb_2O_3 particles, promoting adhesion to the substrate and suppressing fracture formation and propagation. Furthermore, this chameleon behavior has been demonstrated upon transitioning between different temperature conditions [6].

The concept of **self-healing coatings** uses sliding-induced stresses to initiate chemical or physical reactions within the material, ultimately restoring its original characteristics and functionality. In this regard, self-healing tribological materials can be designed in the form of composites containing microcapsules or microchannels with healing agents, such polymers or lubricants released during sliding, or layers of films, which cross-react through reversible oxidation or reduction reactions upon local heating. Notably, the encapsulated lubricants can be in solid form, such as noble metals distributed in a hard ceramic matrix. The use of a ceramic matrix is preferred to maintain the substantial hardness and elastic modulus values while limiting wear. Upon sliding, the applied stresses and temperature increase promote the diffusion of metal ions into the damaged regions, repairing the produced cracks and wear. Usually, these systems only require less than 20 at.-% of the healing component as demonstrated for systems like TiC-Ag, YSZ-Ag, CrN-Ag, and MoCN-Ag [15]. In this context, silver is often the material of choice due to its comparatively low cost and notable mobility at temperatures as low as 300°C–500°C. Another system comprised niobium oxide and silver deposited as alternating layers [16], see Figure 4.22. The tribological stress collective in combination with elevated temperatures (~ 945°C) promoted the formation of a lubricious ternary $AgNbO_3$ phase

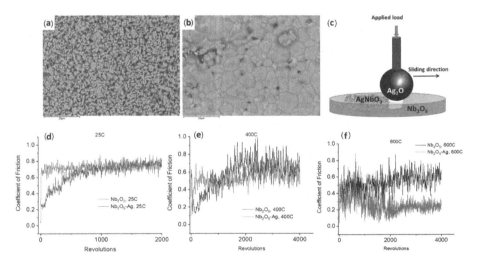

FIGURE 4.22 Preparation of the bulk niobium oxide by (a) pressing and (b) sintering of the pellet. The sample was further tested regarding its tribological performance (c). Tribological test of Nb_2O_5 with and without the presence of Ag at (d) 25°C, (e) 400°C, and (f) 600°C, respectively. (f) Results indicate a reduction in the resulting COF in the case of silver presence at 600°C. (Reprinted from Ref. [16] with permission.)

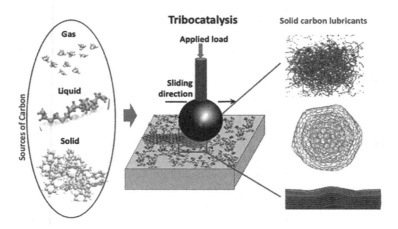

FIGURE 4.23   Schematic of the concept of tribocatalysis to in-situ generate lubricating carbon films. (Reprinted from Ref. [17] with permission.)

at the sliding interfaces that not only helped to repair the damage but also notably decreased friction.

In Section 1.9.3, we discussed that tribochemistry/tribocatalysis phenomena can be one of the possible consequences of the tribological processes [17]. Tribocatalysis can be used to create and promote advanced solid lubrication solutions through the incorporation of tribocatalytically reactive phases in protective coatings. These **tribo-catalytic coatings** are designed to cooperate with the environment and repair damage induced by sliding through tribocatalytically driven formation of protective carbon films on top of the coating (Figure 4.23).

The traditional process of carbon catalysis requires three major elements to proceed, which include access to the catalyst, the availability of carbon precursor, and an activation energy. Many tribological systems, especially those used in industrial settings, have already the two last components, such as carbon precursor (e.g., oil lubricants, which are made from hydrocarbon molecules) and the necessary activation energy (high contact stresses and tribologically-induced heating at the contact interfaces). The remaining catalysts can be meanwhile supplied through the incorporation of catalytically active materials in coatings. In this regard, the presence of the catalysts is essential for the chemical transformation of the carbon source to lubricious carbon films, lowering the activation energy for the required reactions. Recent research has made significant progress in showcasing the tribocatalytic potential of diverse material systems, not only studied in the lab-scale setting but also employed in applied systems. The presence of Cu and Ni in metal nitride (MeN) [18–20] and CoP-based [21] matrices enabled the formation of protective DLC-like films from liquid sources, such as oils, alcohols, and alkanes, and gaseous sources, such as methane gas and ethanol vapor. Alternatively, the use of Pt in the form of Pt-Au films promoted the growth of graphitic carbon films with the

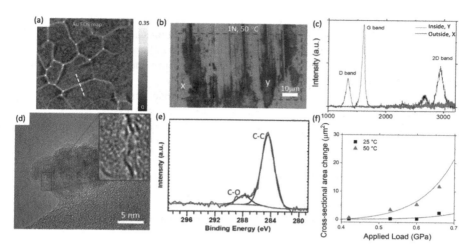

FIGURE 4.24 Tribocatalytic formation of carbon films on a Pt-Au coating from ethanol environment. (a) Structure of the Pt-Au coating with a Pt core and Au segregating along the grain boundaries. (b) Optical micrograph of the wear track formed during tribological testing with (c) Raman signature indicating the formation of a carbon-based tribo-film. (d) TEM analysis confirms the graphitic nature of the tribo-film, which was further supported by (e) XPS. (f) The rate of tribo-film growth increases with temperature and load conditions. (Reprinted from Ref. [22] with permission.)

growth rate increased by higher contact stresses and applied temperature [22], see Figure 4.24. From the application standpoint, the catalytic activity of a thin Pd layer has been confirmed on conventional steel milling equipment [23]. The formation of graphitic layers from internal body fluids has also been detected in metal-on-metal hip replacements [24].

4.4 CHECK YOURSELF

- What is the primary goal of surface engineering, and how does it contribute to the material's suitability for various applications?
- Name three categories of surface engineering methods and briefly describe each.
- How does surface texturing differ from surface finishing, and what role does it play in enhancing material performance?
- Provide examples of mechanical surface finishing methods and their applications in different industries.
- Explain the advantages of surface texturing in reducing friction and wear, considering different lubrication regimes.
- In what situations does surface texturing help generate pressure-induced tribo-films, and how does this contribute to reducing wear?

- What factors influence the choice of surface texturing technique, and how do they vary in terms of lateral texture size and fabrication speed?
- What are the potential drawbacks of imparting enhanced hardness to the entire component, and why is it desirable to limit increased hardness to specific surface layers?
- Describe the process of shot peening and its impact on the surface properties of a material, including the induced compressive stresses.
- Compare flame hardening and induction hardening, highlighting their respective advantages and disadvantages in the context of surface hardening.
- Explain the concept of case hardening and how it differs from bulk hardening, emphasizing the benefits of retaining a softer core with a hardened surface layer.
- Differentiate between intercalation and substitutional solid solutions in the context of diffusion processes for surface hardening, providing examples of each.
- What are the advantages of using coatings as opposed to surface hardening methods for improving the mechanical/tribological properties of materials?
- Name four key parameters that govern the tribological behavior in the contact of two coated surfaces.
- Describe the process of CVD and highlight its advantages and potential drawbacks.
- Describe the process of PVD and highlight its advantages and potential drawbacks.
- How does IBAD influence the properties of a coating during the deposition process?
- What are the considerations that need to be taken into account when selecting appropriate coatings and their deposition strategy?
- Explain the differences between oxide, carbide, nitride, diamond-like carbon (DLC), and polymer-based coatings, including their typical applications.
- How does the structure of polytetrafluoroethylene PTFE contribute to its excellent friction characteristics during sliding against another material?
- Briefly explain the concepts of multilayer coatings, gradient coatings, and disperse coatings, highlighting their respective advantages in specific applications.
- What are some examples of 2D layered structures used as solid lubricant coatings?
- How do solid lubricant coatings, such as graphite and MoS_2, reduce friction and wear in mechanical systems?
- How does humidity affect the shearing capability of graphite, and what is the consequence of the coefficient of friction?
- What is the impact of water molecules on the friction behavior of MoS_2, and how does it compare to graphite?
- What is the advantage of incorporating solid lubricants into composite coatings, and what materials are commonly used for reinforcement?
- Briefly describe the concept of chameleon coatings and how they adapt to changes in environment and temperature conditions.

- How do self-healing coatings use sliding-induced stresses to restore the original characteristics and functionality of the material?
- How do tribocatalytic coatings contribute to advanced solid lubrication solutions, and what is their mechanism for in-situ formation of lubricating carbon films?

4.5 SOLUTIONS FOR THE EXERCISES

EXERCISE 4.1

In the first approach, the underlying concept of 2D materials is based on the prior discussion of the adhesive mechanisms of friction and wear. As indicated in the first module of this book, the COF in simple terms can be described as

$$\mu \sim \frac{\tau}{H},$$

where τ is the shear modulus of the surface material and H stands for the hardness. Therefore, the solid lubrication solutions are specifically designed to support easy shearing via low shearing strength while providing the hardness support of the underlying substrate.

REFERENCES

1. M. Marian, A. Almqvist, A. Rosenkranz, M. Fillon, Numerical micro-texture optimization for lubricated contacts-A critical discussion, *Friction* 10 (2022) 1772–1809.
2. E. Maleki, O. Unal, A. Amanov, Novel experimental methods for the determination of the boundaries between conventional, severe and over shot peening processes, *Surfaces and Interfaces* 13 (2018) 233–254.
3. Induction/flame hardening, Manupedia, https://www.open.edu/openlearn/science-maths-technology/engineering-technology/manupedia/induction/flame-hardening, 2018.
4. V. Moorthy, B.A. Shaw, P. Mountford, P. Hopkins, Magnetic Barkhausen emission technique for evaluation of residual stress alteration by grinding in case-carburised En36 steel, *Acta Materialia* 53 (2005) 4997–5006.
5. K. Holmberg, A. Matthews, H. Ronkainen, Coatings tribology-contact mechanisms and surface design, *Tribology International* 31 (1998) 107–120.
6. A. Shirani, T. Joy, A. Rogov, M. Lin, A. Yerokhin, J.-E. Mogonye, A. Korenyi-Both, S.M. Aouadi, A.A. Voevodin, D. Berman, PEO-Chameleon as a potential protective coating on cast aluminum alloys for high-temperature applications, *Surface and Coatings Technology* 397 (2020) 126016.
7. R. Gonzalez, H. Ashrafizadeh, A. Lopera, P. Mertiny, A. Mcdonald, A review of thermal spray metallization of polymer-based structures, *Journal of Thermal Spray Technology.* 25 (2016) 897–919.
8. K. Kato, N. Umehara, K. Adachi, Friction, wear and N2-lubrication of carbon nitride coatings: A review, *Wear* 254 (2003) 1062–1069.

9. A. Erdemir, O.L. Eryilmaz, G. Fenske, Synthesis of diamondlike carbon films with superlow friction and wear properties, *Journal of Vacuum Science & Technology A* 18 (2000) 1987–1992.

10. S. Biswas, K. Vijayan, Friction and wear of PTFE-a review, *Wear* 158 (1992) 193–211.

11. M. Marian, D. Berman, A. Rota, R.L. Jackson, A. Rosenkranz, Layered 2D nanomaterials to tailor friction and wear in machine elements-a review, *Advanced Materials Interfaces* 9 (2022) 2101622.

12. H. Khare, D. Burris, The effects of environmental water and oxygen on the temperature-dependent friction of sputtered molybdenum disulfide, *Tribology Letters* 52 (2013) 485–493.

13. J. Zabinski, J. Bultman, J. Sanders, J. Hu, Multi-environmental lubrication performance and lubrication mechanism of MoS2/Sb2O3/C composite films, *Tribology Letters* 23 (2006) 155–163.

14. A.A. Voevodin, J.S. Zabinski, Supertough wear-resistant coatings with 'chameleon' surface adaptation, *Thin Solid Films* 370 (2000) 223–231.

15. S.M. Aouadi, J. Gu, D. Berman, Self-healing ceramic coatings that operate in extreme environments: A review, *Journal of Vacuum Science & Technology A: Vacuum, Surfaces, and Films* 38 (2020) 050802.

16. A. Shirani, J. Gu, B. Wei, J. Lee, S.M. Aouadi, D. Berman, Tribologically enhanced self-healing of niobium oxide surfaces, *Surface and Coatings Technology* 364 (2019) 273–278.

17. D. Berman, A. Erdemir, Achieving ultralow friction and wear by tribocatalysis: Enabled by in-operando formation of nanocarbon films, *ACS Nano* 15 (2021) 18865–18879.

18. A. Shirani, Y. Li, O.L. Eryilmaz, D. Berman, Tribocatalytically-activated formation of protective friction and wear reducing carbon coatings from alkane environment, *Scientific Reports* 11 (2021) 20643.

19. K. Jacques, A. Shirani, J. Smith, T.W. Scharf, S.D. Walck, S. Berkebile, O.L. Eryilmaz, A.A. Voevodin, S. Aouadi, D. Berman, MoVN-Cu coatings for in situ tribocatalytic formation of carbon-rich tribofilms in low-viscosity fuels, *ACS Applied Materials & Interfaces* 15 (2023) 25, 30070–30082.

20. G. Ramirez, O.L. Eryilmaz, G. Fatti, M.C. Righi, J. Wen, A. Erdemir, Tribochemical conversion of methane to graphene and other carbon nanostructures: Implications for friction and wear, *ACS Applied Nano Materials* 3 (2020) 8060–8067.

21. A. Shirani, R. Al Sulaimi, A.Z. Macknojia, M. Eskandari, D. Berman, Tribocatalytically-active nickel/cobalt phosphorous films for universal protection in a hydrocarbon-rich environment, *Scientific Reports* 13 (2023) 10914.

22. A. Shirani, Y. Li, J. Smith, J. Curry, P. Lu, M. Wilson, M. Chandross, N. Argibay, D. Berman, Mechanochemically driven formation of protective carbon films from ethanol environment, *Materials Today Chemistry* 26 (2022) 101112.

23. M. Wohlgemuth, M. Mayer, M. Rappen, F. Schmidt, R. Saure, S. Grätz, L. Borchardt, From inert to catalytically active milling media: Galvanostatic coating for direct mechanocatalysis, *Angewandte Chemie International Edition* 61 (2022) e202212694.

24. Y. Liao, R. Pourzal, M.A. Wimmer, J.J. Jacobs, A. Fischer, L.D. Marks, Graphitic tribological layers in metal-on-metal hip replacements, *Science* 334 (2011) 1687–1690.

5 Applied Tribology
Case Studies

Tribology plays a critical role in many applications. In previous chapters, we have already briefly discussed some of them and provided the typical considerations that should be made to address the challenges of friction and wear control. In this module, we will look further at some representative case studies.

5.1 NANOTRIBOLOGY/MEMS

Among different tribological applications, special attention should be given to nanotribology, which resembles sub-micrometric tribological contacts. This scale has a critical importance for nano- and micro-scale devices, a wide range of which are now operational in research laboratory environments and commercial production. In this regard, pressure sensors, accelerometers, actuators, optical switches, and nano- or microelectromechanical systems (NEMS, MEMS) have become a mainstay in electronics, cell phones, and other wireless technology products (Figure 5.1). Nowadays, medical MEMS are used *in vivo* for implants and surgical procedures, while the

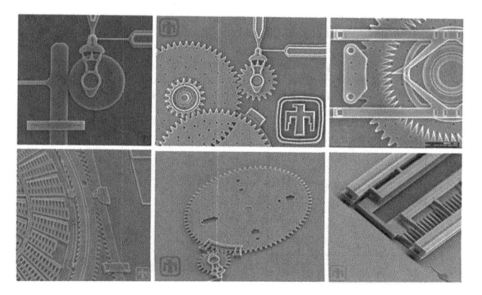

FIGURE 5.1 Examples of MEMS devices: (top, from left to right) dynamometer, gears, and indexing motor, and (bottom, from left to right) torsional ratcheting actuator, optical encoder, and linear rack. (Adapted from Sandia National Laboratories: MicroElectroMechanical Systems (MEMS). Available at https://www.sandia.gov/mesa/mems.)

DOI: 10.1201/9781003397519-5

automotive industry has widely embraced the use of MEMS through crash airbag accelerometers. Nevertheless, a host of MEMS and NEMS applications remain stalled by issues associated with surface science and tribology phenomena originating from high surface-to-volume ratios. In this context, special attention should be given to the nano-scale solid surface contacts, the close proximity of microstructures, and innumerable other device complications that are associated with devices' stiction, adhesion, friction, and wear characteristics associated with the sliding contact of rough contacting asperities.

Stiction in nano- and micro-scale devices refers to the unintentional adhesion of free-standing structures to the substrate or neighboring microstructures in proximity. This can occur during the fabrication process when the restoring forces are not sufficiently high to overcome interfacial forces, such as capillary, van der Waals, and electrostatic forces.

To avoid and/or reduce release-related stiction, different approaches, such as roughening the surface to reduce the contact area, altering the solution-substrate wetting properties, the usage of self-assembled monolayers (SAMs), supercritical drying, and freeze sublimation drying can be considered. The problem of stiction can also arise during device operation, even after their successful release. In-use stiction can occur due to severe operations including large electrostatic forces associated with electric potential differences and/or over-heating induced welding of materials, or due to friction and wear issues. Related to friction, a wide range of studies have been conducted to evaluate the frictional forces at the nano-scale [1]. Numerous studies have made use of AFM to analyze nano-tribological characteristics with the possibility of mimicking a single-asperity contact. In this case, the single-asperity contact's models introduced in Chapter 1 are important to understand the contact geometry and the evolution of the contact interfaces.

In these nano-scale studies, an interesting tribological phenomenon, the puckering effect, has been discovered for graphene (and other 2D materials). The effect recognizes the ability of suspended or poorly adhered to the substrate 2D membranes to experience a notable out-of-plane flexibility. This results in an increased contact area with the AFM tip thus inducing a notable friction increase (Figure 5.2). This effect

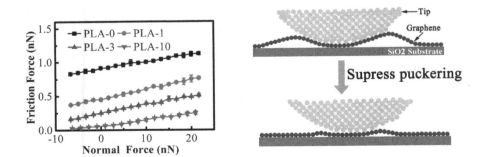

FIGURE 5.2 Friction force between graphene and silicon dioxide substrate decreased due to the realization of a plasma treatment step for durations of 1 (PLA-1), 3 (PLA-3), and 10 minutes (PLA-10), respectively, as compared to the original graphene on silicon dioxide (PLA-0). The observed phenomenon was attributed to the changes in the puckering effect. (Reproduced with permission from Ref. [2].)

can be reduced by improving the adhesion between the 2D material and the substrate by modifying the substrate surface energy [2] or decreasing the rigidity of the films by increasing number of layers [3].

The evolution of single-asperity contacts is largely affected by the substrate. This is critically important for the design of nano-scale devices since the contact resistance depends on the established contact area. The substrate effect on graphene's conductivity has been explored by measuring the conductance between a nano-scale probe and a single layer of graphene with three different levels of substrate support: no substrate (free-standing graphene film), a soft substrate, and a hard substrate. These three systems were studied using conductive AFM experiments complemented by MD simulations. In both experiments and simulations, at a given normal force, the current was shown to increase in the following order: hard substrate < soft substrate < free-standing film (Figure 5.3). Therefore, the influence of a substrate on graphene's conductivity is two-fold. On the one hand, the dominant factor connects with the effective elasticity of the contact, which determines contact size. On the other hand, the secondary factor is the effect of the substrate on atom-atom distances in the contact, which determines resistivity [4].

While the selection of the materials that exhibit low-friction and low-wear characteristics is a possible solution to address the failure of the MEMS/NEMS devices, it can significantly limit their mechanical, electrical, optical, or magnetic properties. Many applications impose their own requirements for the material composition and characteristics that can prevent the use of tribologically effective tribo-pairs. There is still a possibility for the use of an alternative approach. To address the frictional effect, the focus is shifted toward the use of SAMs. Their lubrication concept is similar to traditional oil lubricants but relies on the selective chemisorption of reactive groups of the SAM molecules to the substrate in need of protection [5]. SAMs can be deposited by liquid- or vapor-phase deposition, depending on their structure and targeted substrate material. Their liquid-phase deposition does not

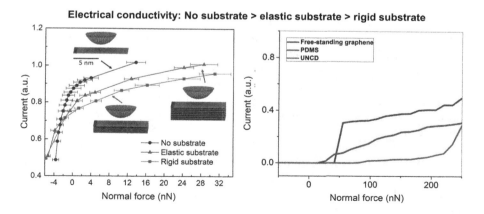

FIGURE 5.3 Computational and experimental studies demonstrating the effect of the underlying substrate on the evolution of graphene/metal contacts. (Adapted with permission from Ref. [4].)

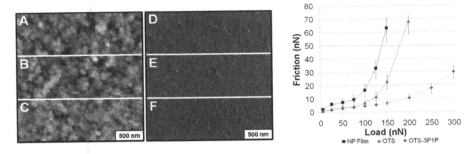

FIGURE 5.4 Topographic (A–C) and the corresponding friction force (D–F) maps of coatings consisting of different nanoparticles (left): Unfunctionalized silica film (A and D), OTS-coated film (B and E), and multicomponent OTS/3P1P film (C and F). Friction versus load results for unfunctionalized (squares), OTS-functionalized (triangles), and OTS/3P1P multicomponent films (circles) at a sliding velocity of 0.1 µm/s (right). (Reprinted from Ref. [6] with permission.)

require a specific equipment, which makes it an interesting approach. However, it has significant limitations for complex structures by not penetrating inside them. Moreover, liquid deposition consumes large amounts of the respective liquid and additional stages such as rinsing and drying of the coated surfaces become necessary. In contrast, vapor-phase deposition offers an excellent diffusivity under high vacuum but is only suitable for molecules with low surface tension. The use of SAMs allows to significantly reduce friction and thus stiction in nano-scale contacts. For instance, Jones et al. [6] studied how two different SAMs (octadecyltrichlorosilane (OTS) and OTS/3-Phenyl-1-proponal (3P1P)) helped to reduce friction between silica nanoparticles and the AFM probe made of Si (Figure 5.4).

5.2 HIGH-TEMPERATURE TRIBOLOGY

What is high temperature? For a practitioner, a suitable definition for high temperature is that the range or precise point beyond which the thermal response of materials causes significant engineering or design problems, which are otherwise not of concern at lower temperatures. Therefore, high-temperature tribology is a fascinating and crucial field that delves into the interactions between surfaces in relative motion under extreme thermal conditions. In these systems, high temperature at the sliding interfaces can be a result of ambient conditions, flash events, or system sources. For instance, turbines used in jet engines and power generation operate at high temperatures, speeds, and contact pressures and heavily rely on an efficient lubrication at these high-temperature conditions. On a tonnage scale, steelmaking, glass manufacturing, or other smelting operations, all use high-temperature greases in moving molten or red-hot materials. On a relative scale with a very short service life, fail-proof performance at high temperatures and extreme pressures is demanded in extra-terrestrial applications such as missiles, spacecraft, and rockets. Aerospace applications call for the most versatility of lubricant materials ranging from low to high temperature

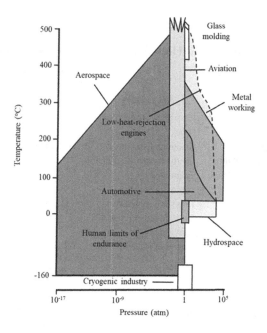

FIGURE 5.5   A map of application temperatures and pressures of various tribological systems. The limits of human endurance are shown in the center for reference. (Redrawn from Ref. [7].)

and alternating pressure cycles. The temperature and pressure regimes relevant for specific applications, which fall under the umbrella of high-temperature tribology are highlighted in Figure 5.5.

Under these requirements, most liquid lubricants are not capable to induce good friction and wear characteristics due to nonreversible changes in the structure and composition of the oil molecules. Certain restrictions are also applied to bulk materials and solid lubricants since high temperature may accelerate oxidation, and corrosion, as well as tend to induce softening and even melting of metals. It is not the temperature alone, but also the applied loads, environment, and dwell time, or duration for which elevated temperatures occur, that are important aspects guiding the design of tribologically efficient high-temperature materials. The need to find solutions for high-temperature applications became the major driving force for developing the area of solid lubrication (reviewed in Chapter 4). Since these applications mostly inquire low friction and wear, the focus is usually on the materials that can provide good lubrication at elevated temperatures (Table 5.1).

It should be noted that the high-temperature conditions usually require the material to provide an efficient thermal management to suppress temperature-induced damages to structural and functional components. For instance, in the aerospace industry, in addition to high temperatures, systems are exposed to erosion processes caused by the environment and minerals (calcium–magnesium–alumina–silicate (CMAS)).

TABLE 5.1

Summary of Commonly Used High-Temperature Lubrication Approaches

Lubrication Approach	Material Examples	Advantages	Operation Temperature (in Air)	Challenges at High-Temperature Applications
Layered materials	MoS_2, WS_2, Graphite	Easy application, low cost	< 300°C	Oxidation at higher temperatures
Diffusion of soft metals	Ag and Au encapsulated in oxide and nitride ceramic matrices	Stable under contact with oxygen, regulated by temperature	300°C–500°C	Fast diffusion to the surface depletes metal lubricant reservoirs
Thermally activated formation of lubricious oxides	Magnéli phases: V_2O_5, MoO_3, TiO_2, WO_3, PbO, ZnO	Provide easy shearing at high temperatures, environment supplies oxygen	500°C–1000°C	Abrasion at low temperatures, lubricant extrusion from contact by the load

CMAS are external siliceous minerals from dust, sand, and volcanic ash that can be ingested by the air intake and pass through the engine when aircraft operate at low altitudes. The melting temperature of CMAS varies between 1190°C and 1260°C depending on composition, while the inlet temperature of high-pressure gas turbine blades is typically around 1300°C–1600°C [8]. Therefore, when the temperature is below the melting point of CMAS, the debris can cause erosive wear by impact damage, and when exceeding the CMAS melting point, siliceous debris melt, deposit, and infiltrate the surfaces. To protect the blades in gas and jet turbines and to allow for a safe operation at elevated temperatures in a highly corrosive environment, a potential solution is to use thermal barrier coatings (TBCs). Traditional TBCs are made from yttria-stabilized zirconia (YSZ), which are not ideal as they still allow wetting and penetration of CMAS through cracks and pores. Prior studies [9] introduced $YSZ-Al_2O_3-SiC$ self-healing composite overlayers as a potential solution. Another strategy that was employed related to precipitation-induced self-healing. A self-healing $YSZ-Al_2O_3-Nb_2O_5$ system provided a 33-fold improvement in CMAS resistance over YSZ.

5.3 LOW-TEMPERATURE TRIBOLOGY

Cryogenic wind tunnels, liquid fuel rockets, space infrared telescopes, superconducting devices, and planetary exploration are examples of high-tech applications that demand effective lubrication for moving parts operating in cryogenic environments as low as 4 K, encompassing cryogenic liquids, gases, and vacuum settings [10]. In the case of liquid lubricants that may be considered to be used at low temperatures, their main challenge relates to the reduced mobility of molecules, which leads to an increased viscosity and reduced fluidity, thus suppressing the ability to form protective films on the involved sliding surfaces. In the case of solid materials, low

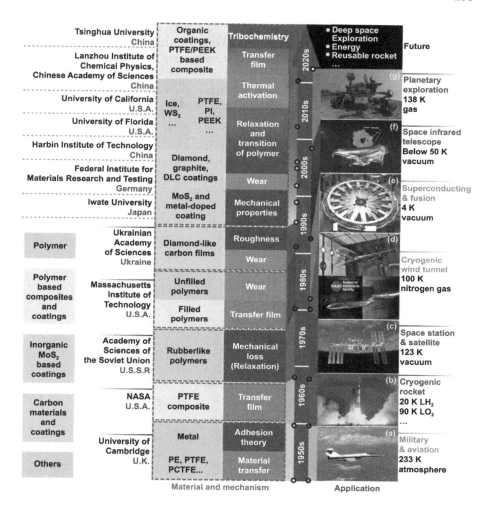

FIGURE 5.6 History and progress on tribological research considering solid lubrication at low temperatures. (Reprinted from Ref. [10] with permission by CC BY 4.0.)

temperatures tend to cause changes in the mechanical properties of materials, which go hand in hand with an increased brittleness and decreased ductility thus affecting the load-bearing capacity of the contacting surfaces. Different materials considered over the last years in low-temperature applications are summarized in Figure 5.6.

The solution usually comes from the use of coatings, while the respective characteristics are precisely tailored for the low-temperature operation regime. These coatings are expected to withstand low temperatures and maintain their integrity as well as adhesion to the substrate upon thermal cycling that may cause dimensional changes to the substrates. Among appropriate choices for these coatings are polymer- (such as epoxy or polyurethane), ceramic- (such as alumina, zirconia, or titanium nitride), or metallic-based coatings (such as zinc or aluminum). Since the exposure of functional components to the cryogenic regime may be undesirable, coatings need to offer a low thermal conductivity, which acts as a thermal barrier or insulator to minimize the heat transfer between the coated component and the surrounding environment. In this case, special attention needs to be given to traditional TBCs used for high-temperature applications, such as yttria-stabilized zirconia. As discussed earlier, the adhesive friction coefficient can be defined as

$$\mu_a = \frac{\tau \cdot A_r}{F_N}. \tag{5.1}$$

According to the contact mechanics models, in the case of an elastic contact, the real area of contact is

$$A_r = c\left(\frac{F_N}{E}\right)^{\frac{2}{3}}, \tag{5.2}$$

thus

$$\mu_a = c_2 \frac{\tau}{\sqrt[3]{F_N \cdot E^2}}. \tag{5.3}$$

In the case of a plastic contact,

$$A_r = \frac{F_N}{H}, \tag{5.4}$$

thus

$$\mu_a = \frac{\tau}{H}. \tag{5.5}$$

With a decreasing temperature, the elastic modulus E and hardness H tend to increase, leading to a decrease in the coefficient of friction (COF) (the shear strength also increases but at a lower rate). It should be noted that the COF does not only depend on the mechanical properties of materials but is also largely affected by accompanying processes taking place during sliding, such as tribo-chemistry. For instance, the influence of temperature on the friction behavior of diamond and DLC films connects with their relative grain size and hydrogen content. Consequently, the COF of microcrystalline diamond films was reported to be high and independent of temperature, while ultra-nanocrystalline diamond films exhibited a significant increase in friction from 220 to 120 K due to the inhibition of the thermally activated hydrogen transport during sliding [11]. Moreover, the tribological behavior of graphite, MoS_2, DLC, and PTFE materials showed a more complex dependence on temperature (Figure 5.7).

The study of cryogenic conditions became important not only from the point of applications but also allowed to capture interesting tribological phenomena. One example relates to the experimental demonstration of electronic friction, an energy

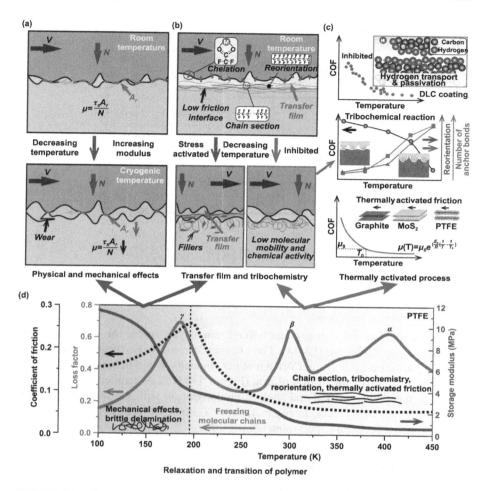

FIGURE 5.7 Summary of friction and wear mechanism for solid lubrication at low temperatures, including (a) physical and mechanical effects, (b) transfer film and tribochemistry, (c) thermally activated process, and (d) relaxation and transition of the used polymer. (Reprinted from Ref. [10] with permission by CC BY 4.0.)

dissipation phenomenon related to electron interactions. In the presence of an interface, when a conducting material is involved in sliding, electrons inside the material can be dragged along. These charge carriers subsequently interact with material defects, thus dissipating energy through ohmic losses. In this regard, it is important to mention that if a material at the interface is electrically insulating, electrons can become trapped, resulting in electrostatic forces affecting friction. The theory of electronic friction suggests that when the material resistivity approaches zero, as seen in superconductors below their critical temperature (T_c), electronic friction is expected to disappear. Several experiments used changes in the resonant frequency of the QCMs coated with lead or niobium films to probe the interactions between the

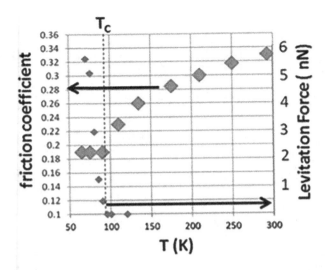

FIGURE 5.8 Tip-sample levitation and friction coefficient μ versus temperature. The $\mu(T)$ curve abruptly changes slope near $T_c = 92.5$ K. (Reproduced from Ref. [15] with permission.)

coatings and adsorbed nitrogen and helium monolayers below the superconductivity transition temperature [12–14]. The phenomenon has been further observed for yttrium barium copper oxide (YBCO) superconductor when being in contact with an AFM tip coated with 100 nm of iron (Figure 5.8) [15]. At temperatures below T_c, the COF remained constant and showed no correlation with the strength of superconducting levitation forces. In contrast, above T_c, the COF gradually increased and nearly doubled between T_c and room temperature.

5.4   HIGH-FRICTION MATERIALS

Many aspects of our daily life connect with friction and wear processes. In a general sense, these characteristics are usually to be reduced to enhance the resulting energy efficiency, reduce the involved fuel consumption, and/or make the components last longer (reduced maintenance costs). As already outlined, different approaches include modern lubrication concepts (oil, greases, solid lubricants, among others), advanced surface engineering (surface hardening, surface texturing, coatings, etc.), material and design changes, among others. Despite the general quest for low-friction and low-wear characteristics, a closer look at the real world unambiguously demonstrates that friction is highly essential for many fundamental processes. Without friction, human beings would not be able to walk, or they would not be able to stop again after having initiated any motion. Many machine components, such as brakes and gears, rely on certain minimum levels of friction, if not even high friction, to induce and/or maintain their overall functionality while at the same time offering sufficient resistance to wear. When aiming at increasing the resulting COF, the proper design of the involved surface topography coupled with the used material pairing is the key strategy.

Related to the first aspect, in the case of soft and rather elastic contacts, an increase of the underlying surface roughness/topography by appropriate surface texturing can help to increase the contact area thus increasing adhesion and, hence, friction. For tribological contacts with a pronounced tendency for plastic deformation, an intentionally designed, rougher topography can help to increase the deformation component of friction. Additionally, by precisely tailoring and adjusting the involved surface roughness and topography of all involved rubbing surfaces, mechanical interlocking between both surfaces can be achieved, which is an interesting approach to increase static and kinetic friction. Regarding the appropriate design of the underlying surface topography, the use of ultra-short pulse laser with pulse durations in the pico- and femtosecond regime has been shown to be promising due to the possibility to generate Laser-Induced Periodic Surface Structures (LIPSS). These periodic surface features, commonly also referred to ripples, tend to have sub-wavelength periodicities, are aligned parallelly or perpendicularly to the laser polarization, and offer aspect ratios (depth versus periodicity) of about and even greater than 1:1. LIPSS fabricated on steel surfaces have shown to increase friction, while the overall magnitude of the increase depended on the relative orientation of the sliding direction with respect to the underlying features (Figure 5.9). As can be seen in Table 5.2, the usage of precisely designed LIPSS has been capable to notably increase friction up to factors up to 4 for metallic substrates and silicon surfaces.

Apart from LIPSS, other techniques can be used for surface modifications to increase friction. In this regard, Dunn et al. [20] created high-friction surfaces (COF > 0.6) on steel by pursuing different texturing strategies by varying pulse overlap and pulse energy (Figure 5.10).

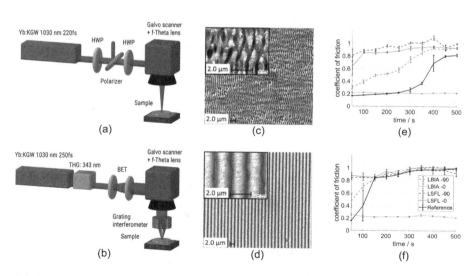

FIGURE 5.9 (a) Low spatial frequency LIPSS setup and (b) laser beam interference ablation setup as well as (c and d) the resulting surface textures. COF for textured surfaces against 100Cr6 ball without lubrication for a load of (e) 50 and (f) 200 mN, respectively. (Reprinted from Ref. [16] with permission by CC BY 4.0.)

TABLE 5.2
Summary of the Conducted LIPSS Studies with the Type of Texture and Structural Dimensions (Periodicity and Depth) Leading the Maximum Increase of Friction

Material	Type of Texture	Periodicity/ nm	Depth/ nm	Tribological Testing	Sliding Direction	Friction Increase Factor	Reference
Crystalline silicon	Periodic grooves	750	150 ± 50	Ball-on-disk	Perpendicular to LIPSS	Max. 1.6	[17]
100Cr6 steel	Periodic grooves	900	200 ± 30	Ball-on-disk	Perpendicular to LIPSS	Max. 4	[16]
CoCrMo alloy	Single- and multi-scale grooves	800	230	Ball-on-disk	Perpendicular to LIPSS	Max. 3	[18]
⟨111⟩ single crystal silicon (p-doped)	Periodic grooves	730		Ball-on-disk	Perpendicular to LIPSS	Max 3.5	[19]

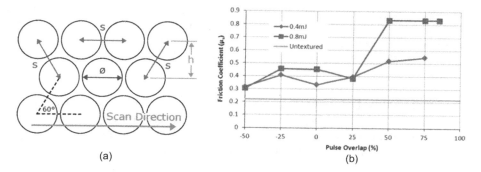

(a)
(b)

FIGURE 5.10 (a) Schematic of "hexagonal" laser pulse used for texturing, where d represents pulse diameter, s is the pulse separation, and h stands for the hatch distance. (b) Observed COF values for the textured surfaces. (Reprinted from Ref. [20] with permission by CC BY 4.0.)

Bio-inspired concepts to artificially increase friction can be also copied from nature. Specific animals including insects, spiders, and lizards have surfaces with defined features and patterns, which are tailor-made to fine-tune adhesion thus enabling a switchable attachment and detachment ability. This discovery has inspired and boosted the creation of polymers with super-adhesive properties induced by surface texturing. The overall idea is based on the approach of contact splitting, for which a larger contact area is split up into many, small-scale contact areas, thus allowing for tunable adhesion characteristics.

Increasing friction can also be very beneficial in mechanical transmission systems such as ultrasonic motors (USMs). Their operation and functionality can also be optimized by the precise use of surface texturing. In this regard, USMs make use of the reverse piezoelectric effect to enable the stator vibration. Under an applied electric field, the motion is elliptical. However, with a superimposed normal load, the motion can be transformed into a rotary motion, thus changing the characteristics and the output torque of the USM. Liu et al. [21] created different textures on polyimide plates (friction material in USMs), which helped to increase the resulting rotational speed, thus enhancing the output torque due to increased friction. Mechanical systems such as clutches and brakes, which require high friction to control the motion between their involved surfaces, have also shown an improved performance and increased friction when creating surface textures.

With respect to material selection, the analysis of material groups and coating candidates reveals different choices. In this regard, high-friction materials are substances or compounds that exhibit a strong resistance to sliding or relative motion when in contact with other surfaces. These materials are designed to increase friction between two objects, resulting in enhanced gripping or braking capabilities. This can be further boosted by the aforementioned surface modifications to enhance their surface energy and adhesive characteristics as well as to induce mechanical interlocking. High-friction materials can be made from various substances, including rubber, polymers, composite materials, or special coatings. These materials are chosen for their ability to generate and maintain high frictional forces while withstanding the associated stresses and wear. At the same time, since high-friction materials are often subjected to significant frictional forces, they need to possess good heat resistance and heat dissipation properties, without sacrificing their frictional characteristics. A few examples of the applications largely in need of high-friction materials are as follows:

- **Automotive Industry**: High-friction materials are used in brake pads and linings to provide effective braking performance. They ensure a strong grip between the brake components, converting the kinetic energy of a moving vehicle into heat through friction, thus enabling efficient deceleration and stopping.
- **Machinery**: High-friction materials are employed in clutches, gears, and couplings to transmit torque effectively and prevent slippage under high loads. This helps maintain the power transfer between rotating components, ensuring smooth operation and preventing loss of energy.
- **Sport Equipment**: High-friction materials are utilized in the manufacturing of athletic shoes, gloves, and sportswear to improve grip and traction. They provide athletes with better control, stability, and maneuverability, enhancing performance and reducing the risk of accidents.
- **Safety Devices**: High-friction materials are utilized in various safety devices, such as anti-slip mats, stair treads, and handrails. These materials offer enhanced traction and grip, reducing the likelihood of slips, trips, and falls, particularly in wet or slippery conditions.
- **Robotics and Automation**: High-friction materials are used in robotic gripping systems, conveyors, and manipulators to ensure secure handling and precise positioning of objects. The high frictional properties enable reliable grasping and movement of items without slippage or loss of control.

5.5 BIOTRIBOLOGY: HUMAN BODY TRIBOLOGY

Among different areas of importance, biotribology is one of the most critical ones. If we look at the human body, it involves various parts and systems, which critically rely on friction and wear. In this regard, the common biotribological systems relate to dental, ocular, skin, orthopedic, cartilages, bones, and spinal disks, which are all by the applied loads (stresses) and relative motion (Figure 5.11).

The human body, as one of the most efficient mechanical systems designed by nature, tries to maintain friction and wear as low as possible. This is possible by proper lubrication of contacts, which is mostly based on water. Most of the contact experience pressures in the range from 1 kPa to 10 MPa, which also connects with the underlying, rather soft surfaces. In this contact, dental contacts can be certainly considered as an exception, which are often fixed by enamel, which is hard as a solution to suppress wear similar to engineering systems.

5.5.1 Ocular Tribology

The eye is a complex organ with multiple moving parts, including the eyelids, cornea, and conjunctiva. These components dynamically interact during blinking and eye

Biotribology

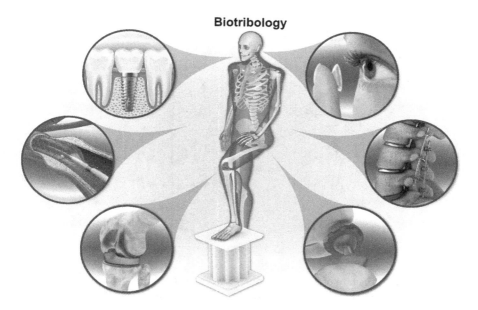

FIGURE 5.11 Biotribological systems of the human body. (Reproduced with permission from Ref. [22].)

movements, experiencing frictional forces that affect the eye's overall function and comfort. To minimize adverse effects, our body uses a tear film, which is a thin layer of fluid that covers the surface of the eye, acting as a lubricant between the cornea and the eyelids. The tear film plays a crucial role in maintaining the health and clarity of the cornea and provides a smooth optical surface for clear vision. Therefore, the composition and rheological properties of the tears are essential to sustain the film stability and prevent large friction as well as irritation on the ocular surface.

The tear film consists of three layers, which include a lipid layer, an aqueous layer, and a mucin layer (Figure 5.12). The lipid layer, secreted by the meibomian glands, serves as a barrier to prevent tear evaporation and provides a smooth surface for the eyelids to slide on during blinking. The aqueous layer, produced by the lacrimal glands, nourishes the cornea and helps wash debris away. The mucin layer, generated by the goblet cells, allows the tear film to adhere to the ocular surface and maintain stability. During the normal eye movements and blinks, the tear film spreads across the cornea, creating a uniform and smooth optical surface. Any disruption or irregularity in the tear film can lead to dry spots and/or abrasions on the cornea, causing discomfort and visual disturbances. Certain eye conditions, such as dry eye disease, can disrupt the tear film dynamics, leading to increased friction and irritation on the ocular surface. By studying ocular tribology, researchers and eye care professionals can develop more effective treatments and interventions to alleviate symptoms and improve the patients' ocular health.

The lubrication in the ocular system can be described by a Stribeck-like curve (Figure 5.13). The pressure of the eye lid to the cornea is on the order of 1–5 kPa. The velocity of movement may significantly vary depending on the activity and person. In

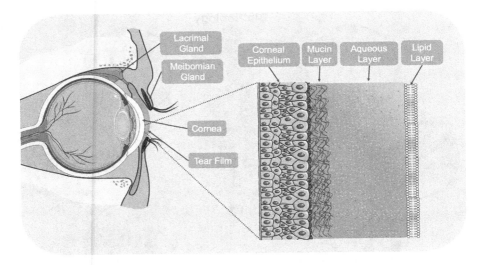

FIGURE 5.12   Structure of a tear-base film between the eyelids responsible for lubrication during eye blinking. (Reprinted and adapted from Ref. [23] with permission by CC BY 4.0.)

the case of blinking, the typical velocity is ~100 mm/s. The viscosity of the tear film also varies from person to person, while for a healthy person, the viscosity is ~1–5 cP (0.001–0.005 Pa s) (for comparison, the viscosity of water is 1 cP or 0.001 Pa s). In the case of dry eye conditions, the viscosity can increase up to 20 cP (0.02 Pa s). In dry eye conditions, the lack of sufficient tears can lead to increased friction between the eyelids and the ocular surface, thus causing eye irritation, a gritty sensation, burning, redness, and even blurred vision. In severe cases, it can damage the surface of the eye, leading to corneal abrasion or other complications. Considering the fluid film lubrication theory, the thickness of the lubricating films (Figure 5.13) can be estimated using

$$h = \frac{\eta \cdot V}{P}. \tag{5.6}$$

In case of healthy condition, the thickness of the tear film during blinking would be

$$h_{\text{healthy}} = \frac{1.5 \cdot 10^{-3}\ \text{Pa} \cdot \text{s} \cdot 100 \cdot \dfrac{10^{-3}\ \text{m}}{\text{s}}}{(1-5) \cdot 10^3\ \text{Pa}} = (0.3 - 1.5) \cdot 10^{-7}\ \text{m}$$

$$= 0.03 - 0.15\ \mu\text{m}, \tag{5.7}$$

which can be connected with the EHL regime. It should be noted that the real thickness of the tear film in the eye is notably increased and on the order of 1–2 μm. This is needed to ensure proper lubrication of the eyes even after evaporation of the fluid after prolonged times without blinking.

If we assume the eye to work like a bearing with lubrication provided by the tear film, a rough friction estimation can be calculated from journal bearing

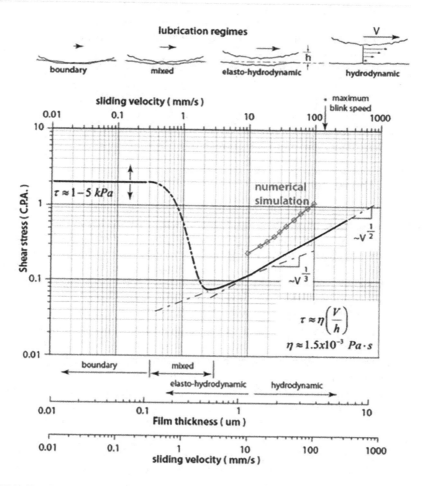

FIGURE 5.13 Summary of the eye lubrication mechanism. (Adapted with permission from Ref. [24].)

representation (Section 3.1.2). In this case, we assume a simple model of two concentric cylinders (the distance a between the eye and the eyelid remains the same and equal to h, while the area of the contact is A). Then, friction can be estimated as

$$F_f = \frac{A \cdot \eta \cdot V_0}{h}.$$ (5.8)

In the case of full eye contact (the eyelid completely covers the eye bulb, $A \sim 100\,\text{mm}^2$), we obtain

$$F_f = \frac{10^{-4}\,\text{m}^2 \cdot 1.5 \cdot 10^{-3}\,\text{Pa·s} \cdot 100 \cdot 10^{-3}\,\text{m/s}}{1 \cdot 10^{-6}\,\text{m}} = 10^{-2}\,\text{N}.$$ (5.9)

For comparison, prior studies (using computer modeling or *ex-vivo* porcine eye measurements) reported the frictional force between the cornea and eyelid to range between 0.01 and 1.0 mN. The difference can be attributed to the simplified assumptions of our model assumptions. Also, the average velocity of blinking changes during the duration of the blink and can be lower than 100 mm/s.

A detailed understanding of ocular tribology is also important for the development of contact lenses, since they are designed to interact with the eye's surface, and their performance relies on minimizing friction and ensuring proper lubrication. Contact lenses, when placed on the eye, can also introduce additional frictional forces that affect their comfort and wearability.

5.5.2 DENTAL TRIBOLOGY

Dental tribology focuses on the study of friction, wear, and lubrication related to oral health, dental materials, and the interactions between dental surfaces. Teeth are exceptionally durable structures, consisting of the visible portion within the mouth, known as the dental crown, and the portion embedded within the bone, referred to as the root. The outermost layer of the dental crown, which is made of enamel resembling one of the hardest tissues in the human body, is the major contributor to the tribological characteristics of the teeth. It is composed of 92%–96% hydroxyapatite $Ca_5PO_4\cdot3OH$, 1%–2% proteins (organic material), and 3%–4% water [25,26]. Dental wear refers to the loss of tooth structure over time due to mechanical actions, such as attrition (occurs when teeth come into contact with each other during chewing or grinding), abrasion (caused by external mechanical actions, such as using a toothbrush with excessive force or abrasive toothpaste), and erosion (loss of tooth structure due to chemical actions, such as exposure to acidic substances). These three processes are often superimposed and combined over time. In this regard, the loss of protective enamel due to erosion can make teeth more susceptible to abrasion and attrition.

We apply the basic concepts of friction and wear when brushing our teeth, using soft-bristled toothbrushes being effective at removing plaque and debris without causing excessive wear on the tooth enamel. In contrast, hard bristles may generate more friction and lead to increased wear on the tooth enamel and potential damage. In this context, the used toothpaste largely affects the brushing process, since it contains mild abrasives that help to increase the cleaning effectiveness by gently scrubbing away stains and surface deposits. Additionally, the toothpaste often contains active ingredients such as fluoride, which upon reaction with hydroxyapatite forms fluorapatite. This is an effective process since fluorapatite is less prone to degradation in an acid environment, thus aiding in strengthening tooth enamel and protecting against tooth decay.

The friction and wear characteristics of human teeth also largely depend on age. Zheng and Zhou [27] studied the wear of human teeth from different age groups (8, 18, 35, and 55) against a titanium alloy in artificial saliva using a reciprocating apparatus. Their analysis demonstrated that the overall age largely influences the enamel hardness and the arrangement of enamel rods, which resemble two critical factors influencing the tribological behavior of human teeth. As a result, more pronounced wear, in the form of delamination and plowing, was observed in the teeth of older participants. In comparison, the permanent teeth of younger and middle-aged participants exhibited enhanced wear resistance (Figure 5.14). Moreover, the enamel rods were shown to be perpendicularly aligned to the surface and smaller in size for young and middle age teeth. Meanwhile, for older age, the enamel rods start to lose alignment while increasing in size.

Since there are still potential decay issues, a detailed understanding of the tribological characteristics of human teeth is important to design materials to repair the functionality of natural enamel. Requirements for these materials are that the material should not wear enamel and that the overall processing and tooth integrity should be easily achievable. Dental implants are commonly utilized to replace damaged or missing teeth (Figure 5.15, left). The materials used must possess biocompatibility, stability, and resistance against fretting and wear. In the pursuit of aesthetic appeal, ceramics or (ceramic or resin) composites are predominantly chosen

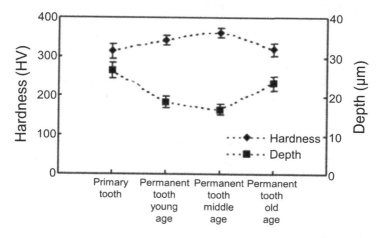

FIGURE 5.14 Variations of wear depth and hardness for teeth at four ages: Primary tooth and permanent tooth at young, middle, and old age. (Redrawn from Ref. [27] with permission.)

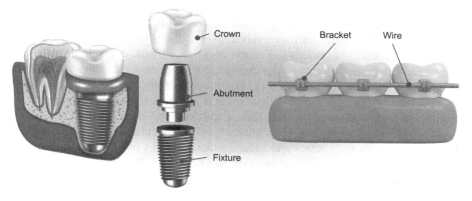

FIGURE 5.15 Structure of dental implants (left) and braces/retainers (right).

for restoration purposes, as polymers exhibit low wear resistance and metals may have an undesirable color. Beyond the visible components of dental implants, such as the crown, metallic alloys based on titanium, zirconium, or tantalum are frequently employed for fixtures, screws, and abutments. While these materials boast good biocompatibility, issues arise from wear caused by micro-movements and associated tribo-corrosion. This wear can lead to the release of toxic elements like aluminum or vanadium, resulting in inflammatory reactions, decreased stability, or even implant loosening. In addressing these concerns, coatings can be utilized and customized through doping elements to enhance bone ingrowth/osseointegration, prevent bacterial growth, and notably, improve mechanical stability, as well as wear and corrosion resistance. Generally, the fast and strong adhesion of the dental implant to the jawbone is of utmost importance. When coatings are applied, they have to withstand the high stresses during screwing without cohesive or adhesive fracturing and spalling in order to avoid particles in the bone and biological immune responses. Typically, achieving rapid and robust adhesion of a dental implant to the jawbone is crucial. When applying coatings, it is imperative that they can endure the substantial stresses incurred during the screwing process without experiencing cohesive or adhesive fracturing and spalling. This resilience is necessary to prevent the generation of particles in the bone and subsequent biological immune responses.

Other dental tribology applications are braces or retainers (Figure 5.15, right). In this context, DLC coatings have been investigated when employed on orthodontic brackets or wires of braces/retainers. Thereby, nickel-titanium (Ni-Ti) alloys are frequently employed due to their superelasticity and shape memory effect. However, there is a growing prevalence of allergy against nickel, which can be released due to Ni-Ti tribo-corrosion when immersed in oral saliva. Therefore, nowadays, alternative materials or coatings are increasingly being studied.

5.5.3 TRIBOLOGY OF JOINTS

Natural joints, such as knee, hip, and shoulder joints, are intricate structures, where bones, cartilage, ligaments, and synovial fluid work together to enable smooth and pain-free movement. Joints are the body's bearings that allow for mobility with low friction and low wear over decades of use. In joints, frictional forces and wear need to be minimized to allow for a smooth movement without causing damage or discomfort. Various adaptations have evolved over millions of years. The presence of cartilage, a specialized smooth and elastic connective tissue, covering the articulating surfaces of bones in joints is important to provide cushion and allow the surfaces to glide smoothly against each other. The lubrication in natural joints is primarily governed by the synovial fluid that contains nutrients and oxygen to support the health and maintenance of the joint tissues and nourishes the cartilage. The curved and congruent surfaces of the joints reduce local contact pressure and distribute the load evenly across the contact area. The contact pressures of joints are higher than for example for eye tribology and can vary in a range of 3–10 MPa depending on the activity. Under these pressures, the body fluids are squished outside the contact. To replenish the lubrication of the contact, additional routes for lubrication are needed. Porous structure of the cartilage allows to hold the synovial fluid so that it can move through and replenish the contact as needed. Unfortunately, with aging, natural functionality of materials suppresses leading to the failure of the joints.

Osteoarthritis (OA) is a chronic degenerative joint disease that is a major contributor to pain and loss of functionality in elderly individuals [28–34]. Early-stage OA is commonly manifested by the degeneration of articular cartilage concomitant with lubrication deficiency [35] and synovium inflammation [36], presenting a complex interaction of mechanical, biochemical, and cellular processes within the joint [37–40]. Impaired lubrication leads to weakened and overloaded cartilage and adverse cellular responses which, in turn, further compromise the lubrication. Specifically, due to inferior lubrication, the production of degradative enzymes by the chondrocytes and the inflamed synovium is greatly enhanced [41], resulting in collagen digestion in the cartilage and reduction in its mechanical properties [41,42]. Weakened cartilage is unable to withstand mechanical loading which further increases wear and damage. Existing treatment approaches include oral medication [43,44], intra-articular (IA) injection of lubricating fluids [45,46], or interventional micro-fracturing to repair local cartilage defects [47,48]. Oral administration of anti-inflammatory or chondroprotective drugs [43,44] is accompanied by multiple systemic side effects and adverse consequences from interactions with other medications commonly prescribed for elderly patients [49]. Currently used intra-articular injections practice hyaluronic acid [50] as a basic component replenishing the lubricating fluid in the joint [45]. The injections are usually administered in series, requiring ~3–4 injections for a 6-month period [51], thus demonstrating their limited effectiveness. Injections with corticosteroids administered at the same time or in parallel are usually used to relieve pain and reduce swelling; if used repeatedly, they also produce side effects [52].

The onset of OA among the population is amplified by growth in life expectancy and poor diet choices [53]. The functional impairment of OA caused by the degradation of the weight-bearing joints (hip, knee) mainly involves restricted daily activities due to pain and limited mobility for most of their lifetime [54]. That necessitated the need for the total joint replacements as a solution to permanent failure of natural joints (Figure 5.16) [55–58]. The first artificial joint replacement was performed in the 1950s [59]. For this purpose, PTFE, as a corrosion resistant and biocompatible material was used. Only after implantation in three hundred patients, two primary drawbacks of PTFE were revealed. Firstly, the PTFE displayed elevated in vivo wear rates, reaching up to 0.5 mm per month. Secondly, the wear debris from PTFE triggered a notably intense immune system reaction, called osteolytic cascade. The immune cells upon recognizing the wear particles try to break them down and eliminate from the body. Upon continuous production of wear debris, the immune system overreaction may further lead to swelling and attacking the bone cells.

After PTFE's failure, the focus was shifted to ultra-high molecular weight polyethylene (UHMWPE), shown to have superior wear properties compared to PTFE. UHMWPE resembles highly aligned molecules, which provide good tribological characteristics when tested in one sliding direction, but failed when tested in out-of-plane motion. To address this issue, the cross-linking of polymer chains by irradiation can be introduced, as it helps to reinforce the UHMWPE, though deteriorating the mechanical strength compared to well-aligned polymer chains.

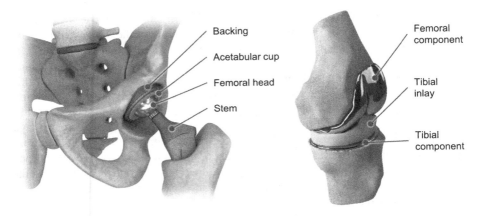

FIGURE 5.16   Total hip (left) and total knee replacement (right).

However, UHMWPE may also fail over time. Various studies have explored the feasi-
bility to eliminate polymers by using metal-on-metal or ceramic-on-ceramic contacts.
Unfortunately, in the case of ceramic-on-ceramic, issues like shuttering can lead to
the emergency surgery. Metal-on-metal contacts are also challenging due to high friction and wear in the absence of lubricant (if the body fails to provide such).

5.5.4   Skin Tribology

The skin's frictional properties play a vital role in our daily activities, such as grasping objects, walking, and performing various movements [60]. Skin tribology also
guides some advanced technological applications involving creating artificial skin
that mimics the material properties and sensory functions (such as tactile feedback)
of human skin. This innovation aims at empowering robotic hands and prostheses to
handle objects with precision and skill, resembling the capabilities of natural human
touch.

The skin's outermost layer, known as the stratum corneum, plays a crucial role in
reducing friction and wear. It is composed of dead skin cells that are tightly packed
and overlaid by lipids. The COF between the skin and different surfaces can vary
significantly, depending on factors such as (1) changes in surface roughness and texture of materials in contact with the skin, (2) presence of natural skin oils, sweat, and
applied lotions or creams (acting as lubricants), (3) exposure to surrounding moisture and temperature (defining mechanical properties of the skin), and (4) the forces
applied to press skin against surfaces and the speed at which the skin moves against
a surface.

In the case of dry skin, the resulting COF is influenced by two primary factors at
the skin-material interface: adhesion due to attractive surface forces and deformation
(manifested as hysteresis and plowing) within the supple, viscoelastic skin tissue. The
interplay of these elements is contingent upon variables such as contact circumstances,
the presence of fluids or lubricants (like sweat, water, and sebum), as well as the relationship between film thickness and the surface roughness of the engaged materials.

It can give rise to various lubrication effects, including boundary lubrication, mixed lubrication, and EHL. In moist or wet conditions, the COF tends to increase. This observation can be attributed to several effects such as the swelling of the outer layer of the skin, which becomes softer. Moreover, capillary forces may act thus sticking the skin and surface sticking together. Additionally, viscous shearing of the liquid layers can play an important role in the resulting COF. Finally, a glue-like layer can form when skin oils and proteins mix with a thin layer of water or sweat.

Skin is naturally equipped with sebaceous glands that produce sebum, an oily substance that lubricates the skin's surface. Sebum helps to reduce friction and provides a protective layer that prevents excessive drying and damage. Additionally, sweat glands produce sweat, which acts as a lubricant and helps to regulate our body temperature. The combined action of sebum and sweat contributes to the lubrication and smooth movement of the skin. However, if the skin hydration level continues to increase, the friction coefficient tends to increase as well (Figure 5.17). This phenomenon was attributed to the observed significant reduction in the elastic modulus of the stratum corneum as its hydration state increases as well as change in the adhesion forces. Meanwhile, with an increase in the apparent contact pressure, friction for dry and wet skin tends to reduce [60].

Skin friction is often accompanied by wear occurring over time due to repeated frictional interactions. Wear in the context of skin tribology refers to a gradual loss of skin cells due to friction and mechanical stresses. This wear on human skin can be challenging to quantify accurately due to its biological nature initiating the process of self-adaptation and repair. Under normal conditions, skin has a natural ability to regenerate and repair itself. However, in certain situations, such as prolonged friction or excessive pressure, wear can result in the formation of blisters. With a decrease in skin thickness and weakened viscoelastic recovery, aging skin becomes more prone and susceptible to wear as well as injuries like abrasions and pressure ulcers. In these cases, friction and shear forces are considered significant contributing factors to the risk.

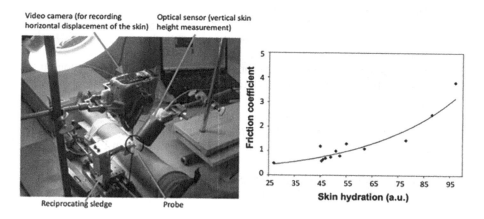

FIGURE 5.17 Apparatus used for real-time monitoring of skin friction and the measured COFs as a function of skin hydration. (Reprinted from Ref. [61] with permission.)

5.6   CHECK YOURSELF

- What is the significance of nanotribology in the context of nano- and micro-scale devices?
- List some examples of tribologically relevant nano- and micro-scale devices widely used in electronics and wireless technology.
- What is stiction in the context of nano- and micro-scale devices, and how can it occur during the fabrication process?
- Explain the puckering effect observed in graphene during tribological studies, and how it relates to friction increase.
- What are self-assembled monolayers (SAMs), and how do they contribute to reducing friction in nano-scale contacts?
- Describe the concept of high-temperature tribology and provide examples of applications where it is crucial.
- Why are most liquid lubricants not suitable for high-temperature tribological systems, and what challenges do they face?
- Discuss the different approaches to high-temperature lubrication, including layered materials, metal encapsulation, and thermally activated formation of lubricious oxides.
- What are some examples of applications that demand effective lubrication solutions for moving parts operating in cryogenic environments?
- Why do liquid lubricants face challenges in low-temperature applications, and how does it relate to the reduced mobility of molecules?
- How do coatings contribute to effective lubrication in low-temperature environments, and what are some examples of suitable coating materials?
- Explain the significance of thermal barrier coatings in low-temperature applications, and why low thermal conductivity is essential.
- What is the impact of temperature on the elastic modulus and hardness of materials in relation to the coefficient of friction (COF)?
- Describe the phenomenon of electronic friction and its experimental demonstration in cryogenic conditions.
- How can changes in surface topography, specifically LIPSS, influence friction?
- Discuss the role of bio-inspired concepts in increasing friction, particularly those inspired by animals with tailored adhesion features.
- How can surface texturing be applied to optimize the operation and functionality of mechanical transmission systems like USMs?
- Provide examples of applications where high-friction materials play a crucial role and explain why these materials are chosen for such purposes.
- Name various biotribological systems in the human body.
- How does the human body minimize friction and wear in biotribological systems?
- Describe the components of the tear film and their roles in ocular tribology.
- What are the consequences of disruption or irregularity in the tear film in ocular tribology?
- What are the key components of dental tribology, and how does enamel contribute to the tribological characteristics of teeth?

- Discuss the role of tribology in natural joints, such as the knee, hip, and shoulder joints.
- What challenges and advancements are discussed in the context of total joint replacements, particularly in hip and knee arthroplasty?

REFERENCES

1. D. Berman, J. Krim, Surface science, MEMS and NEMS: Progress and opportunities for surface science research performed on, or by, microdevices, *Progress in Surface Science* 88 (2013) 171–211.
2. X. Zeng, Y. Peng, H. Lang, A novel approach to decrease friction of graphene, *Carbon* 118 (2017) 233–240.
3. S. Li, Q. Li, R.W. Carpick, P. Gumbsch, X.Z. Liu, X. Ding, J. Sun, J. Li, The evolving quality of frictional contact with graphene, *Nature* 539 (2016) 541–545.
4. X. Hu, J. Lee, D. Berman, A. Martini, Substrate effect on electrical conductance at a nanoasperity-graphene contact, *Carbon* 137 (2018) 118–124.
5. C. Holmes, M. Tabrizian, Chapter 14: Surface functionalization of biomaterials, in: A. Vishwakarma, P. Sharpe, S. Shi, M. Ramalingam (Eds.), *Stem Cell Biology and Tissue Engineering in Dental Sciences*, Academic Press, Boston, MA, 2015, pp. 187–206.
6. R.L. Jones, B.L. Harrod, J.D. Batteas, Intercalation of 3-phenyl-1-proponal into OTS SAMs on silica nanoasperities to create self-repairing interfaces for MEMS lubrication, *Langmuir* 26 (2010) 16355–16361.
7. A. Lansdown, High-temperature lubrication, *Proceedings of the Institution of Mechanical Engineers, Part C: Mechanical Engineering Science* 204 (1990) 279–291.
8. A. Ghoshal, M. Murugan, M.J. Walock, A. Nieto, B.D. Barnett, M.S. Pepi, J.J. Swab, D. Zhu, K.A. Kerner, C.R. Rowe, Molten particulate impact on tailored thermal barrier coatings for gas turbine engine, *Journal of Engineering for Gas Turbines and Power* 140 (2018) 022601.
9. J. Gu, S. Joshi, Y.-S. Ho, B. Wei, T. Huang, J. Lee, D. Berman, N. Dahotre, S. Aouadi, Oxidation-induced healing in laser-processed thermal barrier coatings, *Thin Solid Films* 688 (2019) 137481.
10. W. Cui, H. Chen, J. Zhao, Q. Ma, Q. Xu, T. Ma, Progresses on cryo-tribology: Lubrication mechanisms, detection methods and applications, *International Journal of Extreme Manufacturing* 5 (2023) 022004.
11. Y. Iwasa, A.F. Ashaboglu, E.R. Rabinowicz, T. Tachibana, K. Kobashi, Cryotribology of diamond and graphite, *Cryogenics* 37 (1997) 801–805.
12. A. Dayo, W. Alnasrallah, J. Krim, Superconductivity-dependent sliding friction, *Physical Review Letters* 80 (1998) 1690–1693.
13. J. Sokoloff, M. Tomassone, A. Widom, Strongly temperature dependent sliding friction for a superconducting interface, *Physical Review Letters* 84 (2000) 515.
14. M. Highland, J. Krim, Superconductivity dependent friction of water, nitrogen, and superheated He films adsorbed on Pb (111), *Physical Review Letters* 96 (2006) 226107.
15. I. Altfeder, J. Krim, Temperature dependence of nanoscale friction for Fe on YBCO, *Journal of Applied Physics* 111 (2012) 094916.
16. S. Rung, K. Bokan, F. Kleinwort, S. Schwarz, P. Simon, J.-H. Klein-Wiele, C. Esen, R. Hellmann, Possibilities of dry and lubricated friction modification enabled by different ultrashort laser-based surface structuring methods, *Lubricants* 7 (2019) 43.
17. J. Eichstädt, G. Römer, A. Huis, Towards friction control using laser-induced periodic surface structures, *Physics Procedia* 12 (2011) 7–15.

18. S.H. van der Poel, M. Mezera, G.-w.R. Römer, E.G. de Vries, D.T. Matthews, Fabricating laser-induced periodic surface structures on medical grade cobalt-chrome-molybdenum: Tribological, wetting and leaching properties, *Lubricants* 7 (2019) 70.
19. I. Alves-Lopes, A. Almeida, V. Oliveira, R. Vilar, Influence of laser surface nanotexturing on the friction behaviour of the silicon/sapphire system, *Optics & Laser Technology* 121 (2020) 105767.
20. A. Dunn, J.V. Carstensen, K.L. Wlodarczyk, E.B. Hansen, J. Gabzdyl, P.M. Harrison, J.D. Shephard, D.P. Hand, Nanosecond laser texturing for high friction applications, *Optics and Lasers in Engineering* 62 (2014) 9–16.
21. X. Liu, G. Zhao, J. Qiu, Improving the performance of ultrasonic motors in low-pressure, variable-temperature environments, *Tribology International* 160 (2021) 107000.
22. M. Marian, D. Berman, D. Nečas, N. Emani, A. Ruggiero, A. Rosenkranz, Roadmap for 2D materials in biotribological/biomedical applications-A review, *Advances in Colloid and Interface Science* 307 (2022) 102747.
23. B. Tashbayev, M. Yazdani, R. Arita, F. Fineide, T.P. Utheim, Intense pulsed light treatment in meibomian gland dysfunction: A concise review, *The Ocular Surface* 18 (2020) 583–594.
24. A.C. Dunn, J.A. Tichy, J.M. Urueña, W.G. Sawyer, Lubrication regimes in contact lens wear during a blink, *Tribology International* 63 (2013) 45–50.
25. A. Lanza, A. Ruggiero, L. Sbordone, Tribology and dentistry: A commentary, *Lubricants* 7 (2019) 52.
26. Y. Zheng, K. Bashandeh, A. Shakil, S. Jha, A.A. Polycarpou, Review of dental tribology: Current status and challenges, *Tribology International* 166 (2022) 107354.
27. J. Zheng, Z.R. Zhou, Effect of age on the friction and wear behaviors of human teeth, *Tribology International* 39 (2006) 266–273.
28. X. Ji, Zhang, H., Current strategies for the treatment of early stage osteoarthritis, *Frontiers in Mechanical Engineering* 5 (2019) 57.
29. B. Xia, Di, C., Zhang, J., Hu, S., Jin, H., and Tong, P., Osteoarthritis pathogenesis: A review of molecular mechanisms, *Calcified Tissue International* 95 (2014) 495–505.
30. https://www.cdc.gov/arthritis/data_statistics/arthritis-related-stats.htm.
31. F.M. Maclean, Arthritis and other proliferative joint diseases, *Diagnostic Histopathology* 22 (2016) 369–377.
32. M. Hudelmaier, C. Glaser, J. Hohe, K.H. Englmeier, M. Reiser, R. Putz, F. Eckstein, Age-related changes in the morphology and deformational behavior of knee joint cartilage, *Arthritis & Rheumatism* 44 (2001) 2556–2561.
33. P.G. Passias, J.V. Bono, Total hip arthroplasty in the older population, *Geriatr Aging.* 9 (2006) 535–543.
34. H. Bang, Y.L. Chiu, S.G. Memtsoudis, et al., Total hip and total knee arthroplasties: Trends and disparities revisited, *American Journal of Orthopedics* 39 (2010) E95–E102.
35. M.B. Goldring, M. Otero, Inflammation in osteoarthritis, *Current Opinion in Rheumatology* 23 (2011) 471–478.
36. C.R. Scanzello, S.R. Goldring, The role of synovitis in osteoarthritis pathogenesis, *Bone* 51 (2012) 249–257.
37. S. Raman, U. FitzGerald, J.M. Murphy, Interplay of inflammatory mediators with epigenetics and cartilage modifications in osteoarthritis, *Frontiers in Bioengineering and Biotechnology* 6 (2018) 22.
38. H.L. Stewart, C.E. Kawcak, The importance of subchondral bone in the pathophysiology of osteoarthritis, *Frontiers in Veterinary Science* 5 (2018) 178.
39. F.P. Luyten, M. Denti, G. Filardo, E. Kon, L. Engebretsen, Definition and classification of early osteoarthritis of the knee, *Knee Surgery, Sports Traumatology, Arthroscopy* 20 (2012) 401–406.

40. E. Bonnevie, L.J. Bonassar, A century of cartilage tribology research is informing lubrication therapies, *Journal of Biomechanical Engineering* 142(3) (2020) 031004.
41. J.D. Backus, B.D. Furman, T. Swimmer, C.L. Kent, A.L. McNulty, L.E. Defrate, F. Guilak, S.A. Olson, Cartilage viability and catabolism in the intact porcine knee following transarticular impact loading with and without articular fracture, *Journal of Orthopaedic Research* 29 (2011) 501–510.
42. L.A. Setton, V.C. Mow, F.J. Muller, J.C. Pita, D.S. Howell, Mechanical properties of canine articular cartilage are significantly altered following transection of the anterior cruciate ligament, *Journal of Orthopaedic Research* 12 (1994) 451–463.
43. G. Tiraloche, C. Girard, L. Chouinard, J. Sampalis, L. Moquin, M. Ionescu, A. Reiner, A.R. Poole, S. Laverty, Effect of oral glucosamine on cartilage degradation in a rabbit model of osteoarthritis, *Arthritis & Rheumatism* 52 (2005) 1118–1128.
44. W. Zhang, W.B. Robertson, J. Zhao, W. Chen, J. Xu, Emerging trend in the pharmacotherapy of osteoarthritis, *Frontiers in Endocrinology (Lausanne)* 10 (2019) 431.
45. R. Bannuru, N. Natov, U. Dasi, C. Schmid, T. McAlindon, Therapeutic trajectory following intra-articular hyaluronic acid injection in knee osteoarthritis-meta-analysis, *Osteoarthritis and Cartilage* 19 (2011) 611–619.
46. L. Liu, R. Spiro, Collagen-polysaccharide matrix for bone and cartilage repair, Google Patents, 1999.
47. H. Chen, J. Sun, C.D. Hoemann, V. Lascau-Coman, W. Ouyang, M.D. McKee, M.S. Shive, M.D. Buschmann, Drilling and microfracture lead to different bone structure and necrosis during bone-marrow stimulation for cartilage repair, *Journal of Orthopaedic Research* 27 (2009) 1432–1438.
48. R. Mardones, C.M. Jofré, J.J. Minguell, Cell therapy and tissue engineering approaches for cartilage repair and/or regeneration, *International Journal of Stem Cells* 8 (2015) 48.
49. J. Steinmeyer, F. Bock, J. Stove, J. Jerosch, J. Flechtenmacher, Pharmacological treatment of knee osteoarthritis: Special considerations of the new German guideline, *Orthopedic Reviews (Pavia)* 10 (2018) 7782.
50. K.L. Goa, P. Benfield, Hyaluronic acid, *Drugs* 47 (1994) 536–566.
51. T.P. Stitik, S.M. Issac, S. Modi, S. Nasir, I. Kulinets, Effectiveness of 3 weekly injections compared with 5 weekly injections of intra-articular sodium hyaluronate on pain relief of knee osteoarthritis or 3 weekly injections of other hyaluronan products: A systematic review and meta-analysis, *Archives of Physical Medicine and Rehabilitation* 98 (2017) 1042–1050.
52. https://www.anikatherapeutics.com/products/orthobiologics/cingal/.
53. https://www.niddk.nih.gov/health-information/health-statistics/overweight-obesity.
54. R.C. Corti MC, Epidemiology of osteoarthritis: Prevalence, risk factors and functional impact, *Aging Clinical and Experimental Research* 15 (2003) 359–363.
55. J.A. Singh, S. Yu, L. Chen, J.D. Cleveland, Rates of total joint replacement in the United States: Future projections to 2020–2040 using the national inpatient sample, *The Journal of rheumatology* 46 (2019) 1134–1140.
56. G. Labek, M. Thaler, W. Janda, M. Agreiter, B. Stöckl, Revision rates after total joint replacement: Cumulative results from worldwide joint register datasets, *The Journal of Bone and Joint Surgery. British Volume* 93 (2011) 293–297.
57. J.E. Mueller, A.C.T. van Duin, W.A. Goddard, Development and validation of ReaxFF reactive force field for hydrocarbon chemistry catalyzed by nickel, *The Journal of Physical Chemistry C* 114 (2010) 4939–4949.
58. Z. Zhou, Z. Jin, Biotribology: Recent progresses and future perspectives, *Biosurface and Biotribology* 1 (2015) 3–24.
59. P.F. Gomez, J.A. Morcuende, A historical and economic perspective on Sir John Charnley, Chas F. Thackray Limited, and the early arthroplasty industry, *The Iowa Orthopaedic Journal* 25 (2005) 30.

60. S. Derler, L.-C. Gerhardt, Tribology of skin: Review and analysis of experimental results for the friction coefficient of human skin, *Tribology Letters* 45 (2012) 1–27.

61. M. Kwiatkowska, S.E. Franklin, C.P. Hendriks, K. Kwiatkowski, Friction and deformation behaviour of human skin, *Wear* 267 (2009) 1264–1273.

6 Beyond Traditional Tribology

As we discussed earlier, friction and wear are an integral part of our life not only affecting the performance of traditional mechanical systems but also playing a significant role in everyday activities. In this module, we review several examples addressing tribological issues and needs beyond traditional friction-, wear-, and lubrication-related phenomena. The considered examples demonstrate the importance of a detailed tribological understanding of various manufacturing processes, designs and applied problems as they open new opportunities for these applications.

6.1 FRICTION STIR-WELDING AND FRICTION STIR-PROCESSING

Friction stir-welding (FSW) is an alternative to traditional welding and uses the friction-induced heating and wear of materials (usually metals) in the contact to initiate material intermixing and accelerated diffusion for the formation of the joints (Figure 6.1, top) [1]. Furthermore, this process can be used to precisely induce materials modification allowing to locally manipulate their structural and functional properties thus even creating new alloys and composites, which has been named friction stir-processing (FSP).

The basic concept of FSW or FSP is very similar. The method uses a cylindrical or conical tool with a profiled threaded probe (nib) rotating at a constant speed to enter the joint or modification area until the probe shoulder blocks further penetration of the tool. The parts to be joined together are clamped onto a backing bar to prevent their joint faces from being forced apart by the process.

Once the process has started, high friction and wear are created between the welding tool shoulder, the nib, and the parts being joined. Consequently, frictional heat accompanied by mechanical stresses induces an overall temperature close to or above the melting temperature, thus softening the metals/alloys. During rotation, material adjacent to the tool softens facilitating interdiffusion and mechanical intermixing of the materials. Ideally, the tool is nonconsumable and made of wear-resistant materials, while all material intermixing is generated only for the joining structures. During the process, the tool is moved in the transversal direction along the weld line in a plasticized tubular shaft of metal.

As the melting temperature of metals can range in a large diapason, FSW requires a precise adjustment of the process to select suitable parameters for specific materials involved. Specifically, overheating of the contact should be prevented to avoid damage to the resulting joint structure. The complexity of the process is emphasized by the process-induced structural modifications as they define the quality and

DOI: 10.1201/9781003397519-6

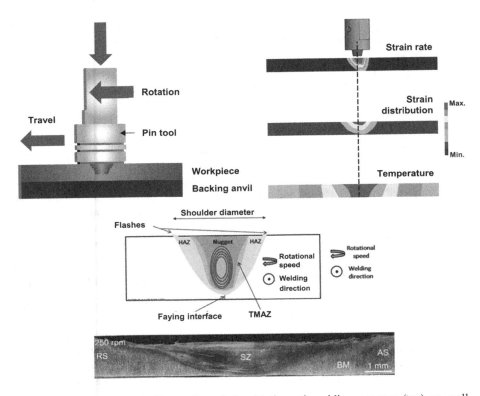

FIGURE 6.1 Schematic illustration of the friction stir-welding process (top) as well as formation of the respective structural zones with the base material, the stir zone, the thermo-mechanically affected zone, and the heating-affected zone (bottom). (Adapted from Ref. [1–3] with permissions by CC BY 4.0 and from Elsevier [3].)

mechanical strength of the joint. Upon pressure release, different zones reinforcing the joint strength are created from the base material (BM) such as the stir zone (SZ), the thermo-mechanically affected zone (TMAZ), and the heating-affected zone (HAZ) (Figure 6.1, bottom). The SZ is the region, where the pin is placed during processing, and it is usually connected with roughly equiaxed grains creating several concentric ring shapes. The TMAZ, which resembles the zone next to the SZ, is formed under lower temperature and strain, resulting in a microstructure with reduced grain size compared to the SZ region. The final affected zone, HAZ, does not directly experience the deformation, but some heating dissipated from SZ and TMAZ, which results in thermally unstable microstructures.

To understand the physics of FSW, it is important to qualitatively evaluate the friction-induced heating. The basic parameters that affect the process connect with the rotational speed of the nib (the major contributor to the heat input), the travel speed of the tool (defines how quickly the process can be completed), the plunge load or plunge position, the tool lead angle, the tool design, and geometry (thickness and composition of the plates determine the needed adjustment in the processing parameters). The heat generation during FSW is primarily caused by two sources: friction at the surface of the tool and the deformation of the material near the tool. It is usually thought that the majority of this generation happens beneath the shoulder due to its greater area of contact with the workpiece. This contact is often represented by sliding friction with a COF μ and an interfacial pressure P. In a simplified model, the torque M rotating the shaft of radius R is responsible for the heat generation. It can be found as

$$M = \int_0^{M_R} dM = \int_0^R r \cdot \mu \cdot P(r) \cdot 2 \cdot \pi \cdot r \, dr, \tag{6.1}$$

with the tool rotational speed ω and the tool radius R. If $P(r)$ is constant in the contact, then

$$M = \frac{2}{3} \pi \cdot \mu \cdot P \cdot R^3. \tag{6.2}$$

When assuming that the shearing is fully converted into heat, the average heat power input per unit time for this process can be described as

$$q_0 = \int_0^{M_R} \omega \, dM = \int_0^R \omega \cdot r \cdot \mu \cdot P \cdot 2 \cdot \pi \cdot r \, dr = \frac{2}{3} \pi \cdot \mu \cdot P \cdot \omega \cdot R^3, \tag{6.3}$$

where ω is the tool rotational speed. This equation suggests that an increase in the radius, pressure, or rotational speed increases the heat input. The resulting temperature increase can be expressed through the heat flux

$$\dot{q} = \frac{q_0}{\pi \cdot R^2} \tag{6.4}$$

as (see Section 1.6.3)

$$\Delta T = \frac{\dot{q} \cdot a}{k} = \frac{q \cdot R}{\pi \cdot R^2 \cdot k} = \frac{2 \cdot \pi \cdot \mu \cdot P \cdot \omega \cdot R^4}{3 \cdot \pi \cdot R^2 \cdot k} = \frac{2}{3} \cdot \mu \cdot P \cdot \omega \cdot \frac{R^2}{k}, \tag{6.5}$$

where a is the contact size and k is the thermal conductivity. As the process occurs near the melting point. The temperature difference ΔT can be defined by the difference in melting temperature of the material used and the processing temperature. For instance, in the case of FSW of aluminum alloys, the temperature in the contact reaches up to 750°C [4] resulting in a difference of $\Delta T \sim 725$°C compared to room temperature.

Several important points related to FSW:

- It should be noted that the melting of the materials affects the COF as the new frictional system becomes lubricated with a liquid layer of the molten material. Consequently, the heating in the contact decreases, and so does the actual contact area, leading to non-uniform joints. Therefore, the COF during the process tends to be rather unstable and depends on various processing parameters such as pressure, velocity, thermal conductivity, etc.

- The rotational speed should not be too high (~800–1500 min⁻¹) to ensure good uniformity of the contact and to avoid the mechanical removal of the material.
- For sufficient welding accuracy, the pin diameter is usually ~6–10 mm.
- The welding speed is usually controlled to ensure the uniformity of the process and to limit the processing time. The welding speed is typically about ~100 mm/min.
- As the process involves abrasive wear, there are specific requirements for the pins (Figure 6.2). The pin material should demonstrate high strength and ductility, good fatigue life, high fracture toughness, low coefficient of thermal expansion, thermal and chemical stability, high wear resistance, and high melting point. The commonly used pin materials are WC-C, steel (mostly martensitic with the inclusion of TiC), WC-Co, SiC, and diamond.

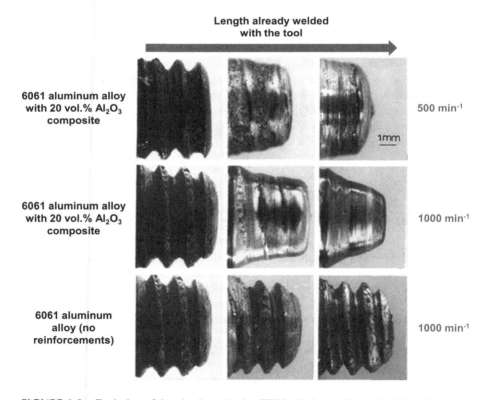

FIGURE 6.2 Evolution of the pin shape during FSW with increasing welded length dependent on rotational velocity and material. Magnifications are the same irrespective of the macrograph. (Reprinted and adapted from Ref. [5] with permission.)

The structure of the joint and the resulting mechanical properties are highly affected by the process parameters:

- Welding speed ↑ – tensile strength ↑ due to a lower heat generation in the local spot, which reduces material diffusion and limits grain growth.
- Rotational speed ↑ – hardness ↓ owing to higher heat input and more melting during joint formation.
- Pin diameter ↑ – strength of the joint ↑ due to the larger area being involved in the process and thus larger joint formation.

Moreover, FSW has a number of advantages over other fusion-based welding techniques. In this regard, this type of joining process is safer and more environmentally friendly as it does not release toxic fumes or splatter. No consumables are required to produce the weld, whereas the process can be automated, which makes it well-suited for industrial purposes and upscaling (the process can be incorporated using simple milling machines). In contrast to traditional welding, FSW does not create a well-known welding pool and can be performed in all positions and for various geometries of the joints. At the same time, the welds are more uniform and require less post-processing. The mechanical characteristics of the joints, which are often even thinner, are good since FSW eliminates common problems such as solidification cracking, porosity, and solute redistribution. Furthermore, the defect density encountered in FSW is highly reduced. Finally, FSW is more cost-effective and requires less specialized training. Some disadvantages of FSW include the presence of an exit hole in the joint when removing the tool and the need to use heavy-duty clamps to hold the materials in place during joining. Additionally, potential issues are insufficient weld temperatures, long tunnel-like defects, poor continuity between the materials from each side of the weld, and the formation of bonds that are hard to detect without X-ray or ultrasonic testing.

Current challenges associated with FSW are as follows:

- The joining/welding of thicker plates is difficult.
- It is mostly used for joining metals, but there is the possibility of joining polymers and ceramics [6].
- Due to differences in thermal expansion, joining different materials is challenging.

Some of these challenges are tried to be overcome by modifications to the conventional FSW process, such as, for instance, the addition of powders or the use of stationary shoulders (only pin rotates) to make the weld smoother and more uniform.

6.2 TRIBOELECTRICITY

Since tribological systems always involve energy dissipation, multiple efforts have been devoted to the concepts of accumulation and reuse of frictional energy. An interesting application of sliding-induced surface modifications relates to triboelectric systems that convert frictional energy into electrical energy through electrostatic charging.

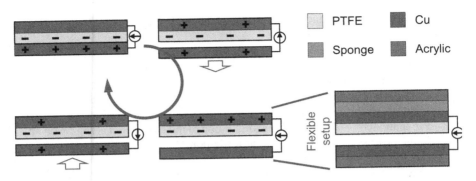

FIGURE 6.3 Design and working principle of a simple TENG in contact separation mode and its extension by a soft foam.

Previously, the charge generation between two contacting surfaces has been considered a negative effect in the electrical and electronic industry causing damage and safety issues to electronic devices [7]. However, the effect found its application niche in triboelectric generators or nanogenerators (TENGs) incorporated into energy harvesting and sensing devices [8–10].

A simple TENG can be made of two metal electrodes (e.g., from Cu) with one being covered with a dielectric film such as PTFE (Figure 6.3). Upon making contact, the Cu film becomes positively charged, while PTFE becomes negatively charged as a result of electron transfer. When the films are separated, the charge transfer is initiated as a result of the induced potential difference. Consequently, the electrode attached to the PTFE becomes positively charged, thus creating a current flow through the externally connected circuit. Once the separation distance starts to reduce and the electrodes are brought back together, the electrical current flow reverses. The connection of the system to a circuit results in a current flow. This is usually referred to as **contact separation mode**.

In the presented example by Wang et al. [11], the simple TENG device had lateral dimensions ranging from 1 to 5 cm, while the thickness of the Cu and PFTE films varied from 200 to 600 μm. This parameter variation allowed the achievement of current densities up to 50 μC/m² in air and 200 μC/m² in high vacuum for the hard contact system, while values up to 142 μC/m² in air and 1003 μC/m² in high vacuum were achieved by incorporating the foam cushion (Figure 6.4). The improvement observed for the soft contact can be attributed to the better distribution of the mechanical stresses at the contacting interface which promotes the charge transfer capability of the device.

For TENGs to be commercialized, the power density, which is related to the triboelectric charge density, is a critical property to control [12]. The gap voltage between the surfaces can be defined as

$$V_{\text{gap}} = \frac{t \cdot \sigma \cdot d}{\varepsilon_0 \left(t + d \cdot \varepsilon_r \right)}, \tag{6.6}$$

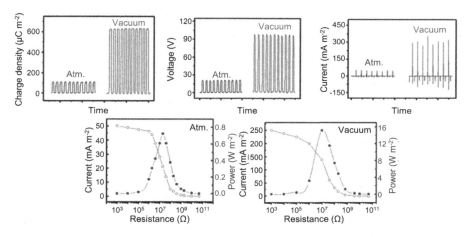

FIGURE 6.4 Output of a soft Cu/PTFE-based TENG contact with a foam cushion at ambient pressure and under high vacuum. (Reprinted from Ref. [11] with permission by CC BY 4.0.)

where t is the thickness of the insulating film (PTFE above), σ relates to the triboelectric charge density, and ε_r and ε_0 are the relative permittivity for the insulating film (PTFE in the described case) and the vacuum permittivity, respectively. Previous studies have achieved high-output charge densities by the injection of ions or the usage of fragmental contact structures. However, the output charge density is mostly limited by the air breakdown voltage described by Paschen's law as [11,12]

$$V_b = \frac{A \cdot P \cdot d}{\ln(P \cdot d) + B},\qquad(6.7)$$

for which A and B are constants describing the composition and pressure of the gas, in which the device is operated. For air at atmospheric pressure, $A = 2.87 \times 10^5$ V/(atm·m), and $B = 12.6$. P is the gas pressure and d resembles the gap distance. To avoid the voltage breakdown, the gap potential difference should be smaller than the breakdown voltage at any operational gap distance. The maximum surface charge density σ is

$$\sigma_{max} = \min\left(\frac{A \cdot P \cdot \varepsilon_0 \left(t + d \cdot \varepsilon_r\right)}{t \cdot \left(\ln(P \cdot d) + B\right)}\right).\qquad(6.8)$$

The theoretical maximum surface charge density allowed without air breakdown at different air pressures and the predicted gap distance for the air breakdown are summarized in Figure 6.5.

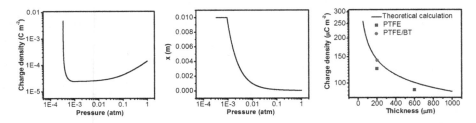

FIGURE 6.5 Theoretical maximum surface charge density allowed without air breakdown at different air pressures (left), corresponding gap distance at which air breakdown is about to occur with a PTFE film of 200 μm (middle), and relationship between the maximum surface charge density and the thickness of the insulating film. (Reprinted and adapted from Ref. [11] with permission by CC BY 4.0.)

In addition to electron transfer, possible sources for the charge transfer can be ion transfer and material transfer from wear debris produced upon contact closure:

- **Electron Transfer**: the driving force connects with phonon–electron interactions assisted by frictional heat. This mechanism is dominant for metal/insulator tribo-pairs and is believed to be impossible for insulator–insulator contacts.
- **Ion Transfer**: the mechanism originates from the tribo-contact-induced transfer of loosely bonded atoms with one polarity, while strongly bonded atoms with the opposite polarity remain on the surface, thus resulting in a charge accumulation.
- **Material Transfer**: charging originates from the random distribution of the surface charges on the wear debris. Because of this, the potential energy transfer is believed to be lower. In this regard, similar to other charging mechanisms, the origin of the electrification from the tribological point of view can be connected to adhesive wear. Prior studies demonstrated that the charge accumulation and the sign of the charge depend on the particle size transferred in the contact zone. Specifically, it has been shown that smaller particles tend to be negatively charged, while larger ones tend to be positively charged. The effect of particle size on the charging density has been investigated by demonstrating the increase in the charge transfer capability for smaller particles [13]: 335–550 μm: 10 nC/g, 250–255 μm: 28 nC/g, 125–150 μm: 50 nC/g, and 90–125 μm: 65 nC/g.

However, the process is self-regulatory and the resulting surface charge density may differ from the maximum possible tribo-generated charge due to charge backflow. The outcomes of different charging mechanisms are summarized in Table 6.1.

The TENG design can be further modified to improve the charge transfer efficiency. One example is the involved charge transfer process of **sliding-based TENGs** as schematically depicted in Figure 6.6.

TABLE 6.1

Summary of Possible Sources of Charge Transfer in Triboelectric Surfaces [14]

Possible Source	Mass	Charge
Electron	10^{-31}kg	1 e
Ion	10^{-20}–10^{-27}kg	1-10 e
Nanoparticle (atom + ion surface charge)	$>10^{-12}$kg	1-10 e

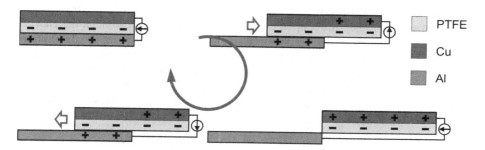

FIGURE 6.6 Charge transfer process of the presented sliding-mode TENG.

This model can be mathematically described using simple assumptions. In the case of the reciprocating motion with displacement l, maximum area of the contact S, applied force F_n, and constant speed of v, the electrical current equals the total charge generated at the contact. Let us assume that all frictional energy is dedicated to the extraction work of N charged particles, $W_{fr} = W_{out}$, where W_{out} is the work of charge extraction. For one particle, the transferred charge Q_i is

$$Q_i = \delta_i S_i, \tag{6.9}$$

where δ_i is the surface charge and S_i is the surface area of the particle. If we assume that during the half cycle of sliding, N number of particles can be extracted ($N = S/S_i$), the total potential charge transfer becomes

$$Q = NQ_i = \delta_i S. \tag{6.10}$$

If we multiply the parts of the equation by the work done during sliding and then substitute it with the frictional energy ($W_{fr} = \mu \cdot F_n \cdot l$):

$$W_{out} \cdot Q = W_{out} \cdot \delta_i \cdot S = W_{fr} \cdot \delta_i \cdot S = \mu \cdot F_n \cdot l \cdot \delta_i \cdot S. \tag{6.11}$$

Then, the total current generated will be equal to

$$I = \frac{Q}{t} = \frac{\mu \cdot F_n \cdot l \cdot \delta_i \cdot S}{W_{out} \cdot t} = \frac{\mu \cdot F_n \cdot \delta_i \cdot S \cdot v}{W_{out}}. \qquad (6.12)$$

This simple model suggests that the generated electrical current will increase with COF, applied load, surface area, and sliding velocity and decrease with the energy needed to extract the charges. The process is a little more complicated as not all the possible charges can be extracted and brought to the counterpart surface, while the mechanism of the preferability of one surface over another for charge accumulation is still debatable [15].

It has been established that materials with higher electron affinity (better capability of trapping electrons) upon contact tend to acquire more negative charges [16–18]. The results of empirical studies combining different materials exposed to tribo-electrical contact are summarized in Figure 6.7. In the case of polymers, highly polar (higher amount of functional oxygen groups) and hydrophilic polymers (such as polyester) tend to charge more positively than less polar ones (such as polypropylene) [16]. Further important aspects to consider for the triboelectricity effect:

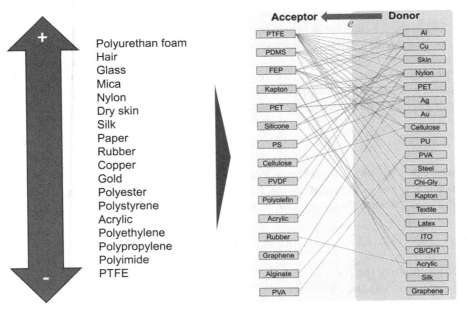

FIGURE 6.7 Triboelectric materials in series following a tendency to easily lose (donors) and gain (acceptors) electrons (left) as well as common material pairings (right). (Reprinted and adapted from Ref. [18] with permission by CC BY 4.0.)

- Materials involved in the contact: microstructure, composition, morphology, wear resistance, electrical characteristics, among others.
- Environment effect: air vs vacuum, humidity (may cause charge leakage), temperature.
- Load conditions and stress distribution during contact.

The triboelectricity concept has been further explored for integration with different concepts, such as thermoelectric generators (for instance, using the power generation of the human body), photonics (combined with solar cells), and biomedical energy harvesting (implantable nanogenerators), among others.

6.3 SEISMIC ACTIVITY

Earthquakes are some of the most destructive natural disasters, and their prediction has been a priority for scientists since the early 20th century [19]. Researchers have been using a variety of methods to attempt to accurately predict earthquakes, such as measuring strain levels (Richter magnitudes), monitoring seismic activity, and using satellite imagery. Since earthquakes are caused by sudden slips between the Earth's tectonic plates due to strain energy release, the fundamental mechanisms accompanying the events can be described using tribological models. This strain energy release can be continuous and slow (aseismic, ~cm/year displacement), inducing creep phenomena of the fault. In the case of earthquakes, the energy release is discrete and fast (seismic, reaching ~m/s displacement). The occurrence of seismic vs. aseismic activity is defined by the friction generated at the slip events. Three general approaches are used to predict and model earthquakes, which are based on

- pure empirical data [20], or
- rock friction analysis [21], and
- stress correlation function to describe pseudo-elastic behavior [22].

The empirical approach uses known data from previous earthquakes and builds a phenological model to predict the duration of the earthquake [23]. This is done through empirical data regarding peak amplitudes, frequency, and energy carried. This can also be used to develop vector predictions of duration, which enables estimations of other parameters such as magnitude [20].

Since the Earth's crust is made up of rocks and minerals, the second approach investigates rock sliding in lab-scale testing. In this case, all combinations of rocks potentially participating in the slip process should be considered. Some rocks, such as igneous coming from the magma beneath the surface, are homogeneous formations. Others, e.g., sedimentary rocks, tend to form a conglomeration of different ceramics. Many additional parameters, which are usually not aligned with real values, should be considered in the lab-scale models. Among them are the porosity and distribution of the fluid pressure, the effect of strain build-up time on the COF, the sliding velocity, and displacement. The last approach, which connects with the elasto-dynamic model, uses stress correlation functions for shear and fracture zone predictions. This is all based on an assumption that the Earth's crust is either in a

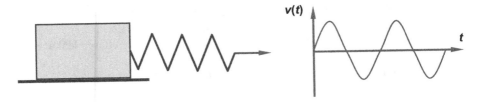

FIGURE 6.8 Simple model representing the stick-slip behavior occurring during an earthquake (left) and the resulting velocity as a function of time for the underlying stick-slip behavior (right).

self-organized critical state or acts as a critical point system trying to retreat from a critical point. In this case, the major event is preceded by a set of small earthquakes along the fault lines that occur and redistribute stress locally. Once the major event happens, the long-range stress distribution from the smaller quakes is lost and the cycle can repeat itself. A simple model of stick-slip dynamics can be used to describe sudden slip events between the involved plates, see Figure 6.8.

The equation describing the movement (position x as a function of time t, velocity v as a function of time t, and acceleration $\ddot{x}$) of the block of mass m (normal load N) attached to the spring with spring constant k and experiencing friction $\Delta\mu_0$ is given by

$$m \cdot \ddot{x} + k \cdot x = \Delta\mu_0 \cdot N, \tag{6.13}$$

$$x(t) = \frac{\Delta\mu \cdot N}{k}\left(1 - \cos(\alpha \cdot t)\right), \tag{6.14}$$

$$v(t) = \frac{\Delta\mu \cdot N}{\sqrt{k \cdot m}}\left(\sin(\alpha \cdot t)\right), \tag{6.15}$$

where

$$\alpha = \sqrt{\frac{k}{m}} \tag{6.16}$$

$\alpha \cdot T = \pi$, where T is the slip duration in the stick-slip event. In this simple model, we consider discrete differences between static and dynamic friction. In this regard, static friction μ_s depends on the history of the contact, particularly how long the contact has been static under a specific load, and can be described as

$$F_s = A + m \cdot T^n \left(\text{or sometimes written as } F_s = A + m \cdot \ln(t)\right), \tag{6.17}$$

where A is the force of adhesion and t reflects the time of the contact. This implies that over time more and more energy is required to move a fault (thus a higher energy build-up occurs), which results in earthquakes with larger magnitudes.

Moreover, dynamic friction depends on the sliding velocity and the nature of the contact. Therefore, in the case of earthquakes, it is sensitive to the rock types involved in the contact and temperature.

$$F_d = A + \mu \cdot N. \tag{6.18}$$

If the difference $\Delta\mu = \mu_s - \mu_d < 0$, the system is stable, no earthquake nucleation occurs as a dynamic rupture is arrested. In the case of $\Delta\mu > 0$, the system is conditionally unstable, and earthquakes may nucleate, causing the dynamic rupture between the surfaces to propagate and stress release. Experimental lab-scale observations demonstrated that the transition between static and dynamic friction is not abrupt, but rather follows the slip-weakening friction effect due to a sliding velocity increase. In this case, μ evolves to the new steady state over a characteristic slip distance D_c (Figure 6.9). The corresponding shear stresses increase until the slip occurs and decrease during the actual slip.

The weak side of this slip-weakening friction model is that it does not account for the resetting of the static friction strength. Therefore, further adjustment has been proposed to include time-dependent friction due to the growth of the asperity contact and frictional healing effect, i.e., slip is needed for healing the surface to reach the new frictional state of static friction, see Figure 6.10.

In this case, friction can be described as

$$\mu = \mu_s + a \ln\left(\frac{V}{V_0}\right) + b \ln\left(\frac{\theta \cdot V_0}{D_c}\right), \tag{6.19}$$

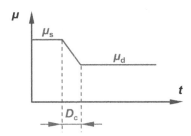

FIGURE 6.9 Change in the resulting coefficient of friction over time with the definition of the characteristics slip distance D_c.

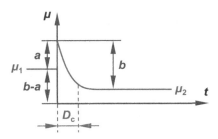

FIGURE 6.10 Characteristic friction response for stick-slip faults to illustrate the correlation between sliding velocity, friction, magnitude, effect, and displacement. (Redrawn from Ref. [23].)

where a is the magnitude of the direct effect and b represents the evolution effect. In the case of $a > b$, strengthening of friction occurs, while in the opposite effect $(a < b)$, weakening of friction is observed leading to unstable rupture. Thus, the second term accounts for the direct effect, and the third term for the evolution effect of friction. The evolution only occurs with slip and depends on the dynamic state variable θ.

The potential parameters affecting slip are numerous thus making the mathematical models complex and requiring lots of empirical data analysis. Among the parameters to be considered as particle size, interactions between the particles, shear stresses, clay contact, roughness, porosity, fluid content, fluid chemistry, temperature, and local normal stresses, among others.

6.4   TRIBOLOGY IN MUSIC

Another interesting example of the use of frictional processes in everyday life relates to musical instruments, such as the violin. The sound is created from the friction between the bow and the strings. Upon initial movement of the bow, the string deflects with the bow. During this process, the friction force is equilibrated by the string's stretching resistance (Figure 6.11, left).

The force generated in response to the strain in the string can be described by Hooke's law, which depends on the change in the length of the string:

$$F_f = 2 \cdot T \cdot \sin \alpha = 2 \cdot T \frac{x}{\sqrt{x^2 + \left(\dfrac{l}{2}\right)^2}} \tag{6.20}$$

For very small deflections of the string ($x \ll l$, then $x^2 + (1/2)^2 \sim (1/2)^2$):

$$F_f = \mu \cdot F_N = \frac{4 \cdot T \cdot x}{l} \tag{6.21}$$

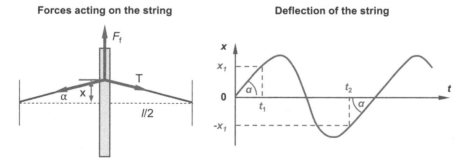

Forces acting on the string   **Deflection of the string**

FIGURE 6.11   Summary of forces experienced by string due to the movement of the bow (left) and deflection of the string as a function of time (right).

This static friction force reaches its maximum value and then the slip between the bow and the string occurs (at time t_1 in Figure 6.11, right), leading to a decrease in the acting friction force. Let's assume that at this moment the minimum contact between the string and the bow exists, leading to free movement of the string (vibrations). Upon the contact release, the velocity of the middle of the string equals the velocity of the bow. Consequently, it continues the movement for some time with a negative acceleration from the strain, reducing this velocity. After reaching zero, the string will move in the opposite direction until reaching the largest negative deflection and then back until being picked up by the bow (at time t_2 in Figure 6.11, right). This movement of the string, also known as "Helmholtz motion" [24], is periodic and involves a single slip in each cycle of stick-slip vibration.

As the deflection of the string and, thus, the energy of the vibrations depends on the COF, the bow is on purpose covered with rosin to increase the resulting frictional forces and improve sticking between the string and the bow [25]. Harder rosin produces more friction, which is beneficial for creating a crisp sound, while softer rosin produces less friction, resulting in a warmer and mellower sound. It should be noted that the rosin is a natural glassy material with a transition temperature near room temperature. As friction is accompanied by heating, the tribological properties of the rosin are largely affected by the bow/string sliding process. If we further elaborate the model of the stick-slip behavior in this system:

- Assume a normal force of 1 N applied through the bow on the string. The Hertzian contact area estimates for steel/glass (representing rosin on string) contact are ~1000 μm² [25]. The average contact pressure becomes 1 GPa.
- The rosin layer is usually 1–2 μm thick.
- For 660 Hz, sound pitch, the period of the string vibrations is 1.5 ms.
- For a typical violin playing, moving at 3 m/s speed slips last for 100 μs per period of vibration.
- The resulting strain rate for the string is about ~107.

As a consequence, the shear strain rate of the rosin can be described by Eyring model [26]:

$$\dot{\gamma} = K \cdot T \cdot e^{-\frac{E+P\cdot\varphi}{R\cdot T}} \cdot \sinh\left(\frac{\tau}{\tau_0}\right), \tag{6.22}$$

where T stands for temperature, P represents the pressure, τ reflects shear stress, and τ_0 is the Eyring stress. As the sliding velocity increases due to traction between the string and the bow, the temperature of the contact and the rosin layer rises (typically reaching temperatures around 30°C). As the rosin begins to soften due to the temperature reaching its glass transition temperature, the motion of the layer tends to become concentrated in a shear band of about 100 nm. This leads to a decreased viscosity and/or shear yield strength of the rosin layer, resulting in a reduced resistance to sliding, thus creating the self-regulatory tribological behavior of the system and enabling a clearer sound.

In this summary, we did not account for the role of the tribological characteristics of the bow and string, which are also important parts of the system. For instance, if the bow is worn or not properly cared for, friction will be inconsistent, resulting in an inconsistent sound. Additionally, the type of strings used can also affect friction and, subsequently, the sound produced.

While the violin presents a most representative example of the usage of the tribological concepts used in music, the tribology is used in other areas, for example for music records, when friction and wear are generated by scanning a diamond stylus over wax, shellac, or vinyl substrate to create the grooves that are then used for reproducing the sound [27,28].

6.5 TRIBOLOGY IN SPORTS

Ice skating and slope skiing are classic winter sports that rely on a unique set of tribological principles [29,30]. Efficient skating is possible when pressure and friction generate enough heat to melt a thin layer of ice and create a thin layer of water between the blade and the ice. This layer of water experiences the squeezing flow supporting the weight of the skater (similar to the hydrodynamic lubrication discussed in Section 3.2.1) and acts as a lubricant, significantly reducing friction and enabling smooth gliding. The process becomes self-regulating since the reduction of friction leads to a decrease in heat production and prevents further melting of ice which in turn again increases friction. In the case of skiing, the mechanism is quite similar. To understand the involved physical processes, we should refer back to melt wear described earlier in Section 1.7.2. In this case, the temperature at the interface is kept constant at 0°C.

The simple model of the slider is presented in Figure 6.12.

For simplicity, let's assume that all the frictional energy generated during the shearing of the slider is contributing to heat. In this case, the heat flux generated by the shearing water is

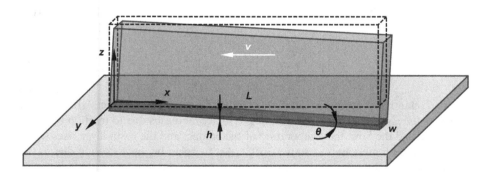

FIGURE 6.12 Schematic of a flat slider on a flat ice substrate. A meltwater layer with thickness h, is created between the slider and the ice substrate. Meltwater layer squeezing is realized by the slider rotating with velocity v. The unshaded rectangle represents the original slider position whereas the shadowed rectangle represents the slider which has been rotated along the leading contact point with an angle θ. (Redrawn from Ref. [30].)

$$d\dot{Q} = \frac{\eta \cdot v^2}{h}. \tag{6.23}$$

This heat flux is used to melt the top layer of ice and to dissipate through conduction into the ice substrate:

$$d\dot{Q} = \frac{H_f \cdot \dot{V}_{melt}}{L \cdot w} + d\dot{Q}_{conduction} = H_f \cdot v_{melt} + d\dot{Q}_{conduction}, \tag{6.24}$$

where H_f is the heat of fusion, $L{\cdot}w$ is the area of the slider contact, v_{melt} is the melt rate. As the thickness of the meltwater layer is much smaller than the contact area, we can use the classic Reynolds equation to describe the flow:

$$\frac{d}{dy}\left(h_x^3\left(\frac{dP}{dy}\right)\right) = 6 \cdot \eta \cdot v \frac{dh_x}{dx} - 12 \cdot \eta \cdot v_s, \tag{6.25}$$

where v_s is the squeezing rate of the meltwater layer. If the slider rotates for an angle θ, the squeezing rate can be written as $v_s = v \tan \theta$. In this case, the change in the thickness of the meltwater layer is

$$\frac{dh_x}{dt} = v_{melt} - v_s = \frac{1}{H_f}\left(d\dot{Q} - d\dot{Q}_{conduction}\right) - v \cdot \tan\theta, \tag{6.26}$$

and then,

$$\frac{dh_x}{dx} = \frac{dh_x}{v \cdot dt} = \frac{1}{H_f}\left(\frac{\eta \cdot v}{h} - \frac{d\dot{Q}_{conduction}}{v}\right) - \tan\theta. \tag{6.27}$$

The pressure of the meltwater layer can be found by integrating the Reynolds equation twice (with boundary conditions of $P=0$ at $y=\pm w/2$):

$$P(x,y) = \frac{3 \cdot \eta \cdot v}{4 \cdot h^3}\left(2 \cdot \tan\theta - \frac{dh}{dx}\right)(w^2 - 4y^2). \tag{6.28}$$

This equation indicates that the maximum pressure happens along the midline of the slider while changing along the rotating angle θ. If the thickness of the meltwater layer is constant and the angle is very small ($\tan\theta = \theta$), the equation can be simplified as

$$P(y) = \frac{3 \cdot \eta \cdot v}{4 \cdot h^3}(2 \cdot \theta)(w^2 - 4y^2). \tag{6.29}$$

At the same time, assuming minimal conduction of heat, Eq. 6.28 will be simplified as

$$0 = \frac{1}{H_f}\left(\frac{\eta \cdot v}{h} - \frac{d\dot{Q}_{\text{conduction}}}{v}\right) - \tan\theta = \frac{1}{H_f}\frac{\eta \cdot v}{h} - \theta, \tag{6.30}$$

and then,

$$h = \frac{\eta \cdot v}{H_f \theta}. \tag{6.31}$$

When substituting Eq. 6.31 in Eq. 6.29, the pressure of the meltwater layer can be found as

$$P(y) = \frac{3 \cdot H_f^3 \cdot \theta^4}{2 \cdot \eta^2 \cdot v^2}\left(w^2 - 4y^2\right). \tag{6.32}$$

This pressure is used to balance the weight of the skater $W = P_{\text{average}} \cdot L \cdot w$

$$W = \int_{-\frac{w}{2}}^{\frac{w}{2}} P(y) L \, dy = \int_{-\frac{w}{2}}^{\frac{w}{2}} \frac{3 \cdot H_f^3 \theta^4}{2 \cdot \eta^2 \cdot v^2}\left(w^2 - 4y^2\right) L \, dy$$

$$= \frac{3 \cdot H_f^3 \cdot L\theta^4}{2 \cdot \eta^2 \cdot v^2} \int_{-\frac{w}{2}}^{\frac{w}{2}} \left(w^2 dy - 4y^2 dy\right)$$

$$= \frac{3 \cdot H_f^3 \cdot L\theta^4}{2 \cdot \eta^2 \cdot v^2}\left\{w^3 - \frac{4}{3}\left[\left(\frac{w}{2}\right)^3 - \left(-\frac{w}{2}\right)^3\right]\right\} = \frac{3 \cdot H_f^3 \cdot L\theta^4}{2 \cdot \eta^2 \cdot v^2}\left(w^3 - \frac{1}{3}w^3\right)$$

$$= \frac{H_f^3 \cdot L \cdot \theta^4 \cdot w^3}{\eta^2 \cdot v^2}. \tag{6.33}$$

Then, the angle needed for supporting the weight of the skater is

$$\theta = \left(\frac{W \cdot \eta^2 \cdot v^2}{H_f^3 \cdot L \cdot w^3}\right)^{\frac{1}{4}}. \tag{6.34}$$

The resulting thickness of the meltwater layer is

$$h = \frac{\eta \cdot v}{H_f \theta} = \left(\frac{\eta^2 \cdot v^2 L \cdot w^3}{W \cdot H_f}\right)^{\frac{1}{4}}, \tag{6.35}$$

which leads to the friction experienced by the slider

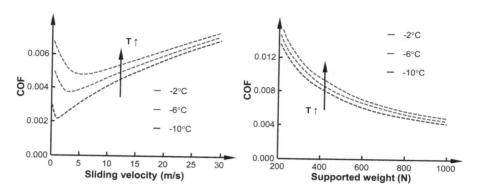

FIGURE 6.13 Coefficient of friction as a function of sliding velocity for a supported load of 800 N (left) and a supported weight at a sliding velocity of 15 m/s (right) considering various ice temperatures. (Redrawn from Ref. [30].)

$$\mu = \dfrac{\tau}{\dfrac{W}{L \cdot w}} = \dfrac{\dfrac{\eta \cdot v}{h}}{W} = \left(\dfrac{\eta^2 \cdot v^2 \cdot H_f L^3 w}{W^3} \right)^{\frac{1}{4}}. \tag{6.36}$$

Again, in this estimation, we neglected the heat dissipation through conduction in ice, which will start playing a major role at lower velocity values (Figure 6.13).

6.6 CHECK YOURSELF

- What is the basic concept of FSW and how does it differ from traditional welding methods?
- Explain the role of the welding tool in the FSW process, and how it induces material intermixing and accelerated diffusion.
- Describe the structural zones formed during the FSW process, including the stir zone, the thermo-mechanically affected zone, and the heating-affected zone.
- How does the rotational speed of the welding tool impact the FSW process, and what considerations should be taken into account to prevent overheating?
- Calculate the processing regimes (pressure times rotational speed) for the FSW of aluminum alloys. Can you use a simple drilling tool as the FSW tool?
- How do TENGs convert frictional energy into electrical energy, and what is the basic design of a simple TENG?
- What are the key parameters influencing the power density of a TENG, and how is the gap voltage between surfaces defined in the context of TENGs?

- Explain the possible sources of charge transfer in triboelectric surfaces, including electron transfer, ion transfer, and material transfer from wear debris.
- How can the design of a TENG be modified to improve charge transfer efficiency, and what is the significance of surface charge and area in this context?
- What factors influence the triboelectricity effect, and why is it important to consider the materials involved in contact, environmental conditions, and load conditions in TENG applications?
- Think about the systems and encounter the everyday tribological sliding and the approaches for reusing this energy for the benefits. Good examples: self-powered highways with the energy supplied by the friction between the tires and the pavement, self-illuminated corridors with the light being turned on when people walk, and self-generation of the power connected to the clothing. What are the potential advantages and disadvantages of such systems?
- How is the sound produced in a violin, and what role does friction play in the interaction between the bow and the strings?
- Describe the stick-slip behavior in the bow-string interaction in a violin, and how does the choice of rosin affect the sound produced?
- In the context of the violin example, how do the temperature and tribological properties of rosin influence the stick-slip behavior and sound quality?
- Why is the consistency of friction between the bow and strings crucial for the sound quality in a violin, and how can factors like bow condition and string type impact this consistency?
- Explain the tribological principles involved in winter sports like ice skating and slope skiing, focusing on the formation of a meltwater layer and its impact on friction.

REFERENCES

1. A. Heidarzadeh, S. Mironov, R. Kaibyshev, G. Çam, A. Simar, A. Gerlich, F. Khodabakhshi, A. Mostafaei, D.P. Field, J.D. Robson, A. Deschamps, P.J. Withers, Friction stir welding/processing of metals and alloys: A comprehensive review on microstructural evolution, *Progress in Materials Science* 117 (2021) 100752.
2. U. Chadha, S.K. Selvaraj, N. Gunreddy, S. Sanjay Babu, S. Mishra, D. Padala, M. Shashank, R.M. Mathew, S.R. Kishore, S. Panigrahi, A survey of machine learning in friction stir welding, including unresolved issues and future research directions, *Material Design & Processing Communications* 2022 (2022) 1–28.
3. M.-N. Avettand-Fènoël, A. Simar, A review about friction stir welding of metal matrix composites, *Materials Characterization* 120 (2016) 1–17.
4. S.S.O. Dadi, C. Patel, B. Appala Naidu, Effect of friction-stir welding parameters on the welding temperature, *Materials Today: Proceedings* 38 (2021) 3358–3364.
5. R.A. Prado, L.E. Murr, D.J. Shindo, K.F. Soto, Tool wear in the friction-stir welding of aluminum alloy 6061+20% Al2O3: A preliminary study, *Scripta Materialia* 45 (2001) 75–80.

6. A. Zafar, M. Awang, S.R. Khan, Friction stir welding of polymers: An overview, in: M. Awang (Ed.), *2nd International Conference on Mechanical, Manufacturing and Process Plant Engineering*, Springer Singapore, Singapore, 2017, pp. 19–36.

7. S. Li, J. Wang, W. Peng, L. Lin, Y. Zi, S. Wang, G. Zhang, Z.L. Wang, Sustainable energy source for wearable electronics based on multilayer elastomeric triboelectric nanogenerators, *Advanced Energy Materials* 7 (2017) 1602832.

8. R. Dharmasena, K. Jayawardena, C. Mills, R. Dorey, S. Silva, A unified theoretical model for Triboelectric Nanogenerators, *Nano Energy* 48 (2018) 391–400.

9. S. Pan, Z. Zhang, Fundamental theories and basic principles of triboelectric effect: A review, *Friction* 7 (2019) 2–17.

10. Y. Zi, S. Niu, J. Wang, Z. Wen, W. Tang, Z.L. Wang, Standards and figure-of-merits for quantifying the performance of triboelectric nanogenerators, *Nature Communications* 6 (2015) 8376.

11. J. Wang, C. Wu, Y. Dai, Z. Zhao, A. Wang, T. Zhang, Z.L. Wang, Achieving ultrahigh triboelectric charge density for efficient energy harvesting, *Nature Communications* 8 (2017) 88.

12. S. Wang, Y. Xie, S. Niu, L. Lin, C. Liu, Y.S. Zhou, Z.L. Wang, Maximum surface charge density for triboelectric nanogenerators achieved by ionized-air injection: Methodology and theoretical understanding, *Advanced Materials* 26 (2014) 6720–6728.

13. G. Rowley, Quantifying electrostatic interactions in pharmaceutical solid systems, *International Journal of Pharmaceutics* 227 (2001) 47–55.

14. S. Naik, Experiment and Discrete Element Based Modeling of Granular Flow: Tribocharging and Particle Size Reduction in Pharmaceutical Manufacturing, Doctoral Dissertations. 640, University of Connecticut - Storrs. https://digitalcommons.lib.uconn.edu/dissertations/640 (2014).

15. H. Zou, Y. Zhang, L. Guo, P. Wang, X. He, G. Dai, H. Zheng, C. Chen, A.C. Wang, C. Xu, Quantifying the triboelectric series, *Nature Communications* 10 (2019) 1427.

16. S.D. Cezan, A.A. Nalbant, M. Buyuktemiz, Y. Dede, H.T. Baytekin, B. Baytekin, Control of triboelectric charges on common polymers by photoexcitation of organic dyes, *Nature Communications* 10 (2019) 276.

17. A. Yu, Y. Zhu, W. Wang, J. Zhai, Progress in triboelectric materials: Toward high performance and widespread applications, *Advanced Functional Materials* 29 (2019) 1900098.

18. R. Zhang, H. Olin, Material choices for triboelectric nanogenerators: A critical review, *EcoMat* 2 (2020) e12062.

19. V.G. Kossobokov, A.K. Nekrasova, Earthquake hazard and risk assessment based on Unified Scaling Law for Earthquakes: Greater Caucasus and Crimea, *Journal of Seismology* 22 (2018) 1157–1169.

20. J.J. Bommer, P.J. Stafford, J.E. Alarcón, Empirical equations for the prediction of the significant, bracketed, and uniform duration of earthquake ground motion, *Bulletin of the Seismological Society of America* 99 (2009) 3217–3233.

21. T.E. Tullis, Rock friction and its implications for earthquake prediction examined via models of Parkfield earthquakes, *Proceedings of the National Academy of Sciences* 93 (1996) 3803–3810.

22. P. Mora, D. Place, Stress correlation function evolution in lattice solid elasto-dynamic models of shear and fracture zones and earthquake prediction, Earthquake Processes: Physical Modelling, *Numerical Simulation and Data Analysis Part II* (2002) 2413–2427.

23. N.A. Abrahamson, W.J. Silva, Empirical ground motion models, Report to Brookhaven National Laboratory (1996).

24. H.L. Helmholtz, *On the Sensations of Tone as a Physiological Basis for the Theory of Music*, Cambridge University Press, Cambridge, 2009.

25. J.H. Smith, J. Woodhouse, The tribology of rosin, *Journal of the Mechanics and Physics of Solids* 48 (2000) 1633–1681.

26. J. Woodhouse, T. Putelat, A. McKay, Are there reliable constitutive laws for dynamic friction? *Philosophical Transactions of the Royal Society A: Mathematical, Physical and Engineering Sciences* 373 (2015) 20140401.

27. M. Evans, *Vinyl: The Art of Making Records*, Simon and Schuster, New York, 2022.

28. D. Sarpong, S. Dong, G. Appiah, 'Vinyl never say die': The re-incarnation, adoption and diffusion of retro-technologies, *Technological Forecasting and Social Change* 103 (2016) 109–118.

29. J. Hjelle, Winter Sports Tribology-An Experimental Approach To Understanding Kinetic Friction and Equipment Performance in Speed Skating, NTNU, 2017.

30. F. Du, Analytical theory of ice-skating friction with flat contact, *Tribology Letters* 71 (2023) 5.

Index